SEP 17 '84

S0-APP-350

DOBIS NO= 31601 MASTER NO= 31598
QD506 Kitahara, Ayao
E381   Electrical phenomena at
1984   interfaces, fundamentals,
c.3   measurements and
       applications.

WITHDRAWN

IBM
ALMADEN RESEARCH LIBRARY   031601

QD506
E381
1984, C. 3

KITAHARA, A.
    ELECTRICAL PHENOMENA AT INTERFACES;
FUNDAMENTALS, MEASUREMENTS

TIECKELMANN
2013

Electrical Phenomena at Interfaces

SURFACTANT SCIENCE SERIES

CONSULTING EDITORS

MARTIN J. SCHICK
Diamond Shamrock Corporation
Process Chemicals Division
Morristown, New Jersey

FREDERICK M. FOWKES
Department of Chemistry
Lehigh University
Bethlehem, Pennsylvania

Volume 1: NONIONIC SURFACTANTS, edited by Martin J. Schick

Volume 2: SOLVENT PROPERTIES OF SURFACTANT SOLUTIONS, edited by Kozo Shinoda *(out of print)*

Volume 3: SURFACTANT BIODEGRADATION, by Robert D. Swisher *(out of print)*

Volume 4: CATIONIC SURFACTANTS, edited by Eric Jungermann

Volume 5: DETERGENCY: THEORY AND TEST METHODS *(in three parts)*, edited by W. G. Cutler and R. C. Davis

Volume 6: EMULSIONS AND EMULSION TECHNOLOGY *(in two parts)*, edited by Kenneth J. Lissant

Volume 7: ANIONIC SURFACTANTS *(in two parts)*, edited by Warner M. Linfield

Volume 8: ANIONIC SURFACTANTS–CHEMICAL ANALYSIS, edited by John Cross

Volume 9: STABILIZATION OF COLLOIDAL DISPERSIONS BY POLYMER ADSORPTION, by Tatsuo Sato and Richard Ruch

Volume 10: ANIONIC SURFACTANTS–BIOCHEMISTRY, TOXICOLOGY, DERMATOLOGY, edited by Christian Gloxhuber

Volume 11: ANIONIC SURFACTANTS–PHYSICAL CHEMISTRY OF SURFACTANT ACTION, edited by E. H. Lucassen-Reynders

Volume 12: AMPHOTERIC SURFACTANTS, edited by B. R. Bluestein and Clifford L. Hilton

Volume 13: DEMULSIFICATION: INDUSTRIAL APPLICATIONS, by Kenneth J. Lissant

Volume 14: SURFACTANTS IN TEXTILE PROCESSING, by Arved Datyner

Volume 15: ELECTRICAL PHENOMENA AT INTERFACES: FUNDAMENTALS, MEASUREMENTS, AND APPLICATIONS, edited by Ayao Kitahara and Akira Watanabe

OTHER VOLUMES IN PREPARATION

Electrical Phenomena at Interfaces
Fundamentals, Measurements, and Applications

edited by

Ayao Kitahara
Science University of Tokyo
Tokyo, Japan

Akira Watanabe
Nara Women's University
Nara, Japan

MARCEL DEKKER, INC. New York and Basel

Library of Congress Cataloging in Publication Data
Main entry under title:

Electrical phenomena at interfaces.

  (Surfactant science series; 15)
  Includes indexes.
  1. Surface chemistry. 2. Electric double layer.
I. Kitahara, Ayao [date]. II. Watanabe, Akira
[date]. III. Series
QD506.E44   1984   541.3'453   84-7835
ISBN 0-8247-7186-9

Copyright © 1984 by Marcel Dekker, Inc. All rights reserved.

Neither this book nor any part may be reproduced or transmitted in any form or by any means, electronic or mechanical, including photocopying, microfilming, and recording, or by any information storage and retrieval system, without permission in writing from the publisher.

Marcel Dekker, Inc.
270 Madison Avenue, New York, New York 10016

Current printing (last digit):
10 9 8 7 6 5 4 3 2 1

Printed in the United States of America

In remembrance of Prof. Akira Watanabe.

Progress of the Series

Since its start 17 years age, 20 books have appeared in this series. Using an interdisciplinary approach, these monographs and collective volumes have covered many facets of surfactant science and have been written by outstanding experts drawn from all over the world. We intend to continue this approach in reviewing current knowledge of surfactants. Since, over the span of many years, some volumes have become outdated, second updated editions of some volumes will appear to keep abreast with the most recent developments in surfactant science.

Since an understanding of the physical chemistry of surfactant action requires knowledge of surface science, volumes dealing with the fundamentals of surface science will also be included in this series. The present volume, *Electrical Phenomena at Interfaces: Fundamentals, Measurements, and Applications,* is the first of these volumes dealing with an important area of surface science.

In conclusion, we hope that the books in this series will present to the reader a well-balanced account of our knowledge of surfactants in the areas of fundamentals, evaluation methods, and major applications. The final objective, as stated in Volume 1, is to generate an international exchange of views and stimulate new research.

Martin J. Schick
Frederick M. Fowkes

Preface

This book is written for scientists and engineers who want to study the fundamentals of and current developments in interfacial electric phenomena and their relation to colloid stability. Graduate students interested in surface and colloid science are also encouraged to use this book. The study of electrokinetic phenomena has evolved considerably since the first studies were carried out by Quincke, Helmholtz, and Lipman in the last century.

Zeta-potential obtained from the measurements of electrokinetic phenomena is not only an important property in interfacial science, but it is also profoundly related to colloid science in the sense that it controls the stability of colloids as dispersions, emulsions, and foams. Nevertheless, many studies now show that it is very difficult to obtain reproducible and reliable values of zeta-potential of colloid particles. Zeta-potential seems to be very sensitive to changes in the interface.

In 1966, a group of Japanese surface and colloid chemists which had engaged in the study of electrokinetic phenomena and/or colloid stability, formed a committee under the Division of Surface Chemistry in the Japan Oil Chemists' Society. This group measured, compared, and discussed zeta-potentials of such samples as titanium oxide, microcapsule, and silica or silver iodide. The values of the zeta-potential of the silver iodide were very similar among those tested and were reproducible. The committee cooperatively published a textbook in Japanese, *Kaimen Denki Genshō—Riron, Sokutei, Ōyō (Interfacial Electric Phenomena: (Theories, Measurements, and Applications)*, edited by A. Kitahara and A. Watanabe, Kyōritsu Shuppan Co., Tokyo

(1972). This was an introductory textbook intended to orient young students and engineers in Japan to interfacial electric science.

As such a book had not yet appeared internationally, Prof. Watanabe appealed to the members of the group to publish a book on interfacial electric phenomena in English. He intended to bring the contents of the Japanese version up to date, and approached Marcel Dekker, Inc. with the plan to publish the book. He further consulted me and we agreed to edit and promote the book jointly. To our great loss, however, Prof. Watanabe died soon after. In publishing this book, we hope to keep alive the memory and wishes of Prof. Watanabe, and to show our deep gratitude not only for the inspiration he gave in the planning and publishing of the book, but also for the help and support he gave.

In the course of preparing this book, two other related books were published. One is *Electrokinetic Phenomena,* by S. S. Duhkin and B. V. Derjaguin, in the series on surface and colloid science edited by Prof. E. Matijevic. The other is *Zeta Potential in Colloidic Science,* by R. T. Hunter, in the series on colloid science edited by Profs. R. H. Ottewill and R. L. Rowell. Both books are excellent, and each book has its own characteristic. I believe our book maintains its own character separable from the other two and that they will all benefit each other.

The style of this book is similar to the Japanese version. The content is different, however, being more advanced and up to date than the Japanese. Several new authors have contributed and new chapters have been added to the Japanese version. The present book is divided into three parts. Part I contains the fundamentals of electrical phenomena at interfaces in aqueous and nonaqueous systems. Here readers will obtain a fundamental understanding of this field. Part II is composed of necessary experimental methods. The methods are described in detail, which readers will find useful. Part III outlines a wide range of applications in this field. Industrial and practical problems are discussed from the view of interfacial electric science and colloid stability.

Finally, we would like to gratefully acknowledge the assistance provided by Ms. Patricia Hoagland and Ms. Chimi Daly, residents of Japan, who have spent many valuable hours in correcting the English in this book. We are grateful to Dr. Maurits Dekker of Marcel Dekker, Inc., for his continuing encouragement.

Ayao Kitahara

Contributors

Shoji Fukushima Shiseido Co., Ltd., Yokohama, Japan

Kunio Furusawa, Ph.D. Department of Chemistry, The University of Tsukuba, Ibaraki, Japan

Sei Hachisu, Ph.D.* Institute of Applied Physics, The University of Tsukuba, Ibaraki, Japan

Ken Higashitsuji, Ph.D. Research and Development Department, Marubishi Oil Chemical Co., Ltd., Osaka, Japan

Tetsuya Imamura, Ph.D. Tochigi Research Laboratories, Kao Corporation, Tochigi, Japan

Shuichi Karasawa Material and Device Technology Department, Technical Division, Ricoh Co., Ltd., Tokyo, Japan

Ayao Kitahara, Ph.D. Department of Industrial Chemistry, Science University of Tokyo, Tokyo, Japan

Shigeharu Kittaka, Ph.D. Department of Chemistry, Faculty of Science, University of Science, Okayama, Japan

Isao Kumano Research and Development Laboratories, Toyo Ink Mfg. Co., Ltd., Itabashi, Tokyo, Japan

Tamotsu Kondo, Ph.D. Faculty of Pharmaceutical Sciences, Science University of Tokyo, Shinjuku-ku, Japan

---

*Dr. Hachisu is now Professor Emeritus.

Mutsuo Matsumoto, Ph.D. Laboratory of Surface Chemistry, Institute for Chemical Research, Kyoto University, Kyoto, Japan

Tetsuo Morimoto, Ph.D. Department of Chemistry, Faculty of Science, Okayama University, Okayama, Japan

Kazuo Nishizawa Laboratory of Interfacial Technique, Kyoto, Japan

Masataka Ozaki, Ph.D. Department of Chemistry, Faculty of Literature and Science, Yokohama City University, Yokohama, Japan

Toshiro Suzawa, Ph.D. Department of Applied Chemistry, Faculty of Engineering, Hiroshima University, Higashihiroshima, Hiroshima, Japan

Hisako Tamai Tagaya, Ph.D. Faculty of Education, Shiga University, Ohtsu, Shiga, Japan

Katsuo Takahashi, Ph.D. Department of Inorganic Chemistry, The Institute of Physical and Chemical Research, Wako-shi, Saitama, Japan

Isao Tari, Ph.D. Department of Synthetic Chemistry, Faculty of Engineering, Okayama University, Tsushima-naka, Okayama, Japan

Fumikatsu Tokiwa, Ph.D. Research and Development Division, Kao Corporation, Tokyo, Japan

Shinnosuke Usui, Ph.D. Research Institute of Mineral Dressing and Metallurgy, Tohoku University, Sendai, Japan

Hiroshi Yamada, Ph.D.* Research Laboratory, Sanyo-Kokusaku Pulp Co., Ltd., Tokyo, Japan

---

*Dr. Yamada is now retired.

Contents

Progress of the Series, *Martin J. Schick and Frederick M. Fowkes* iv
Preface v
Contributors vii

PART I: FUNDAMENTALS

1. Surface Electricity 3
   Sei Hachisu

2. Electrical Double Layer 15
   Shinnosuke Usui

3. Interaction of Electrical Double Layers and Colloid Stability 47
   Shinnosuke Usui and Sei Hachisu

4. Electrokinetics 99
   Sei Hachisu

5. Nonaqueous Systems 119
   Ayao Kitahara

PART II: MEASUREMENTS

6. Electrocapillary Measurements   147
   Katsuo Takanaski, Hisako Tagaya, Ken Higashitsuji, and Hisako Tamai Tagaya

7. Electrokinetic Measurements   183
   Shigeharu Kittaka, Kunio Furusawa, Masataka Ozaki, Tetsuo Morimoto and Ayao Kitahara

8. Stability Measurement of Disperse Systems   225
   Kunio Furusawa and Mutsuo Matsumoto

PART III: APPLICATIONS

9. Detergency   269
   Fumikatsu Tokiwa and Tetsuya Imamura

10. Flotation   285
    Shinnosuke Usui

11. Fibers   299
    Toshiro Suzawa

12. Paper   323
    Hiroshi Yamada

13. Electrocapillary Emulsification   339
    Kazuo Nishizawa

14. Pigments and Paints   355
    Isao Kumano

15. Cosmetics   369
    Shoji Fukushima

16. Antirusting   387
    Isao Tari

17. Electrokinetic Phenomena in Biological Systems   397
    Tamotsu Kondo

18. Reproduction in Copying and Electrophoretic Display 413

    Shuichi Karasawa

Author Index 437
Subject Index 457

Electrical Phenomena at Interfaces

PART I: FUNDAMENTALS

1
Surface Electricity

Sei Hachisu* The University of Tsukuba, Sakura, Ibaraki, Japan

1.1. Origin of Surface Electricity 3
1.2. Fundamental Quantities: ϕ, χ, and ψ 5
    1.2.1. Isolated systems 5
    1.2.2. Two-phase systems 7
1.3 Zero Point of Charge and Isoelectric Point 12
References 14

1.1 ORIGIN OF SURFACE ELECTRICITY

In a stably dispersed lyophobic colloid (e.g., a negative AgI sol) each particle carries a negative electric charge due to selective "adsorption"[†] of iodide ions, and is surrounded by an atmosphere of positive charge. This entity, consisting of two layers, one negatively charged and the other positively charged, is called an electrical double layer. When either (or rarely both) of the layers demonstrates a diffuse distribution of the electric charge, the system is referred to as a diffuse double layer. Typical cases are illustrated in Fig. 1.1.

The diffuse part of a double layer has a thickness ranging from 10 Å to 1 μm and plays important roles in colloid and surface phenom-

---

*Currently Professor Emeritus.
†Strictly speaking, this should not be termed adsorption but rather, considered to be a kind of deposition (see Sec. 1.2.2).

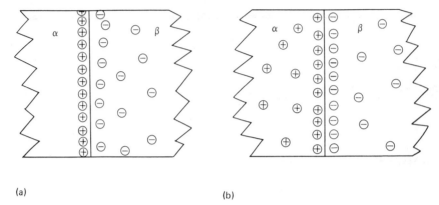

Figure 1.1 Two typical patterns of charge distribution produced by potential-determining ions. (a) Solid-liquid system. (b) Liquid (or semiconductor)-liquid system.

ena. It can move relative to the particle surface to produce electrokinetic phenomena such as electrophoresis, electroosmosis, and surface conductivity. When two particles approach each other, the diffuse double layers interact to create a repulsive force that is responsible for the stability of colloids.

There is another electrical double layer, which is extremely thin, probably one molecule thick, and can be said to be a molecular capacitor. These layers are bound to the surface, and the electric charges neither move nor separate from the surface. They produce a cliff-like drop or rise in the electric potential at the surface or interface.

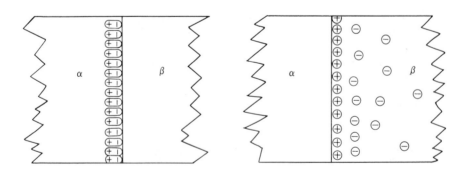

Figure 1.2 (a) Bound double layer or molecular capacitor. (b) Charge distribution caused by specific adsorption of ions.

In contrast to diffuse double layers, they do not directly produce any appreciable effect, but exert an influence on the magnitude of the diffuse double layers.

The origin of the double layers is somewhat complex. Diffuse double layers are the result of an uneven distribution of some type of active free ions. Two classes of active ions are distinguished, according to the mechanism producing the uneven distribution. One includes the potential-determining ions (I ions in AgI sols) and the other includes specifically adsorbing ions (e.g., Ca^{2+} to hematite surface). Bound double layers are produced by the orientation of dipoles at the interface (see Fig. 1.2a) and by some surface polarization, the details of which are not clear in most cases. These origins or mechanisms can interfere with each other, and the total effect is surface electicity.

1.2 FUNDAMENTAL QUANTITIES: ϕ, χ, and ψ

1.2.1 Isolated Systems

When two phases are brought into contact, electric charges appear in the vicinity of the phase boundary. This is shown schematically in Figs. 1.1 and 1.2. In every case, positive and negative charges are aligned opposite each other. The plane of separation of these opposite charges is usually the material boundary of the two phases (Fig. 1.1a and b), but sometimes the two oppositely charged layers are on the same site of the boundary (Fig. 1.2a and b).

In order to introduce some basic concepts concerning surface electricity, a conducting body isolated in empty or nearly empty space is considered. Such a body generally carries some electricity generated by an accidental occurrence. The electric field produced by this charge extends into the space around the body.

The electric potential is shown in Fig. 1.3a. This is the one found in most textbooks on electricity. It is to be noted that in the elementary theory of electricity, the oversimplification is made that the solid body is homogeneous throughout. With a real body of matter this is not the case. Generally, at the surface of a solid (or liquid) body, there is an electrical double layer one molecule thick, as mentioned in Sec. 1.1. This double layer produces a potential distribution as depicted in Fig. 1.3b. In this system, three kinds of electric potentials are distinguished. They are ϕ, the potential inside the body; χ, the potential jump at the surface; and ψ, the potential just outside the surface. ϕ is referred to as the inner potential or Galvani potential and ψ as the outer potential or Volta potential. From Fig. 1.3b we have the following relationship:

$$\phi = \chi + \psi \tag{1.1}$$

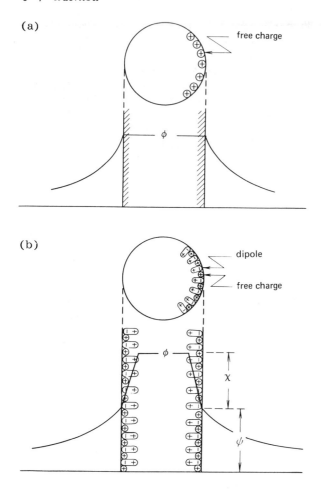

Figure 1.3 Potential distribution in isolated systems. (a) Ideal uniform body. (b) Real substance. At the surface there is a dipole orientation or some other surface polarization.

ψ is the potential produced by a free charge on the body and is measurable in principle (although a sensitive apparatus and great skill are necessary for the actual measurement). In contrast to this, the χ potential cannot be measured. This requires some explanation.

In the theory of electricity, when the measurement of the electric field is discussed, an ideal probe—an imaginary particle that carries a unit electric charge but does not possess any attributes of matter—is considered; it is supposed to be able to penetrate or move through every type of matter without any resistance or interaction since it senses only the electric field. In actual measurements, however, the probe is made of matter that excludes everything from its inherent volume, and it cannot penetrate bodies of matter without disturbing their original states. This situation makes it impossible to measure ϕ or χ potentials that are inside or at the very surface of the body.

The particle that most resembles an ideal probe would be an electron. It carries an electric charge (minus the protonic charge e) but does not possess a finite size, and resides in every kind of matter as the common ingredient. The work that is required to take it out of a substance can be measured by thermal emission or by photoelectric effect. It looks as if we have obtained the quantity $(-e\chi)$; however, this is not true. The work is not equal to $(-e\chi)$ because an electron is not an ideal probe but has some chemical affinity ($\Delta\mu$) with matter. Therefore, the work is to be given by $-e\chi + \Delta\mu$. We have no means to measure one independently of the other.

For these reasons the absolute values of χ, and consequently of ϕ, cannot be known. We can only measure the change of χ through the change of the outer potential ψ, as will be explained in the next section.

1.2.2 Two-Phase Systems

The isolated systems mentioned above are only of theoretical interest. The main body of phenomena of surface electricity is found at the interface between two phases that are in contact.

When two phases are allowed to contact, a migration of electricity takes place from one phase to the other across the boundary, and a potential difference appears between the two phases. The carriers of the electricity are particular species of ions (or electrons, when the phases are metals or semiconductors) that can reside in both phases. They are either common constituents or common solutes to the two phases. These ions are called the potential-determining (PD) ions. For example, they are the metal ions in metal electrode-water systems, Ag and I ions in AgI-water systems. For simplicity, the case of one species of PD ions will be discussed below. The results can be easily extended to the case of two ion species.

The migration or distribution of ions is governed by the law of thermodynamics. Generally, molecules or ions have some escaping tendency from the environment in which they are now residing. When the two phases (α and β) are brought into contact, the PD ions (assumed to be positively charged) migrate from one phase (β), in which the escaping tendency of the ions is larger, to the other phase (α). As the result of this migration, phase α becomes positively charged and phase β becomes negatively charged. The electric field produced by this charging will hinder further migration of the ions. Hence after a certain number of ions have migrated, a state of equilibrium will be reached where the drive for migration caused by the difference

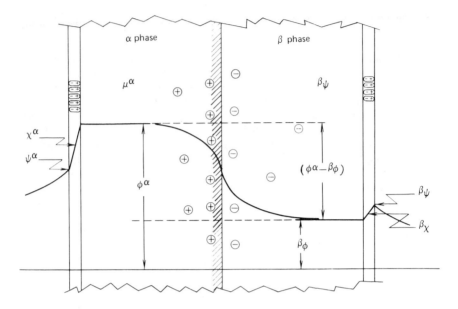

Figure 1.4 Potential distribution in a two-phase system, without considering the potential jump at the interface.

in escaping tendencies is counterbalanced by the electric field that is produced. Thus a difference in electric potential is established between the two phases, as shown in Fig. 1.4.

The thermodynamic concept of the escaping tendency (of i species of ions or molecules in phase α) is the chemical potential $^{\alpha}\mu_i$, which is given by the formula

$$^{\alpha}\mu_i = {^{\alpha}\mu_i^{\circ}} + RT \ln {^{\alpha}C_i} \tag{1.2}$$

where $^{\alpha}C_i$ denotes the molar concentration of i species of ions (or molecules) and $^{\alpha}\mu_i^{\circ}$ is the standard chemical potential. In exact treatments $^{\alpha}C_i$ is to be replaced by the activity $^{\alpha}a_i$. By use of the chemical potentials, the equilibrium condition is written

$$^{\beta}\mu_i - {^{\alpha}\mu_i} = e\nu(^{\alpha}\phi - {^{\beta}\phi}) \tag{1.3}$$

Substituting Eq. (1.2) for $^{\alpha}\mu_i$ and $^{\beta}\mu_i$, Eq. (1.3) becomes

$$\nu e(^{\alpha}\phi - {^{\beta}\phi}) = (^{\beta}\mu_i^{\circ} - {^{\alpha}\mu_i^{\circ}}) + RT \ln \frac{^{\beta}C_i}{^{\alpha}C_i} \tag{1.4}$$

Surface Electricity / 9

This indicates that the potential difference between the two phases is determined by the difference between the PD ion concentrations in the two phases.

Equation (1.3) may be rearranged as

$$^{\alpha}\mu_i + \nu e\,^{\alpha}\phi = \,^{\beta}\mu_i + \nu e\,^{\beta}\phi \tag{1.5}$$

($\mu + e\nu\phi$) is the electrochemical potential; and Eq. (1.5) is another form of the equilibrium condition, the equality of the electrochemical potentials over the two phases. In the equilibrium state, the negative and positive charges that are separated into the two phases attract each other and concentrate near the interface to form a diffuse electrical double layer as shown in Fig. 1.4*. A gradient of the electric potential appears on both sides of the interface. These gradients fade out as the distance from the interface increases. In the bulk of the phases the electric and chemical potentials and the distribution of the charges are uniform. Quantities such as $^{\alpha}\phi$, $^{\beta}\phi$, $^{\alpha}\mu$, and $^{\beta}\mu$ refer to the bulk of the phases.

When the potential distribution at the interface is examined closely, it is not as simple as given by Fig. 1.4. There is a potential jump, produced by the orientation of dipoles or by some polarization at the interface. The phenomenon would be as shown in Fig. 1.5. Thus there are three potentials associated with the interface. They are the potential $^{\alpha}D$ on the α side, the potential $^{\beta}D$ on the β side, and the potential jump. The potential jump may be expressed by $^{\alpha\beta}\chi$, following the convention in the case of a surface exposed to vacuum or gases. $^{\alpha}D$ and $^{\beta}D$, which are produced by free charges of PD ions, are called double-layer potentials. The difference between ϕ and D is one important quantity in colloid studies.

Several potentials associated with the surface and the interface are illustrated in Figs. 1.4 and 1.5. At the surface exposed to vacuum or gas, there are Volta potentials $^{\alpha}\psi$ and $^{\beta}\psi$ just outside the surface, and $^{\alpha}\chi$ and $^{\beta}\chi$ the potential jumps at the surface proper.

---

*This may indicate that the electric charge in a diffuse layer consists solely of PD ions (or of a deficit of PD ions). This is, however, not true in most cases. Generally, in a liquid phase, other ion species coexist with PD ions. They do not take part in the PD reaction and are called indifferent ions. They are also attracted by the electric force to join the diffuse layers. Ordinarily, the concentration of indifferent ions is much higher than that of PD ions. Hence the electric charge in the diffuse layer consists almost entirely of indifferent ions. The readers will see in Chap. 2 that in the theory of electrical double layers, the presence of PD ions is totally neglected. There, the potential difference between the phases (actually, the surface potential of the particle) is a given quantity.

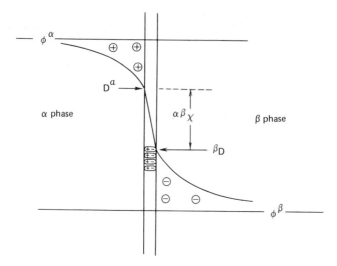

Figure 1.5 Potential distribution near the interface, taking into account the potential jump at the interface.

Several relations at the surface and the interface are

$$^\alpha\phi = {}^\alpha\chi + {}^\alpha\psi \tag{1.6}$$

$$^\beta\phi = {}^\beta\chi + {}^\beta\psi \tag{1.7}$$

$$^\alpha D - {}^\beta D = {}^{\alpha\beta}\chi \tag{1.8}$$

As mentioned earlier, $^\alpha\psi$ and $^\beta\psi$ are measurable quantities in principle, but the measurement needs great skills and high-precision apparatus. In contrast to this, the difference $^\alpha\psi - {}^\beta\psi$ can be obtained rather easily by the use of an electric circuit. This makes the investigation of the χ potential possible in the following way. The difference $^\alpha\psi - {}^\beta\psi$ is given by the relation

$$^\alpha\psi - {}^\beta\psi = ({}^\alpha\phi - {}^\beta\phi) - ({}^\alpha\chi - {}^\beta\chi) \tag{1.9}$$

in which $({}^\alpha\phi - {}^\beta\phi)$ is the characteristic quantity of the system and is a constant. Thus, if one of the two χ's, say $^\alpha\chi$, is altered by some surface treatment (maintaining $^\beta\chi$ constant), a relation

$$\Delta({}^\alpha\psi - {}^\beta\psi) = \Delta{}^\alpha\chi \tag{1.10}$$

holds. Therefore, from the measurements of $({}^\alpha\psi - {}^\beta\psi)$, we can determine the change in the $^\alpha\chi$ potential.

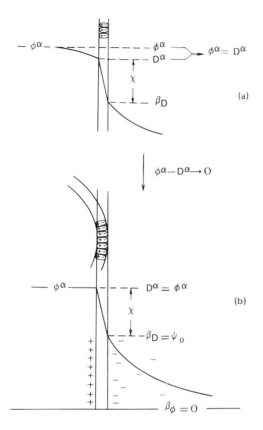

Figure 1.6 Potential distribution inside a colloid particle when $^\alpha\phi = {}^\alpha D$; there is no atmosphere in the particle.

In the treatment thus far, it has been tacitly assumed that the two phases in consideration were liquids for which the PD ions were the common solutes. In most real systems, however, one of the two phases is a solid, of which the PD ions are the constituent. In such cases, the chemical potential of the PD ions in the solid is a constant.

Since ions in the solid phase do not move easily, only ions at the surface take part in the PD reaction. Migration of the ions in this case is from the surface of the solid to the bulk of the liquid, or vice versa. A potential-determining process in this case consists of dissolution or deposition from or to the solid surface. The charge produced in the solid phase is a surface charge.

Colloids are a type of two-phase system in which one phase (say, α) is finely divided and suspended in the other (β), which is called the medium. In the study of colloids, our main concern is with the particle-medium interface. Here the inner potential $^\beta\hat{\phi}$ of the medium is taken as zero; furthermore, the potential distribution in the finely divided phase is considered to be uniform. That is, the double-layer potential $^\alpha D$ is taken to be equal to the inner potential $^\alpha\phi$, as shown in Fig. 1.6. The potential distribution in the system becomes rather simple (see Fig. 1.6b), and Eq. (1.8) becomes

$$^\alpha\phi = {^{\alpha\beta}\chi} + {^\beta D} \qquad \beta_D = \beta_\phi \tag{1.11}$$

There is an analogy between Fig. 1.6 and Fig. 1.3b; the double-layer potential $^\beta D$ in Fig. 1.6 corresponds to the Volta potential ψ in Fig. 1.3b, since both are produced by free charges. The difference is that $^\beta D$ is the result of the partition equilibrium of PD ions, while ψ is due to an accidental charging. The distribution of the potential in the β phase in Fig. 1.6 is governed by Gouy's double-layer theory (see Chap. 2), while in Fig. 1.3b the potential distribution is determined by Coulomb's law.

In the colloid theory, the double-layer potential is conveniently expressed by ψ_0, called the surface potential. Equation (1.11) is written as (with superscripts removed)

$$\phi = \chi + \psi_0 \tag{1.12}$$

Dispite the resemblance of Eq. (1.12) to Eq. (1.1), the double-layer potential should not be confused with the Volta potential.

1.3 ZERO POINT OF CHARGE AND ISOELECTRIC POINT

When the chemical potential of PD ions in the α phase is constant, $^\alpha\mu_i$ is taken as equal to $^\alpha\mu_i^\circ$. Equation (1.4) becomes

$$\nu e(^\alpha\phi - {^\beta\phi}) = (^\beta\mu_i^\circ - {^\alpha\mu_i^\circ}) + RT \ln {^\beta C_i} \tag{1.13}$$

This is the Nernst equation. In the case of colloids, $^\beta\phi$ is taken to be zero; by use of Eq. (1.12), the equation above is rewritten as

$$\psi_0 = \frac{^\beta\mu_i^\circ - {^\alpha\mu_i^\circ} - \chi}{\nu e} + \frac{RT \ln {^\beta C_i}}{\nu e} \tag{1.14}$$

At a PD ion concentration that satisfies the condition

$$0 = \frac{^\beta\mu_i^\circ - {^\alpha\mu_i^\circ} - \chi}{\nu e} + RT \ln (^\beta C_i)_0 \tag{1.15}$$

the surface potential or double-layer potential ψ_0 becomes zero, and at the same time the surface charge is also zero. This concentration of PD ions is called the zero point of charge (ZPC). By use of the ZPC, Eq. (1.14) becomes

$$\psi_0 = \frac{RT}{\nu e} \ln \frac{^\beta c_i}{(^\beta c_i)_0} \qquad (1.16)$$

Hence, if the ZPC is known, the surface potential ψ_0 can be calculated from Eq. (1.16). In view of the difficulty in measuring the surface potential, this conclusion is very important.

The ZPC is determined experimentally by a potentiometric titration of the adsorbed PD ions by the dispersed (or α) phase [1]. This method utilizes the fact that the capacity of electrical double layers depends on the total electrical concentration of the medium. Therefore, as shown in Fig. 1.7, the adsorption curves have different slopes depending on the total electrolyte concentration. Every curve passes the zero point of adsorption, so the ZPC is obtained as the common intersection of these curves. Figure 1.7 shows the case of hemative [2] dispersion where the PD ions are H^+ and OH^- ions.

Another method for ZPC determination is to determine the PD ion concentration where electrokinetic phenomena disappear. Electrophoresis is employed in most cases. It is to be noted that in this

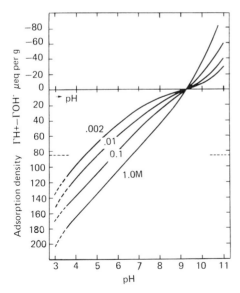

Figure 1.7 $(\Gamma_{H^+} - \Gamma_{OH^-})$ versus pH curves of a hematite suspension at various KCl concentrations. The intersection is the zero point of adsorption or the zero point of charge.

IEP: isoelectric point
the pH at which a particular molecule or surface carries no net electrical charge

method the point actually measured is the zero point of the zeta (ζ) potential, called the isoelectric point (IEP).

As will be mentioned in a later chapter, the ζ potential is the electric potential at the slipping plane, which is located some distance (a few molecule thickness) from the interface. Between the interface (where the surface potential ψ_0 is referred to) and the slipping plane, there can exist ions that are specifically adsorbed onto the interface, and the charge of these ions greatly influences the value of the ζ potential. Therefore, if some ions are specifically adsorbed, the ζ potential may have a finite value to generate some electrokinetic phenomena even at the ZPC; in the opposite situation, at the IEP, where no electrokinetic phenomena are recognized, the system may not be situated at the ZPC. In these cases, the titration method has to be used. However, the ZPC from titration may not give the true value because the electric field produced by the specifically adsorbed ions affects the equilibrium condition of the PD ion adsorption [3].

In addition, the value of the ZPC is known to be affected by the adsorption of some polar ions. An example is the shift of the ZPC of an AgI sol by the addition of a small amount of acetone [4]: the ZPC moves from pAg 5.5 to pAg 2.5 (pAg = $-\log C_{Ag}$).

REFERENCES

1. J. Lyklema and J. Th. G. Overbeek, *J. Colloid Interface Sci.*, 16, 595 (1961).
2. R. J. Atkinson, A. M. Posner, and J. P. Quirk, *J. Phys. Chem.*, 71, 550 (1967).
3. J. Lyklema and A. Breeuwsma, *J. Colloid Interface Sci.*, 48, 437 (1973).
4. J. Th. G. Overbeek, in *Colloid Science* (H. R. Kruyt, ed.), Vol. 1, Elsevier, Amsterdam, 1952, p. 165.

2
Electrical Double Layer

Shinnosuke Usui Research Institute of Mineral Dressing and Metallurgy, Tohoku University, Sendai, Japan

2.1. Thermodynamics of Charged Interfaces 15
    2.1.1. Polarizable interface 15
    2.1.2. Nonpolarizable interface 17
2.2. Electrocapillarities 19
    2.2.1. Electrocapillary curves 19
    2.2.2. Differential capacity 22
2.3. Gouy-Chapman Model 25
2.4. Stern and Grahame Models 30
2.5. Other Models 36
2.6. Oxide-Aqueous Solution Interfaces 42
    References 44

2.1 THERMODYNAMICS OF CHARGED INTERFACES

2.1.1 Polarizable Interface

Consider a mercury electrode in contact with an aqueous KCl solution. An electric potential (E) is applied to the mercury electrode from an external source (potentiometer) against an Ag-AgCl reference electrode (Fig. 2.1). Cu' and Cu" represent copper lead wires. In this case, no common components exist between mercury and the aqueous phase, and accordingly, it is assumed that there is no current passing through the mercury-solution interface. Thus the mercury electrode acts as an ideally polarized electrode. By applying the Gibbs adsorption equation [Eq. (2.1)] to the mercury-aqueous solution interface and including electrons in the mercury phase, Eq. (2.2) is obtained,

Cu'|Ag|AgCl|KCl, H_2O|Hg|Cu"

Figure 2.1 Polarizable mercury electrode.

in which the chemical potential for a charged component is replaced by the electrochemical potential.

$$-d\gamma = \sum_i \Gamma_i \, d\tilde{\mu}_i \tag{2.1}$$

where γ is the interfacial tension and Γ_i and μ_i are the surface excess and chemical potential of the ith species, respectively.

$$-d\gamma = \Gamma_{Hg^+} d\tilde{\mu}_{Hg^+} + \Gamma_e \, d\tilde{\mu}_{e(Hg)} + \Gamma_{K^+} d\tilde{\mu}_{K^+} + \Gamma_{Cl^-} d\tilde{\mu}_{Cl^-}$$
$$+ \Gamma_{H_2O} \, d\mu_{H_2O} \tag{2.2}$$

where $\tilde{\mu}$ is the electrochemical potential. After applying the equilibrium conditions for each phase* and rearranging the electrochemical potential terms using μ_{Hg} as a constant, Eq. (2.3) is obtained.

$$-d\gamma = \sigma \, dE_- + \Gamma_{K^+} d\mu_{KCl} + \Gamma_{H_2O} \, d\mu_{H_2O} \tag{2.3}$$

where σ is the surface charge density as defined by

$$\sigma = \Gamma_{Hg^+} - \Gamma_e \tag{2.4}$$

and E_- refers to the electric potential difference measured relative to a reference electrode which is reversible to anions (Cl^-). For an aqueous phase at constant temperature and pressure, the Gibbs-Duhem relation is

$$x_{KCl} \, d\mu_{KCl} + x_{H_2O} \, d\mu_{H_2O} = 0 \qquad d\mu_{H_2O} = -\frac{x_{KCl} \, d\mu_{KCl}}{x_{H_2O}} \tag{2.5}$$

where x refers to the mole fraction.

By eliminating $d\mu_{H_2O}$ from Eq. (2.3) using Eq. (2.5),

$$-d\gamma = \sigma \, dE_- + \Gamma_{K^+(H_2O)} \, d\mu_{KCl} \tag{2.6}$$

is obtained, where

---

*$d\tilde{\mu}_{Hg^+} = d\mu_{Hg} - d\tilde{\mu}_{e(Hg)}$, $d\tilde{\mu}_{Cl^-} = d\tilde{\mu}_{e(Ag)} = d\tilde{\mu}_{e(Cu')}$, $d\tilde{\mu}_{e(Hg)} = d\tilde{\mu}_{e(Cu'')}$, $d(\tilde{\mu}_{e(Cu')} - \tilde{\mu}_{e(Cu'')}) = FdE_-$, and $d\mu_{K^+} = d\mu_{KCl} - d\mu_{Cl^-}$.

$$\Gamma_{K^+(H_2O)} = \Gamma_{K^+} - \frac{x_{KCl}}{x_{H_2O}} \Gamma_{H_2O} \tag{2.7}$$

$\Gamma_{K^+(H_2O)}$ is the relative surface excess of K^+ and is independent of the choice of the dividing plane by which interfacial region is defined [1]. Equation (2.6) is the thermodynamic relation for the system shown in Fig. 2.1. The detailed treatment can be found in the textbook by Delahay [2]. Equation (2.8) is obtained in a similar manner if a reference electrode that is reversible to cations (K^+) is used.

$$-d\gamma = \sigma \, dE_+ + \Gamma_{Cl^-(H_2O)} \, d\mu_{KCl} \tag{2.8}$$

In general, the following formula holds:

$$-d\gamma = \sigma \, dE_\pm + \Gamma_\mp \, d\mu \tag{2.9}$$

From Eq. (2.9) it follows that

$$\left(\frac{\partial \gamma}{\partial E}\right)_\mu = -\sigma \tag{2.10}$$

$d\mu = d\mu_{KCl}$
$\Gamma_- = \Gamma_{Cl^-(H_2O)}$
$\Gamma_+ = \Gamma_{K^+(H_2O)}$

Equation (2.10) is known as the electrocapillary equation or the Lippmann equation, where the subscript (±) of E can be omitted because the solution compositions are kept constant and hence the reference electrode becomes irrelevant. Equations (2.11) and (2.12) may be derived from Eq. (2.9).

$$\Gamma_- = -\left(\frac{\partial \gamma}{\partial \mu}\right)_{E_+} \tag{2.11}$$

$$\Gamma_+ = -\left(\frac{\partial \gamma}{\partial \mu}\right)_{E_-} \tag{2.12}$$

Thus the relative surface excesses, $\Gamma_- (\Gamma_{Cl^-(H_2O)})$ or $\Gamma_+ (\Gamma_{K^+(H_2O)})$, can be obtained. In experimentation, it is not always possible to obtain a reference electrode that is reversible to a particular type of anion or cation. A detailed treatment of this problem is given in Refs. 3-5.

2.1.2 Nonpolarizable Interface

Consider the system shown in Fig. 2.2, which differs from that of Fig. 2.1 in that both the mercury (α) phase and the aqueous (β) phase contain M (e.g., Tl) in common. The Gibbs adsorption equation for the mercury-aqueous solution interface is given by

			(β)		(α)	
Cu'	Ag	AgCl	KCl, MCl,	H_2O	Hg, M	Cu''

Figure 2.2 Nonpolarizable mercury electrode.

$$-d\gamma = \Gamma^\alpha_{M^+} d\tilde{\mu}^\alpha_{M^+} + \Gamma^\alpha_{Hg^+} d\tilde{\mu}^\alpha_{Hg^+} + \Gamma^\alpha_e d\tilde{\mu}^\alpha_e$$
$$+ \Gamma^\beta_{M^+} d\tilde{\mu}^\beta_{M^+} + \Gamma^\beta_{K^+} d\tilde{\mu}^\beta_{K^+} + \Gamma^\beta_{Cl^-} d\tilde{\mu}^\beta_{Cl^-} + \Gamma^\beta_{H_2O} d\mu^\beta_{H_2O}$$

(2.13)

By rearranging Eq. (2.13) using the electrochemical equilibrium conditions for each phase with the additional requirements,

$$d\tilde{\mu}^\alpha_{M^+} = d\tilde{\mu}^\beta_{M^+} \tag{2.14}$$

and

$$d\mu^\beta_{MCl} = d\mu^\alpha_M + F\, dE \tag{2.15}$$

$$-d\gamma = -(\sigma^\beta - F\Gamma^\beta_{M^+(H_2O)})\, dE + (\Gamma^\alpha_{M^+(Hg)} + \Gamma^\beta_{M^+(H_2O)})\, d\mu^\alpha_M$$
$$+ \Gamma^\beta_{K^+(H_2O)}\, d\mu_{KCl} \tag{2.16}$$

is derived, where

$$\sigma^\alpha = F(\Gamma^\alpha_{M^+} + \Gamma^\alpha_{Hg^+} - \Gamma^\alpha_e)$$
$$= -\sigma^\beta = F(\Gamma^\beta_{M^+} + \Gamma^\beta_{K^+} - \Gamma^\beta_{Cl^-}) \tag{2.17}$$

Equation (2.16) is the thermodynamic relation for the nonpolarizable interface illustrated in Fig. 2.2. If the compositions in the α and β phases are kept constant,

$$\left(\frac{\partial \gamma}{\partial E}\right)_\mu = \sigma^\beta - F\Gamma^\beta_{M^+(H_2O)} \tag{2.18}$$

This is the electrocapillary equation for nonpolarizable interfaces. If $a_{MCl} \ll a_{KCl}$, then $\Gamma^\beta_{M^+(H_2O)}$ can be ignored, and

$$\left(\frac{\partial \gamma}{\partial E}\right)_\mu = \sigma^\beta \tag{2.19}$$

This is identical to Eq. (2.10). Detailed derivations may be found in Mohilner's paper [6].

The same type of electrocapillary equation was derived for both polarizable and nonpolarizable interfaces. It should be mentioned, however, that E in Eq. (2.10) is varied by means of a potentiometer, while E in Eq. (2.19) is determined by the activity of the potential-determining species present in both phases.

Investigations on nonpolarizable metal-aqueous solution interfaces are seldom done and most of these are carried out on a silver iodide-aqueous solution system. The Lippmann equation for this system is

$$\left(\frac{\partial \gamma}{\partial E}\right)_{\mu,\ \text{except Ag}^+ \text{ and I}^-} = \sigma \tag{2.20}$$

where

$$\sigma = F(\Gamma_{Ag^+} - \Gamma_{I^-}) \tag{2.21}$$

Ag^+ and I^- are the potential-determining ions for AgI. The experiment involves the measurement of σ by potentiometric titration of a colloidal suspension of AgI with $AgNO_3$ and KI in the presence of a supporting electrolyte as a function of pAg. The curve of σ versus E is obtained from this [7,8]. Similar work was done using sulfide- [9,10] and oxide- [11] aqueous solution systems. In the case of oxide, the adsorption density of H^+ and OH^-, the potential-determining ions, was determined by pH measurements. In this way, the nature of the electrical double layer can be studied even on solids that do not serve as electrodes.

2.2 ELECTROCAPILLARITIES

2.2.1 Electrocapillary Curves

Figure 2.3 shows a schematic drawing of an electrocapillarometer. It is virtually identical to the apparatus shown in Fig. 2.1 in cell construction. The interfacial tension (γ) at the mercury-aqueous solution interface in the capillary (A) can support the mercury head (H); thus

$$\gamma = \frac{Rg\rho H}{2} \tag{2.22}$$

where R is the radius of the capillary, ρ the density of mercury, and g the gravitational constant. When the electric potential is changed by an amount ΔE by the potentiometer (P) and B is a reversible reference electrode, ΔE is completely applied at the mercury-aqueous

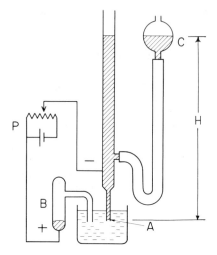

Figure 2.3 Schematic drawing of electrocapillarometer. A, Capillary tip; B, reference electrode; C, mercury pool; P, potentiometer; H, mercury head.

solution interface and it polarizes the mercury. This results in the movement of the mercury level in A due to the change of γ. By adjusting H in such a way that the mercury level in A is brought to a fixed position, γ can be evaluated by Eq. (2.22) as a function of the polarization potential of the mercury(E). The γ versus E curve is called the electrocapillary curve and some examples are shown in Fig. 2.4. Equation (2.10) indicates that the slope of the electrocapillary curve gives the surface charge density (σ), which implies that at a maximum point of the curve, the electrocapillary maximum (ECM), σ becomes zero. This point at which $\sigma = 0$ is also called the point of zero charge (PZC). In the case in which no specific adsorption takes place (e.g., mercury in contact with a NaF aqueous solution) the polarization potential of -0.48 V versus normal calomel electrode (NCE) corresponds to the ECM. The electric potential measured with respect to the ECM (E $-$ E_{ECM}) is called the rational potential [12]. It is seen that γ decreases in both the negative and the positive polarization branches with respect to the ECM. This is caused by the repulsion among surface charges, whereby less work is required for surface expansion. It is also seen that γ decreases significantly in the positive polarization branch for NaBr and KI solutions with the shift of the ECM toward the negative potential, indicating the specific adsorption of halide anions at the mercury-aqueous solution interface. The shift of the ECM is explained as follows. If anions are specifically adsorbed at the PZC, positive charges are induced on the

Electrical Double Layer / 21

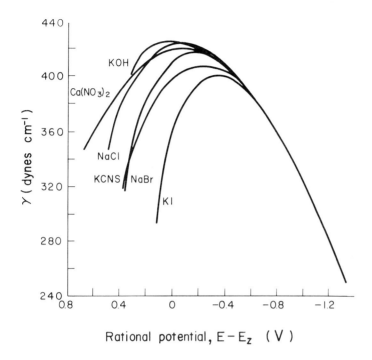

Figure 2.4 Electrocapillary curves of mercury in contact with aqueous solutions of various electrolytes (18°C). Ez: ECM potential (−0.48 V versus NCE) in NaF aqueous solution. (From Ref. 12).

mercury surfaces and additional negative polarization is required to reestablish the state of zero surface charge on the mercury. The relation between the shift of the PZC and the chemical potential of specifically adsorbed ions is referred to as the Esin-Markov effect. In the negative polarization branch each curve is seen to overlap, indicating no specific adsorption of cations in these potential regions. Figure 2.5 shows electrocapillary curves of mercury in 0.5 M Na_2SO_4 aqueous solutions in the absence and in the presence of n-heptyl alcohol. A large depression of γ is seen at the midpotential region, which includes the ECM.

If electrocapillary curves of systems in which the electrolyte composition is varied are plotted, Γ_- and Γ_+ at a particular value of E are obtained using Eqs. (2.11) and (2.12). Figure 2.6 shows Γ_- as a function of the polarization potential (E versus NCE) of mercury in various potassium salt solutions [3]. E = −0.48 V versus NCE corresponds to the ECM for KF, which shows no specific adsorption. It can be seen that specific adsorption of anions other than F^- takes place at all polarization potentials except for large negative values.

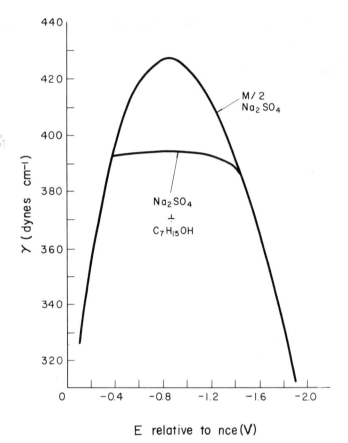

Figure 2.5 Effect of n-heptyl alcohol on an electrocapillary curve of mercury. (From Ref. 12.)

2.2.2 Differential Capacity

The electrical double layer is composed of two charged layers, a surface charge layer on the metal and an ionic layer in the solution which is equal in magnitude to and opposite in charge from the layer in the metal. The major difference between this double layer and an electric condenser is that the capacity of the electrical double layer depends on the electric potential across the double layer, whereas that of the electric condenser shows no potential dependency. In this respect, the differential capacity C, defined by Eq. (2.23), is of significance in the study of the electrical double layer.

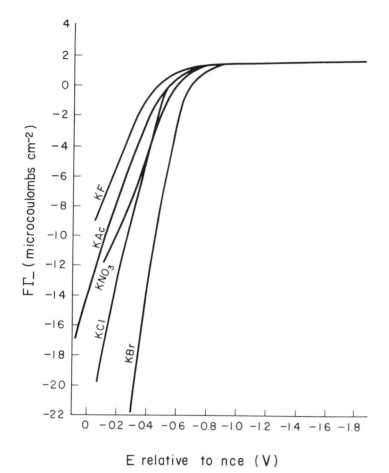

Figure 2.6 Γ_- as a function of the polarization potential of mercury in contact with aqueous solutions of various potassium salts. (From Ref. 3, by permission of American Institute of Physics.)

$$C = \left(\frac{\partial \sigma}{\partial E}\right)_\mu = -\left(\frac{\partial^2 \gamma}{\partial E^2}\right)_\mu \qquad (2.23)$$

This equation indicates that the differential capacity is obtained by double differentiation of the electrocapillary curve. Grahame [13] developed a method for measuring the differential capacity of a dropping mercury electrode using an impedance bridge (Fig. 2.7). This contributed greatly to elucidating the structure of the electrical double layer. Details of the measurement are described in Sec. 6.2.

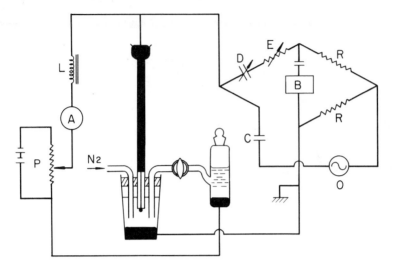

Figure 2.7 Schematic illustration of apparatus to measure the differential capacity on a dropping mercury electrode. P, Potentiometer; A, galvanometer; L, choke coil; D, capacitor decade; E, resistance box; B, amplifier and detector; R, resistor; C, capacitor; O, oscillator. (From Ref. 13. Copyright 1941 the American Chemical Society.)

The differential capacity is used to obtain the surface charge density and interfacial tension by Eqs. (2.24) and (2.25).

$$\sigma = \int_{E_{ECM}}^{E} C \, dE \tag{2.24}$$

$$\gamma = \gamma_{ECM} - \iint_{E_{ECM}}^{E} C \, dE^2 \tag{2.25}$$

The differential capacity is widely used to study the structure of double layers because it gives direct information on the events at the electrode-solution interface.

In some cases, electrocapillary curves are obtained by integration of differential capacity versus potential curves due to their accuracy. When this is done, it is necessary to know the electric potential of the ECM, E_{ECM}, of the system under investigation. Direct measurements of the double-layer capacity of AgI-aqueous electrolyte (KNO_3) solution interfaces using an alternating current were reported [14].

The integral capacity K is defined as

$$K = \frac{\sigma}{E_0} \tag{2.26}$$

where

$$E_0 = E - E_{ECM} \tag{2.27}$$

In general, since σ is not proportional to E_0, K is not constant and is related to C as follows:

$$C = K + E_0 \left(\frac{dK}{dE_0}\right)_\mu \tag{2.28}$$

At $E_0 = 0$, C becomes equal to K.

2.3 GOUY-CHAPMAN MODEL

From Sec. 2.2 it can be seen that double-layer properties such as surface charge density and surface excess can be obtained thermodynamically. It is necessary to set up a model of the electrical double layer in order to elucidate its structure from thermodynamic information.

The double-layer model first proposed was the Helmholtz molecular condenser model, which consisted of a surface charge layer and a counter charge layer. This parallel-plate condenser model predicts a constant capacity, which contradicted the experimental results.

Gouy [15] and Chapman [16] proposed the diffuse double-layer model, taking into account the thermal motion of counterions in the solution phase. Counterions are attracted to the surface charges by coulombic forces and at the same time they tend to move toward the bulk solution because of their thermal motion. Thus an equilibrium distribution of counterions is established (Fig. 2.8).

The electric potential (ψ) in the diffuse layer is analyzed in terms of two basic equations. One is the Poisson equation,

$$\Delta \psi = -\frac{4\pi\rho}{\varepsilon} \tag{2.29}$$

where ρ is the volume charge density at a point where the electric potential is ψ and ε is the dielectric constant of the medium. The other is the Boltzmann equation,

$$n_+ = n_{+0} \exp\left(\frac{-v_+ e\psi}{kT}\right)$$

$$n_- = n_{-0} \exp\left(\frac{v_- e\psi}{kT}\right) \tag{2.30}$$

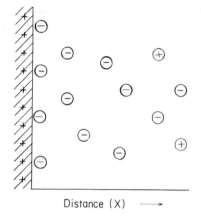

Figure 2.8 Gouy-Chapman model for the electrical double layer.

where $n_{+(-)}$ is the number concentration of ions in the diffuse layer, $n_{+(-)0}$ that of the bulk solution, $v_{+(-)}$ the valence of the ions, e the elementary charge, k the Boltzmann constant, T the absolute temperature, and subscripts + and − refer to the cation and anion, respectively. For a symmetrical electrolyte ($n_+ = n_- = n$, $v_+ = v_- = v$),

$$\rho = ve(n_+ - n_-)$$

$$= -2nve \sinh \frac{ve\psi}{kt} \tag{2.31}$$

From Eqs. (2.29)-(2.31), for a flat doulbe layer the Poisson-Boltzmann equation is obtained as

$$\frac{d^2\psi}{dx^2} = \frac{8\pi nve}{\varepsilon} \sinh \frac{ve\psi}{kT} \tag{2.32}$$

For a small value of ψ ($ve\psi/kT < 1$), Eq. (2.32) simplifies to Eq. (2.33) (Debye-Hückel approximation).

$$\frac{d^2\psi}{dx^2} = \kappa^2 \psi \tag{2.33}$$

where

$$\kappa^2 = \frac{8\pi ne^2 v^2}{\varepsilon kT} \tag{2.34}$$

κ is the Debye reciprocal length parameter. Integration of Eq. (2.32) yields [17]

$$\psi = 2\frac{kT}{ve} \ln \frac{1 + \gamma \exp(-\kappa x)}{1 - \gamma \exp(-\kappa x)} \tag{2.35}$$

with

$$\gamma = \frac{\exp(Z/2) - 1}{\exp(Z/2) + 1} \qquad Z = \frac{ve\psi_0}{kT} \tag{2.36}$$

where ψ_0 is referred to as the surface potential and x is the distance from the surface.

In the case of $\kappa x \gg 1$, ψ is approximated as follows [17]:

$$\psi = 2\frac{kT}{ve} \ln[1 + 2\gamma \exp(-\kappa x)]$$

$$\cong \frac{4kT}{ve} \gamma \exp(-\kappa x) \tag{2.37}$$

The concentration distribution of cations and anions across the diffuse layer and the corresponding electric potential are illustrated schematically in Fig. 2.9 for the case where the surface is positively charged. Counterions (n_-) accumulate around the surface, while ions of the same sign as the surface charge (co-ions, n_+) are excluded from the surface as a result of the electrostatic interaction. The depletion of co-ions and the excess of counterions result in the screening of the surface charge. However, there is a substantial difference in the screening effect between counterions and co-ions; there is no upper limit to the concentration of counterions if the ionic size is neglected, while the lowest concentration of co-ions is limited to zero. Therefore, in the case of large surface potential, the screening effect is brought about predominantly by counterions, and co-ions play no significant role. Thus, for asymmetrical electrolytes, v and n in the foregoing formulas refer to those of the counterions.

The surface charge per unit area, σ, is equal to the negative charge per unit area of diffuse layer, σ_d, and is given by

$$\sigma = -\sigma_d = -\int_0^\infty \rho\, dx = \frac{\varepsilon}{4\pi} \int_0^\infty \frac{d^2\psi}{dx^2} dx$$

$$= -\frac{\varepsilon}{4\pi} \left(\frac{d\psi}{dx}\right)_{x=0} \tag{2.38}$$

Differentiating Eq. (2.32) once setting $\psi = 0$ and $d\psi/dx = 0$ at $x = \infty$ yields

$$\left(\frac{d\psi}{dx}\right)_{x=0} = -\left(\frac{32\pi nkT}{\varepsilon}\right)^{1/2} \sinh\frac{ve\psi_0}{2kT} \tag{2.39}$$

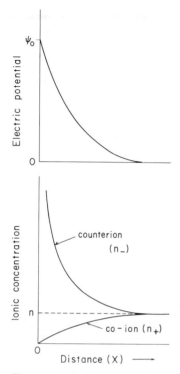

Figure 2.9 Concentration of counterions (n_-) and co-ions (n_+) and corresponding electric potential as a function of the distance from the surface.

The initial negative sign in Eq. (2.39) arises from the fact that ψ decreases as x increases. Thus σ is given by

$$\sigma = \left(\frac{2n\varepsilon kT}{\pi}\right)^{1/2} \sinh \frac{ve\psi_0}{2kT} \qquad (2.40)$$

The operational equation for an aqueous solution at 25°C is

$$\sigma = 11.72(c)^{1/2} \sinh(19.46v\psi_0) \qquad (2.41)$$

where the units of σ, c, and ψ_0 are $\mu C/cm^2$, mol/liter, and volts, respectively.

With the Debye-Hückel approximation, Eq. (2.33) yields

$$\psi = \psi_0 \exp(-\kappa x) \qquad (2.42)$$

The distance $1/\kappa$ at which $\psi = \psi_0/e$ (e = 2.718 ...) is called the thickness of the diffuse layer. Equation (2.43) is obtained from Eqs. (2.38) and (2.42):

$$\sigma = \frac{\varepsilon\kappa}{4\pi}\psi_0 \tag{2.43}$$

This equation indicates that σ is proportional to ψ_0 for a constant value of κ. $\varepsilon\kappa/4\pi$ is the capacity of a parallel-plate condenser with a plate distance of $1/\kappa$, indicating that the thickness of the diffuse layer may be regarded as the distance at which diffuse layer charges accumulate. For aqueous solutions at 25°C, $1/\kappa$ is approximated as

$$\frac{1}{\kappa} = \frac{3 \times 10^{-8}}{v(c)^{1/2}} \text{ cm} \tag{2.44}$$

where c is given in mol/liter. For a 1-1 electrolyte, $1/\kappa = 100$ Å for $c = 1 \times 10^{-3}$ mol/liter and 10 Å for 1×10^{-1} mol/liter.

In the case of spherical particles of radius a, κa is a measure of sphericity. For $\kappa a > 10$, the flat double-layer model will serve as a good approximation, while for $\kappa a < 0.1$ the situation resembles that of point charge system where the Debye-Hückel theory is applicable. For a spherical double layer the Poisson-Boltzmann equation takes the form

$$\frac{d^2\psi}{dr^2} + \frac{2}{r}\frac{d\psi}{dr} = \frac{8\pi nve}{\varepsilon} \sinh \frac{ve\psi}{kT} \tag{2.45}$$

An explicit solution for Eq. (2.45) is not available, but the Debye-Hückel approximation leads to

$$\psi = \psi_0 \frac{a}{r} \exp[\kappa(a-r)] \tag{2.46}$$

and

$$Q = a\varepsilon(1 + \kappa a)\psi_0 \tag{2.47}$$

where Q is the total surface charge of a particle and r is the radial distance.

From Eq. (2.41), the differential capacity (C) of a flat double layer is obtained as

$$C = \frac{d\sigma}{d\psi_0} = 228.5v(c)^{1/2} \cosh(19.46v\psi_0) \tag{2.48}$$

where the units of C, ψ_0, and c are $\mu F/cm^2$, volts, and mol/liter, respectively. Equation (2.48) gives $C > 200$ $\mu F/cm^2$, for even small ψ_0 at c = 1 mmol/liter and v = 1, which is very large compared with the experimental value. Furthermore, the surface concentration calculated from Eq. (2.30) yields values as large as 160 mol/liter for 1-1 electrolyte concentration of 1×10^{-3} mol/liter at $\psi_0 = 300$ mV. These anomalous results arise from the treatment of the ions as point charges. Stern solved this problem by taking into account the size of ions in contact with the surface.

2.4 STERN AND GRAHAME MODELS

Stern [18] considered that ions have finite sizes and there is a plane of closest approach, the Stern plane, within which no ions can exist. The Stern plane is located at a distance δ from the surface, and is assumed to be of the order of several angstrom units in magnitude. The solution part is divided by the Stern plane into two parts: the Stern layer and the diffuse layer. The distribution of ions and the corresponding drop of electric potential in a flat double layer is illustrated schematically in Fig. 2.10. In the Stern layer [i.e., from

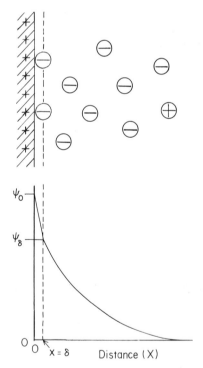

Figure 2.10 Schematic representation of the distribution of ions and corresponding electric potential across the double layer based on the Stern model.

from x = 0 ($\psi = \psi_0$) to x = δ ($\psi = \psi_\delta$)] the electric potential changes linearly because there is no free charge in this layer. The distribution of ions in the diffuse layer is given by the Gouy-Chapman theory. Let n_1 be the number of ions adsorbed at the Stern plane per unit area, n the number concentration of ions in the bulk per unit volume, Z_1 the maximum number of adsorption sites per unit area, Z the maximum number of ions in the bulk solution per unit volume, and $\Delta \overline{G}$ the work necessary to move one mole of ions from the bulk to the surface. Then from the Langmuir theory it follows that

$$\frac{n_1}{n} = \frac{Z_1 - n_1}{Z - n} \exp\left(-\frac{\Delta \overline{G}}{RT}\right) \qquad (2.49)$$

Usually, n << Z holds; then

$$\theta = \frac{n_1}{Z_1} = \frac{1}{1 + (Z/n) \exp(\Delta \overline{G}/RT)} \qquad (2.50)$$

where θ is the surface coverage. Stern identified n/Z as the mole fraction of the ions. In actuality, θ << 1 and

$$n_1 = \frac{Z_1}{Z} n \exp\left(-\frac{\Delta \overline{G}}{RT}\right)$$

$$= 2bn \exp\left(-\frac{\Delta \overline{G}}{RT}\right) \qquad (2.51)$$

where b is the radius of the adsorbed ion. In the derivation above, the ratio of the thickness of the adsorption layer to a cube of the bulk liquid with a volume of 1 cm³ is 2b:1; thus Z_1/Z equals 2b. Equation (2.51) is identical to Grahame's equation [12]. Since $\Delta \overline{G}$ consists of both electrical and chemical work, it is written

$$\Delta \overline{G} = vF(\psi_\delta - \phi) \qquad (2.52)$$

where ψ_δ is the electric potential at the Stern plane and ϕ is the specific chemical adsorption potential of the counterion. The specific adsorption density σ_1 (μC/cm²) in an aqueous solution at 25°C becomes

$$\sigma_1 = 1.93vbc \exp[-38.9v(\psi_\delta - \phi)] \qquad (2.53)$$

where b is given in angstrom units, c in mol/liter, and ψ_δ and ϕ in volts. The principle of electroneutrality requires that

$$\sigma = \sigma_1 + \sigma_d \qquad (2.54)$$

and the surface charge density σ is given by

$$\sigma = K_s(\psi_0 - \psi_\delta) \qquad (2.55)$$

where K_s is the Stern layer capacity. The diffuse layer charge σ_d is obtained from Eq. (2.40), in which ψ_0 is replaced by ψ_δ. Stern succeeded in overcoming the shortcomings of the Gouy-Chapman theory, but the coincidence of the Stern plane at which the ions are specifically adsorbed with the plane from which the diffuse layer initiates was still responsible for an ambiguity.

Grahame [12] further divided the Stern layer into two parts. The specifically adsorbed ions are not hydrated, so that they may approach the surface closer than the Stern plane. The locus of the specifically adsorbed ions is referred to as the inner Helmholtz plane, while the plane at which the diffuse layer initiates was called the outer Helmholtz plane. The distribution of ions and the corresponding electric potential in the case where anions are specifically adsorbed on a negatively charged surface is illustrated schematically in Fig. 2.11. The electric potential ψ_2 at the outer Helmholtz plane corresponds to ψ_δ in

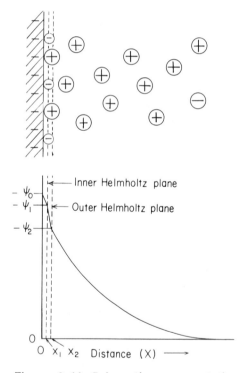

Figure 2.11 Schematic representation of the distribution of ions and corresponding electric potential across the double layer based on the Stern-Grahame model. Anions being specifically adsorbed on negatively charged surface.

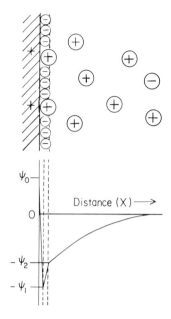

Figure 2.12 Schematic representation of the distribution of ions and corresponding electric potential across the double layer based on the Stern-Grahame model. Anions being specifically adsorbed in excess on positively charged surface.

Fig. 2.10. In the Helmholtz layer (from $x = 0$ to $x = x_1$ and from $x = x_1$ to $x = x_2$), the electric potential changes linearly with respect to distance. In the case where strong specific adsorption of anions takes place on a positively charged surface, the reversal of electric potential occurs as illustrated in Fig. 2.12. The detailed structure of the Helmholtz layer and its relation to the electrode reaction were discussed by Parsons [19].

Specific adsorption is one of the most important topics in the study of electrical double layers. In general, anions exhibit stronger specific adsorption than cations. Grahame [12] reported that at mercury-aqueous solution interfaces, the specific adsorption of halide anions increases with increasing ionic radii (i.e., $F^- < Cl^- < Br^- < I^-$), which also corresponds to the order of solubility of mercury halides. Thus the nature of the specific adsorption was correlated with the covalent forces between the halide ions and mercury. Anderson and Bockris [20] analyzed the free-energy change in specific adsorption from the viewpoint of the energy of interaction between metal-ion, metal-water molecule, and ion-water molecule. They interpreted the order of specific adsorption of halide ions in terms of the hydration energy without invoking covalent forces. The specific adsorption forces are complex and involve coulombic and noncoulombic forces. Devanathan and Tilak [21] stated that dehydration of the ions is the necessary condition and covalent forces is the sufficient condition for specific adsorption.

It has been stated that strong specific adsorption is accompanied by sign reversal in the electric potential ψ_2. Figure 2.13 shows ψ_2 as a function of E versus NCE for mercury in contact with 0.1 N concentrations of various potassium salts (see Fig. 2.6). It is to be noted that ψ_2 increases in negative value with increasing E except with KF. Divalent Ba^{2+} suppresses ψ_2 markedly in the negative potential region compared with K^+.

If the molecular adsorption of an ionic surfactant follows the Langmuir isotherm, ψ_δ can be calculated using Eq. (2.50). Watanabe,

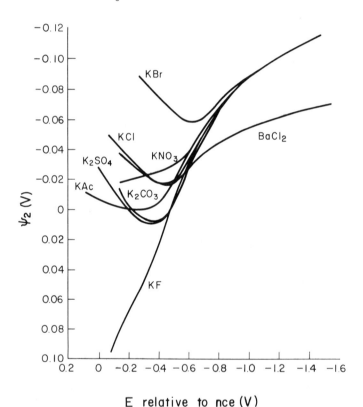

Figure 2.13 Relation between the potential at the outer Helmholtz plane and the polarization potential of mercury in aqueous solutions of various electrolytes (0.1 N) at 25°C. (From Ref. 3, by permission of American Institute of Physics.)

Tsuji, and Ueda [22] derived Eq. (2.56), by which ψ_δ for a solid surface in contact with an aqueous (monovalent) surfactant solution (c, mol/liter) in the presence of a 1-1 type of supporting electrolyte (c_e, mol/liter) can be calculated.

$$\psi_\delta = \frac{2kT}{e} \sinh^{-1}\left[\left(\frac{c_e}{c + c_e}\right)^{1/2} \sinh \frac{e\psi_\delta^*}{2kT} \pm \frac{eKcZ_1}{2Y(1 + Kc)}\right] \quad (2.56)$$

with

$$Y = \left[\frac{\varepsilon RT(c + c_e)}{2000\pi}\right]^{1/2} \quad (2.57)$$

where Z_1 is the maximum number of adsorption site per unit area, R the gas constant, ψ_δ^* refers to c = 0, ± in the brackets refers to cationic or anionic surfactant, and K is the equilibrium constant of the adsorption and is related to $\Delta \bar{G}$ for small values of c as follows:

$$K = \frac{\exp(-\Delta \bar{G}/RT)}{55.6} \quad (2.58)$$

For $e\psi_\delta/kT \ll 1$ (Deybe-Hückel approximation), sinh x = x, and Eq. (2.56) simplifies to

$$\psi_\delta = \left(\frac{c_e}{c + c_e}\right)^{1/2} \psi_\delta^* \pm \frac{KK'c}{1 + Kc} \quad (2.59)$$

with

$$K' = \left[\frac{2000\pi kT}{\varepsilon(c + c_e)N}\right]^{1/2} Z_1 \quad (2.60)$$

For a spherical particle of radius a, the Debye-Hückel approximation leads to

$$\psi_\delta = \left(\frac{c_e}{c + c_e}\right)^{1/2} \psi_\delta^* \pm \frac{KK''c}{1 + Kc} \quad (2.61)$$

with

$$K'' = \frac{4\pi aeZ_1}{\varepsilon(1 + \kappa a)} \quad (2.62)$$

Setting $a = \infty$ in Eq. (2.62), Eq. (2.60) is obtained. For sufficiently high concentrations of supporting electrolytes ($c_e \gg c$), Eq. (2.61) for anionic surfactants becomes

$$\psi_\delta = \psi_\delta^* - \frac{KK''c}{1 + Kc} \quad (2.63)$$

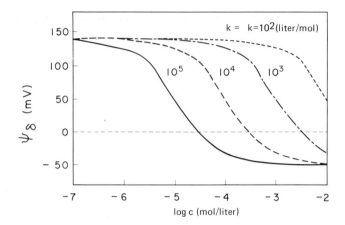

Figure 2.14 Theoretical ψ_δ versus log c curves for adsorption of anionic surfactant. $\psi_\delta^* = 140$ mV, $Z_1 = 1.94 \times 10^{12}$ cm^{-2}, $\kappa a = 1.23$, and a = 100 Å. (From Ref. 22.)

ψ_δ versus log c curves for anionic adsorption calculated using Eq. (2.63) for various values of K are shown in Fig. 2.14, in which $\psi_\delta^* = 140$ mV, $Z_1 = 1.94 \times 10^{12}$ cm^{-2}, $\kappa a = 1.23$, and a = 100 Å are assumed. The results agree with those from ζ potential versus log c curves for a system of silver iodide-sodium alkyl sulfate under corresponding conditions. Equation (2.64) is obtained from Eq. (2.63):

$$\left(\frac{d\psi_\delta}{d \log c}\right)_{c=c_0} = -2.303 \psi_\delta^* \left(1 - \frac{\psi_\delta^*}{K''}\right) \tag{2.64}$$

$$\frac{1}{c_0} = -K\left(1 - \frac{K''}{\psi_\delta^*}\right) \tag{2.65}$$

where c_0 is the concentration at which ψ_δ becomes zero. The point at which electrophoretic mobility becomes zero, $\zeta = 0$, is called the isoelectric point (IEP). Thus K and K'' (K') can be obtained from the slope of ψ_δ versus log c curve at IEP and likewise $\Delta \bar{G}$ and Z_1.

Adsorption of organic compounds at an electrified interface are reviewed by Frumkin and Damaskin [23].

2.5 OTHER MODELS

Devanathan [21,24] analyzed the double-layer structure on the basis of the Stern-Grahame model. Since there is no free charge between the metal surface and the inner Helmholtz plane,

$$\psi_0 - \psi_1 = \frac{\sigma}{K_{m-1}} \tag{2.66}$$

with

$$K_{m-1} = \frac{\varepsilon}{4\pi x_1} \tag{2.67}$$

where K_{m-1} and x_1 are the capacity and the thickness of the inner Helmholtz layer, respectively, and σ is the surface charge. Similarly,

$$\psi_1 - \psi_2 = \frac{\sigma_d}{K_{1-2}} \tag{2.68}$$

with

$$K_{1-2} = \frac{\varepsilon}{4\pi(x_2 - x_1)} \tag{2.69}$$

where ε is the dielectric constant of the Helmholtz layer and x_2 is the thickness of the outer Helmholtz layer. Differentiation of Eq. (2.66) with respect to σ leads to

$$\frac{d\psi_0}{d\sigma} = \frac{1}{C} = \frac{1}{K_{m-1}} + \frac{d\psi_1}{d\sigma} \tag{2.70}$$

After substitution of $d\psi_1/d\sigma$, which is derived from Eq. (2.68) and Eq. (2.54), into Eq. (2.70), Eq. (2.71) is obtained:

$$\frac{d\psi_0}{dq} = \frac{1}{C} = \frac{1}{K_{m-1}} + \left(\frac{1}{K_{1-2}} + \frac{1}{C_d}\right)\left(1 - \frac{d\sigma_1}{d\sigma}\right) \tag{2.71}$$

C is the total differential capacity and is determined experimentally, and C_d ($=d\sigma_d/d\psi_2$) is the differential capacity of the diffuse layer. In the case of no specific adsorption (i.e., $\sigma_1 = 0$ and $d\sigma_1/d\sigma = 0$), it follows that

$$\frac{1}{C} = \frac{1}{K_{m-1}} + \frac{1}{K_{1-2}} + \frac{1}{C_d} \tag{2.72}$$

There is no inner Helmholtz plane because of the absence of specific adsorption, so

$$\frac{1}{K_{m-1}} + \frac{1}{K_{1-2}} = \frac{1}{C_H} \tag{2.73}$$

Then Eq. (2.74) is obtained.

$$\frac{1}{C} = \frac{1}{C_H} + \frac{1}{C_d} \tag{2.74}$$

C_H ($=\varepsilon/4\pi x_2$) is the differential capacity of the Helmholtz layer, which depends only on x_2 and ε, and has been termed the solvent capacity (K_s). In other words, in the absence of specific adsorption, the electrical double layer is regarded as an electric condenser comprising the Helmholtz layer and the diffuse layer connected in series. The capacity C_d is calculated from Eq. (2.48) by replacing ψ_0 with ψ_2:

$$C_d = 228.5v(c)^{1/2} \cosh(19.46v\psi_2) \quad (2.75)$$

C_d is also expressed as a function of σ_d:

$$C_d = 228.5v(c)^{1/2}\left(1 + \frac{\sigma_d^2}{137c}\right) \quad (2.76)$$

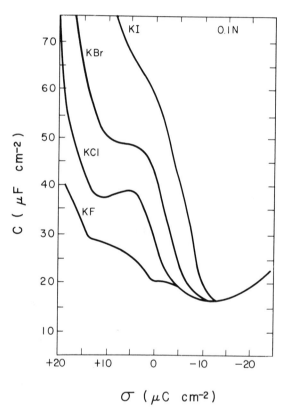

Figure 2.15 Differential capacity of mercury as a function of surface charge density in 0.1 N potassium halide aqueous solutions at 25°C. (From Ref. 24.)

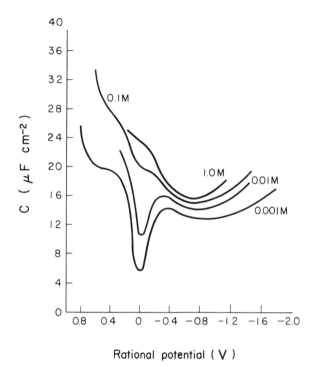

Figure 2.16 Differential capacity versus potential curve of mercury in contact with NaF aqueous solutions. (From Ref. 12.)

In Fig. 2.15 the differential capacities of mercury in contact with potassium halide aqueous solutions (0.1 N) are shown as a function of σ. It can be seen that $C = 16.2$ $\mu F/cm^2$ at $\sigma = -13$ $\mu c/cm^2$, where no specific adsorption takes place. $C_d = 270$ $\mu F/cm^2$ is calculated from Eq. (2.76), and $C_H = 17.2$ $\mu F/cm^2$ is calculated using Eq. (2.74). Using the value of $x_2 = 3.7$ Å obtained by assuming the hexagonal close packing of water molecules on mercury surfaces, Devanathan calculated the magnitude of ε of the Halmholtz layer as 7.2 from the relationship $C_H = \varepsilon/4\pi x_2$. This value of ε is rather small when compared to $\varepsilon = 80$ of bulk water, indicating that the water dipoles attain saturation orientation on the metal surfaces. Equation (2.75) indicates that C_d increases with increasing electrolyte concentration. Thus from Eq. (2.74), the total differential capacity C is dominated by C_H at large electrolyte concentrations. On the other hand, C_d dominates C at low electrolyte concentrations. It can be seen from Eq. (2.75) or Eq. (2.76) that C_d is a minimum at $\psi_2 = 0$ ($\sigma_d = 0$). In Fig. 2.16, C exhibits a distinct minimum at ECM (-0.48 V versus NCE) in NaF solutions of low concentration. This provides one method for determining the ECM. In the presence of specific adsorption, the variation of the total differential capacity depends on $d\sigma_1/d\sigma$. The specific adsorption of thiourea on mercury-aqueous solution interfaces was reasonably explained by the Devanathan model [25].

The differential capacity is sensitive to properties such as the dielectric constant of the adsorbing layer and the response of surface charges to changes in the surface potential, which yields direct information about the behavior of adsorbing molecules at the interfaces. In Fig. 2.17, differential capacity versus potential curves are shown for the system identical to that of Fig. 2.5. Alcohol molecules are excluded from the aqueous phase to adsorb at the mercury surface as a result of the hydrophobic nature of the hydrocarbon chain, and the differential capacity decreases markedly in the midpotential region, owing to the low dielectric constant of the molecule. Two peaks appearing at each side of the potential reflect the adsorption of the alcohol

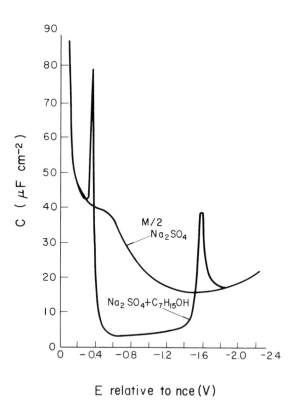

Figure 2.17 Effect of n-heptyl alcohol on differential capacity of mercury in contact with Na_2SO_4 aqueous solution. (From Ref. 12.)

molecules. The desorption is accompanied by the displacement of alcohol molecules by SO_4^{2-} ions (left-hand side) or Na^+ ions (right-hand side). This displacement occurs over a narrow range of potential values, and is revealed as the sharp peaks in the differential capacity curve. The explanation of the fact that the peaks do not occur at exactly the potentials of the breaks in the electrocapillary curve (Fig. 2.5) was attributed to the difference in the temperature between the experiments [12]. Extensive studies on the differential capacity of mercury in contact with aqueous solutions of ionic surfactants were carried out by Eda, Tamamushi, and Takahashi [26].

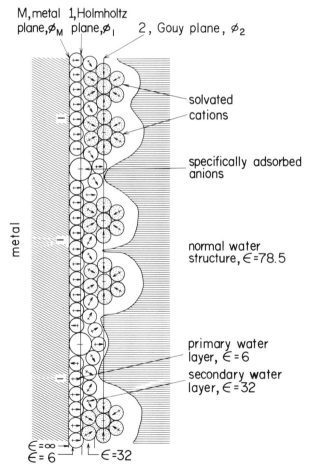

Figure 2.18 Bockris model of the electrical double layer. (From Ref. 27, by permission of the Royal Society.)

42 / Usui

Bockris, Devanathan, and Müller [27] proposed a more detailed model of electrode-solution interfaces as shown in Fig. 2.18, where the outer Helmholtz plane is termed a Gouy plane. In this model, the hydrated counterions are separated from the electrode surface by a monolayer of oriented water molecules.

2.6 OXIDE-AQUEOUS SOLUTION INTERFACES

A metal oxide-aqueous solutions interface is one of the most important systems for the study of the electrical double layer from both a theoretical and a practical aspect. The structure of the electrical double layer at oxide-aqueous solution interfaces can be inferred by the analogy of a silver iodide-aqueous solution interface with a metal-aqueous solution interface.

For AgI, the lattice constituent ions, Ag^+ and I^-, are the potential-determining ions. For oxides, H^+ and OH^- are the potential-determining ions and the surface potential ψ_0 is given by the Nernst-like equation

$$\psi_0 = \frac{RT}{F} \ln \frac{a_{H^+}}{a^\circ_{H^+}} \tag{2.77}$$

where a_{H^+} is the activity of H^+ and $a^\circ_{H^+}$ refers to the PZC. The operational formula for aqueous solutions at 25°C is

$$\psi_0 = 0.059(pH^\circ - pH) \quad \text{volts} \tag{2.78}$$

where pH° refers to the PZC.

The PZC is one of the most important characteristics of oxides because it is a measure of the interaction between oxide surfaces and the aqueous media. Detailed studies on the electrical double layer of oxides have been carried out by many investigators [28-44].

An objection [31] to the application of Eq. (2.77) to metal oxide-aqueous solution interfaces was presented. The major point is that the potential-determining ions (H^+ and OH^-) are not lattice constituents for oxides. Levin and Smith [39] interpreted the deviation of ψ_0 of oxides from Eq. (2.77) in terms of the discreteness charge effect. Healy and White [43] proposed the surface regulation mechanism by which the surface charge or surface potential of oxides is determined. Oxide surfaces are charged positively or negatively in accordance with the following reactions:

$$AH_2^+ \rightleftharpoons AH + H^+ \tag{2.79}$$

$$AH \rightleftharpoons A^- + H^+ \tag{2.80}$$

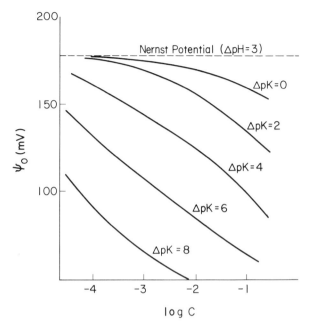

Figure 2.19 ψ_0 as a function of log c for various ΔpK (=$pK_- - pK_+$) at Δ pH (= pH $-$ pH°) = 3. (From Ref. 43.)

The equilibrium constants (K_+ and K_-) for the reactions above are

$$K_+ = \frac{[AH][H^+]_s}{[AH_2^+]} \tag{2.81}$$

$$K_- = \frac{[A^-][H^+]_s}{[AH]} \tag{2.82}$$

where [AH], [A$^-$], and [AH$_2^+$] are surface concentrations (activities, more exactly) of neutral, anion, and cation sites, respectively, and $[H^+]_s$ is the surface concentration of H$^+$ and is given by

$$[H^+]_s = H \exp\left(-\frac{e\psi_0}{kT}\right) \tag{2.83}$$

where H is the bulk concentration of H$^+$. The surface charge density σ is then given by

$$\sigma = eN_s \frac{[AH_2^+] - [A^-]}{[AH] + [AH_2^+] + [A^-]} \tag{2.84}$$

where N_s is the total number of surface sites available and e is the elementary charge. Equation (2.85) is obtained from Eqs. (2.81)-(2.84):

$$\sigma = eN_s \frac{(H/K_+) \exp(-e\psi_0/kT) - (K_-/H) \exp(e\psi_0/kT)}{1 + (H/K_+) \exp(-e\psi_0/kT) + (K_-/H) \exp(e\psi_0/kT)} \tag{2.85}$$

On the other hand, σ is also given by Eq. (2.40) on the basis of the Gouy-Chapman model. Equating Eq. (2.85) to Eq. (2.40), ψ_0 is obtained as a function of K_+, K_-, c, and H. As an example, ψ_0 as a function of log c at ΔpH (=pH - pH°) = 3 and at various ΔpK (=pK_- - pK_+) values is shown in Fig. 2.19. If ψ_0 obeys the Nernst equation, it should be 177 mV (dashed line) irrespective of the electrolyte concentration. It can be seen from Fig. 2.19 that ψ_0 obeys the Nernst equation only for small values of ΔpK and for low electrolyte concentrations.

REFERENCES

1. E. A. Guggenheim, *Thermodynamics*, North-Holland, Amsterdam, 1959, p. 263.
2. P. Delahay, *Double Layer and Electrode Kinetics*, Wiley-Interscience, New York, 1965.
3. D. C. Grahame and B. A. Soderberg, *J. Chem. Phys.*, 22, 449 (1954).
4. M. A. V. Devanathan, *Proc. R. Soc. Lond. A*, 264, 133 (1961).
5. E. Dutkiewiz and R. Parsons, *J. Electroanal. Chem.*, 11, 100 (1966).
6. D. M. Mohilner, *J. Phys. Chem.*, 66, 724 (1962).
7. J. Th. G. Overbeek, in *Colloid Science* (H. R. Kruyt, ed.), Vol. 1, Elsevier, Amsterdam, 1952, p. 115.
8. B. H. Bijsterbosh and J. Lyklema, *Adv. Colloid Interface Sci.*, 9, 147 (1978).
9. W. L. Freyberger and P. L. de Bruyn, *J. Phys. Chem.*, 61, 586 (1957).
10. I. Iwasaki and P. L. de Bruyn, *J. Phys. Chem.*, 62, 594 (1958).
11. G. A. Parks and P. L. de Bruyn, *J. Phys. Chem.*, 66, 967 (1962).
12. D. C. Grahame, *Chem. Rev.*, 41, 441 (1947).

13. D. C. Grahame, *J. Am. Chem. Soc.*, *63*, 1207 (1941).
14. J. H. A. Pieper and D. A. de Vooys, *Electroanal. Chem. Interfacial Electrochem.*, *53*, 243 (1974).
15. L. Gouy, *J. Phys.* (4), *9*: 457 (1910); *Ann. Phys.* (9), *7*: 129 (1917).
16. P. L. Chapman, *Philos. Mag.* (6), *25*: 475 (1913).
17. E. J. W. Verwey and J. Th. G. Overbeek, *Theory of the Stability of Lyophobic Colloids*, Elsevier, Amsterdam, 1948.
18. O. Stern, *Z. Elektrochem.*, *30*, 508 (1924).
19. R. Parsons, in *Advances in Electrochemistry and Electrochemical Engineering*, Vol. 1 (P. Delahay and C. Tobias, eds.), Interscience, New York, 1961, p. 1.
20. T. N. Andersen and J. O'M. Bockris, *Electrochim. Acta*, *9*, 347 (1964).
21. M. A. V. Devanathan and B. V. K. S. R. A. Tilak, *Chem. Rev.*, *65*, 635 (1965).
22. A. Watanabe, F. Tsuji, and S. Ueda, *Denki Kagaku*, *29*, 777 (1961).
23. A. N. Frumkin and B. B. Damaskin, in *Modern Aspects of Electrochemistry* (J. O'M. Bockris, ed.), Vol. 3, Academic Press, New York, 1954, p. 149.
24. M. A. V. Devanathan, *Trans. Faraday Soc.*, *50*, 373 (1954).
25. M. A. V. Devanathan, *Proc. R. Soc. Lond. A*, *264*, 133 (1961).
26. K. Eda, *Nippon Kagaku Zasshi*, *80*, 343, 347, 349, 463, 465, 708 (1959); *81*, 689, 875, 879 (1960); *85*, 828 (1964); K. Eda and B. Tamamushi, *Proc. 3rd Int. Congr. Surf. Activ.*, *B/II*, *1*, 291 (1960); K. Eda, K. Takahashi, and B. Tamamushi, *Proc. 4th Int. Congr. Surf. Activ.*, *B/II*, 19 (1964).
27. J. O'M. Bockris, M. A. V. Devanathan, and K. Müller, *Proc. R. Soc. Lond. A*, *274*, 55 (1963).
28. G. H. Bolt, *J. Phys. Chem.*, *61*, 1166 (1957).
29. G. A. Parks and P. L. de Bruyn, *J. Phys. Chem.*, *66*, 967 (1962).
30. G. A. Parks, *Chem. Rev.*, *65*, 177 (1965).
31. Y. G. Berube and P. L. de Bruyn, *J. Colloid Interface Sci.*, *27*, 305 (1968).
32. T. W. Healy and D. W. Fuerstenau, *J. Colloid Sci.*, *20*, 376 (1965).
33. G. Y. Onoda and P. L. de Bruyn, *Surf. Sci.*, *4*, 48 (1966).
34. L. Block and P. L. de Bruyn, *J. Colloid Interface Sci.*, *32*, 518, 527, 533 (1970).
35. S. M. Ahmed and D. Maksimov, *J. Colloid Interface Sci.*, *29*, 97 (1969).
36. Th. F. Tadros and J. Lyklema, *J. Electroanal. Chem.*, *17*, 267 (1968); *22*, 1 (1969).

37. R. J. Atkinson, A. M. Posner, and J. P. Quirk, *J. Phys. Chem.*, *71*, 550 (1967).
38. J. Yopps and D. W. Fuerstenau, *J. Colloid Sci.*, *19*, 61 (1964).
39. S. Levin and L. Smith, *Discuss. Faraday Soc.*, *52*, 290 (1971).
40. R. P. Abendroth, *J. Colloid Interface Sci.*, *34*, 591 (1970).
41. J. Lyklema, *Croat. Chem. Acta*, *43*, 249 (1971).
42. A. Breeuwsma and J. Lyklema, *Discuss. Faraday Soc.*, *52*, 324 (1971).
43. T. W. Healy and L. R. White, *Adv. Colloid Interface Sci.*, *9*, 303 (1978).
44. J. Westall and H. Hohl, *Adv. Colloid Interface Sci.*, *12*, 265 (1980).

3
Interaction of Electrical Double Layers and Colloid Stability

Shinnosuke Usui Research Institute of Mineral Dressing and Metallurgy, Tohoku University, Sendai, Japan

Sei Hachisu* The University of Tsukuba, Ibaraki, Japan

3.1. Interaction of Double Layers 48
    3.1.1. Potential energy of interaction between flat double layers via the force 48
    3.1.2. Potential energy of interaction between flat double layers via the free energy 50
    3.1.3. Spherical double layers 53
3.2. London-van der Waals Attraction 55
3.3. DLVO Theory 57
3.4. Colloid Stability—Kinetics of Coagulation 61
3.5. Heterocoagulation 65
    3.5.1. Dissimilar double-layer interaction 66
    3.5.2. Constant surface charge model 70
    3.5.3. Regulation model 74
    3.5.4. London-van der Waals interaction between dissimilar substances 77
    3.5.5. Examples of heterocoagulation 78
3.6. Thin Liquid Films 80
3.7. Concentrated Dispersion 83
    3.7.1. Introduction 83
    3.7.2. Ordered structure in monodisperse latices 84
    3.7.3. The mechanism of the ordered formation 87
    3.7.4. Alder transition 89
    3.7.5 Exposition of Alder transition 91
    References 94

*Currently Professor Emeritus.

The Derjaguin-Landau-Verwey-Overbeek (DLVO) theory [1,2] has made it possible to discuss the stability of lyophobic colloids quantitatively. This theory is based on the interaction due to overlapping of electrical double layers and London-van der Waals forces between colloid particles.

3.1 INTERACTION OF DOUBLE LAYERS

3.1.1 Potential Energy of Interaction Between Flat Double Layers via the Force

Consider the force of interaction P between two flat double layers separated by a distance 2h (Fig. 3.1). The dashed line in Fig. 3.1 represents the electric potential that a single double layer would have in the absence of interaction. For simplicity, the double-layer structure is assumed to be the Gouy-Chapman model. The force P consists of two components, P_E and P_0; thus

$$P_E = -\frac{\varepsilon}{8\pi}\left(\frac{d\psi}{dx}\right)^2 \qquad (3.1)$$

$$P_0 = (n_+ + n_-)kT - 2nkT \qquad (3.2)$$

where ε is the dielectric constant, ψ the electric potential, x the distance from the surface, $n_{+(-)}$ the number concentration of cations

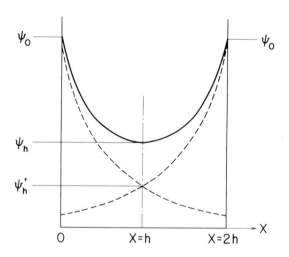

Figure 3.1 Schematic representation of the electric potential between flat double layers.

(anions), n that of the bulk solution, k the Boltzmann constant, and T the absolute temperature. P_E is the electrostatic force and is always an attraction. P_0 is the force resulting from the osmotic pressure due to the accumulation of ions in the diffuse layers and is always a repulsion in the case of two identical double layers. The Poisson-Boltzmann equation, Eq. (2.32), may be integrated to yield Eq. (3.3), from which P_E is calculated.

$$\left(\frac{d\psi}{dx}\right)^2 = \frac{8\pi nkT}{\varepsilon}\left(2\cosh\frac{ve\psi}{kT} + C\right) \tag{3.3}$$

where e is the elementary charge and v is the valence of the counterions. C is an integration constant and is determined from the boundary condition that $d\psi/dx = 0$ at $x = h$; thus

$$C = -2\cosh\frac{ve\psi_h}{kT} \tag{3.4}$$

where ψ_h is the electric potential at $x = h$. By combining Eqs. (3.3), (3.4), and (3.1), P_E is obtained.

$$P_E = -nkT\left(2\cosh\frac{ve\psi}{kT} - 2\cosh\frac{ve\psi_h}{kT}\right) \tag{3.5}$$

Equation (3.6) is derived from Eqs. (2.30) and (3.2) ($n_{+0} = n_{-0} = n$):

$$P_0 = 2nkT\left(\cosh\frac{ve\psi}{kT} - 1\right) \tag{3.6}$$

P_E and P_0 vary with ψ, but $P_E + P_0$ is constant at any point between flat plates, which is a requirement for the system to be in equilibrium. Therefore, the calculation of P results in the calculation of P_0 at $x = h$, where $d\psi/dx = 0$ and hence $P_E = 0$. Thus

$$P = P_{0(x=h)} = 2nkT\left(\cosh\frac{ve\psi_h}{kT} - 1\right) \tag{3.7}$$

For weak interactions, ψ_h is considered to be relatively small and $ve\psi_h/kT < 1$. In that situation, Eq. (3.7) simplifies to

$$P = nkT\left(\frac{ve\psi_h}{kT}\right)^2 \tag{3.8}$$

For weak interactions, it can be assumed that

$$\psi_h = 2\psi'_h \tag{3.9}$$

where ψ'_h is the value of ψ at $x = h$ for a single double layer in the absence of interaction. By using Eq. (2.37) for ψ'_h, Eq. (3.10) is obtained.

$$\psi_h = \frac{8kT}{ve} \gamma \exp(-\kappa h) \tag{3.10}$$

where

$$\gamma = \frac{\exp(Z/2) - 1}{\exp(Z/2) + 1} \qquad Z = \frac{ve\psi_0}{kT} \tag{3.11}$$

and κ is the Debye reciprocal length parameter defined by Eq. (3.12).

$$\kappa = \left(\frac{8\pi n e^2 v^2}{\varepsilon kT}\right)^{1/2} \tag{3.12}$$

After substituting for ψ_h into Eq. (3.8), P becomes

$$P = 64 nkT \gamma^2 \exp(-2\kappa h) \tag{3.13}$$

The potential energy of interaction, V_R, is obtained as follows:

$$V_R = -2 \int_\infty^h P \, dh$$

$$= \frac{64 nkT}{\kappa} \gamma^2 \exp(-2\kappa h) \tag{3.14}$$

3.1.2 Potential Energy of Interaction Between Flat Double Layers via the Free Energy

There are two ways to obtain the free energy of an electrical double layer. Method 1 involves the calculation of the reversible work necessary for the building of a double layer, and method 2 involves the reversible work necessary to diminish the double layer by a discharging process. Method 2 is more general and was used by Verwey and Overbeek to calculate the potential energy for the double-layer interaction. Method 1 is useful when the relationship between σ and ψ_0 is established, as in the case of the linear approximation to the Poisson-Boltzmann equation. It will be used in what follows.

The free energy of a double layer per unit area, G_{dl}, consists of two contributions,

$$G_{dl} = G_{ch} + G_{el}$$

$$= -\sigma\psi_0 + \int_0^\sigma \psi_0' \, d\sigma' \tag{3.15}$$

where ψ_0' and σ' are surface potential and surface charge density, respectively, at each stage of building the double layer. The chemical term, G_{ch} ($= -\sigma\psi_0$), involves the free-energy change associated with the adsorption of potential determining ions to establish a double layer. Since this process occurs spontaneously, G_{ch} is negative in sign. The electrical term, G_{el} [integral term in Eq. (3.15)], involves the electrical work necessary for charging the double layer. This process requires an energy and G_{el} is positive. G_{ch} is always larger in magnitude than G_{el} and hence G_{dl} is always negative, which means that double-layer formation is a spontaneous process. After integration by parts, Eq. (3.15) is written as

$$G_{dl} = -\int_0^{\psi_0} \sigma' \, d\psi_0' \tag{3.16}$$

If ψ_0 is small, the linear approximation [see Eq. (2.43)] may be used and G_{dl} simplifies to

$$G_{dl} = -\frac{1}{2}\sigma\psi_0 \tag{3.17}$$

The potential energy of interaction, V_R, for two flat double layers is the free-energy change in bringing them from infinity to a distance of 2h; thus

$$V_R = 2(G_{dl}^h - G_{dl}^\infty) \tag{3.18}$$

From Eq. (3.17)

$$G_{dl}^h = -\frac{1}{2}\sigma^h \psi_0 \tag{3.19}$$

where σ^h is the surface charge density when two plates are separated by a distance 2h. It should be noted that as the two double layers approach each other, ψ_0 is held constant (constant potential process). In order to calculate σ^h, it is necessary to know $(d\psi/dx)_{x=0}$ at a plate distance of 2h. This is obtained from Eqs. (3.3) and (3.4).

$$\left(\frac{d\psi}{dx}\right)_{x=0} = -\left(\frac{8\pi nkT}{\varepsilon}\right)^{1/2} (2\cosh Z - 2\cosh U)^{1/2} \tag{3.20}$$

where

$$Z = \frac{ve\psi_0}{kT} \quad U = \frac{ve\psi_h}{kT} \tag{3.21}$$

The initial negative sign in Eq. (3.20) arises from the fact that ψ decreases with increasing x. For $Z < 1$ ($U < 1$)

$$2\cosh Z - 2\cosh U = Z^2 - U^2 = Z^2\left(1 - \frac{U^2}{Z^2}\right) \tag{3.22}$$

After using the relationship [2, p. 96]

$$\frac{U}{Z} = \text{sech } \kappa h \tag{3.23}$$

Eq. (3.24) is obtained:

$$\left(\frac{d\psi}{dx}\right)_{x=0} = -\left(\frac{8\pi nkT}{\varepsilon}\right)^{1/2} Z \tanh \kappa h \tag{3.24}$$

Hence

$$\overset{h}{\sigma} = -\frac{\varepsilon}{4\pi}\left(\frac{d\psi}{dx}\right)_{x=0} = \frac{\varepsilon\kappa kT}{4\pi ve} Z \tanh \kappa h \tag{3.25}$$

$$= \frac{\varepsilon\kappa}{4\pi} \psi_0 \tanh \kappa h \tag{3.26}$$

Equation (3.27) follows from Eqs. (3.19) and (3.26).

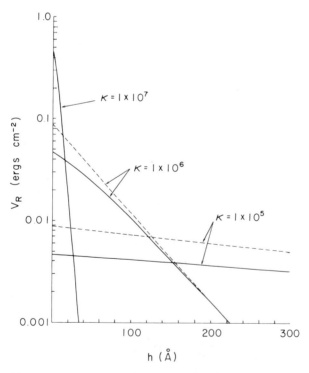

Figure 3.2 V_R as a function of h for various values of κ at $Z = 1$. Solid line, exact calculation; dashed line, approximation [Eq. (3.30)].

$$G_{dl}^{h} = -\frac{\varepsilon \kappa}{8\pi}\psi_0^2 \tanh \kappa h \tag{3.27}$$

and accordingly,

$$G_{dl}^{\infty} = -\frac{\varepsilon \kappa}{8\pi}\psi_0^2 \tag{3.28}$$

Thus, from Eq. (3.18), V_R is given by

$$V_R = \frac{\varepsilon \kappa}{4\pi}\psi_0^2(1 - \tanh \kappa h)$$

$$= \frac{2nkT}{\kappa} Z^2(1 - \tanh \kappa h) \tag{3.29}$$

For large κh (weak interaction), $1 - \tanh \kappa h = 2 \exp(-2\kappa h)$ and using the approximate relation $Z = 4\gamma$ and

$$V_R = \frac{64nkT}{\kappa}\gamma^2 \exp(-2\kappa h) \tag{3.30}$$

which is identical to Eq. (3.14).

Although Eq. (3.30) has been derived for the case of small surface potential and weak interaction, this approximate equation is widely used for practical purposes. Figure 3.2 shows V_R as a function of h for $Z = 1$ and at various values of κ. The solid lines represent the exact calculations, while the dashed lines are the results calculated using Eq. (3.30). It can be seen that the approximation gives higher values at small separations. It should be noted that V_R decreases at large separations and it increases at short separations with increasing κ.

3.1.3 Spherical Double Layers

Consider two spherical particles (radius a) separated by the shortest distance H_0 (Fig. 3.3).

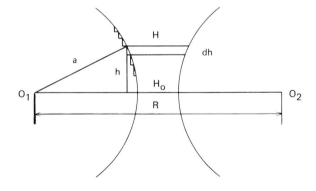

Figure 3.3 Interaction of spherical double layers.

$$H_0 = R - 2a \tag{3.31}$$

where R is the center-to-center distance. The potential energy of interaction between two spherical double layers can be calculated using Derjaguin's integration method [3], which is based on the interaction energy between flat double layers. If it is assumed that the spherical double-layer interaction, $V_{R(sp)}$, is an accumulation of the interactions of infinitesimal parallel rings, then

$$V_{R(sp)} = \int_0^\infty 2\pi h 2(G_H - G_\infty) \, dh \tag{3.32}$$

where $2G_H$ is the free energy per unit area between parallel plates separated by a plate distance H and h is the distance from the axis of symmetry of a ring under consideration. This equation is valid when the range of interaction is much smaller than the particle radius. Since $(H - H_0)/2 = a - (a^2 - h^2)^{1/2}$ and $2h \, dh = a \, dH(1 - h^2/a^2)^{1/2} \cong a \, dH$ for small values of h, Eq. (3.32) becomes

$$V_{R(sp)} = 2\pi a \int_{H_0}^\infty (G_H - G_\infty) \, dH \tag{3.33}$$

By making use of Eq. (3.29) for the interaction energy between plates, Eq. (3.34) is obtained.

$$2(G_H - G_\infty) = \frac{\varepsilon \kappa \psi_0^2}{4\pi} \left(1 - \tanh \frac{\kappa H}{2}\right) \tag{3.34}$$

Thus Eq. (3.33) can be integrated to give

$$V_{R(sp)} = \frac{\varepsilon a \psi_0^2}{2} \int_{H_0}^\infty \left(1 - \tanh \frac{\kappa H}{2}\right) d\left(\frac{\kappa H}{2}\right)$$

$$= \frac{\varepsilon a \psi_0^2}{2} \ln[1 + \exp(-\kappa H_0)] \tag{3.35}$$

If Eq. (3.30) is used for parallel interactions, then

$$V_R(sp) = 4.62 \times 10^{-6} \frac{a\gamma^2}{v^2} \exp(-\kappa H_0) \tag{3.36}$$

It should be mentioned that these equations for $V_{R(sp)}$ have been derived for small surface potential and large κa, and that they give larger values than the true ones. Overbeek [2, p. 139] pointed out that the errors produced by Eq. (3.35) are approximately 5% for $\kappa a = 10$, 10% for $\kappa a = 5$, and 30% for $\kappa a = 2$, respectively. Derjaguin's method gives less accurate results as κH_0 becomes large. Bell, Levine, and McCartney [4] proposed a linear superposition approximation method to remedy this.

3.2 LONDON-VAN DER WAALS ATTRACTION

The London-van der Waals force between two atoms or molecules is a short-range force and its potential energy is expressed as the sixth power of the separation distance r, that is,

$$V_{London} = -\frac{\lambda}{r^6} \tag{3.37}$$

with

$$\lambda = \frac{3}{4} \alpha^2 h\nu_0 \tag{3.38}$$

where α is the polarizability and $h\nu_0$ is a characteristic energy corresponding to the principal specific frequency ν_0 and is sometimes identified as the ionization potential I. Because of the additivity of this force, the London-van der Waals force between macroscopic materials becomes a long-range force. Hamaker [5] calculated the potential energy per unit area, V_A, between two large parallel plates of thickness δ separated by a distance 2h as

$$V_A = -\frac{A}{48\pi}\left[\frac{1}{h^2} + \frac{1}{(h+\delta)^2} - \frac{2}{(h+\delta/2)^2}\right] \tag{3.39}$$

with

$$A = \pi^2 q^2 \lambda \tag{3.40}$$

where q is the number of atoms contained in 1 cm^3 of the substance. A is called the Hamaker constant, which is of the order of magnitude of 10^{-13}-10^{-12} erg. For $h \ll \delta$, V_A is approximated by

$$V_A = -\frac{A}{48\pi h^2} \tag{3.41}$$

The corresponding force, F_A, is given by

$$F_A = -\frac{dV_A}{d(2h)} = -\frac{A}{48\pi h^3} \tag{3.42}$$

The London-van der Waals force is essentially electromagnetic in origin. If the distance between two atoms becomes comparable with the wavelength of the London frequency, a lessening of the London–van der Waals force may be expected. This is called the retardation effect. In this situation, V_A varies with h^{-3} instead of h^{-2}.

For the interaction of two equal spheres of radius a and a distance R between their center, V_A is given by

$$V_{A(sp)} = -\frac{A}{6}\left(\frac{2}{s^2-4} + \frac{2}{s^2} + \ln\frac{s^2-4}{s^2}\right) \qquad (3.43)$$

where $s = R/a = 2 + H_0/a$ (see Fig. 3.3). For $H_0 \ll a$, Eq. (3.43) is approximated as

$$V_{A(sp)} = -\frac{Aa}{12H_0} = -\frac{A}{12(s-2)} \qquad (3.44)$$

For small separation distances between the spheres, the London–van der Waals attraction decreases more slowly than that between flat plates.

For two unequal spheres of radius a_1 and a_2 ($a_1 < a_2$), $V_{A(sp)}$ is given by [5]

$$V_{A(sp)} = -\frac{A}{12}\left(\frac{y}{x^2+xy+x} + \frac{y}{x^2+xy+x+y}\right.$$

$$\left. + 2\ln\frac{x^2+xy+x}{x^2+xy+x+y}\right) \qquad (3.45)$$

where $x = H_0/2a_1$, and $y = a_2/a_1$, respectively. When $x \ll 1$, Eq. (3.45) simplifies to

$$V_{A(sp)} = -\frac{Aa_1 a_2}{6(a_1+a_2)H_0} \qquad (3.46)$$

So far, the London–van der Waals interaction between particles in a vacuum has been considered. If particles embedded in a condensed medium, which is usually the case for practical colloid systems, are considered, the influence of the medium has to be taken into account. For the case where two particles of material 1 are embedded in a medium 3, the Hamaker constant A ($=A_{11/3}$) is given by [6]

$$A_{11/3} = A_{11} + A_{33} - 2A_{13} \qquad (3.47)$$

where A_{11}, A_{33}, and A_{13} refer to the Hamaker constants for particle–particle, medium–medium, and particle–medium interactions in a vacuum, respectively. If it is assumed that $A_{13} = (A_{11}A_{33})^{1/2}$, then

$$A_{11/3} = (A_{11}^{1/2} - A_{33}^{1/2})^2 \qquad (3.48)$$

Thus the London-van der Waals interaction between identical particles in a medium is always attractive.

Fowkes [7] proposed a method to estimate the Hamaker constant by the dispersion component of the surface tension. The Hamaker constant A_{11} is given by

$$A_{11} = 6\pi r_{11}^2 \gamma_1^d \tag{3.49}$$

and

$$A_{11/3} = 6\pi r_{33}^2 [(\gamma_3^d)^{1/2} - (\gamma_1^d)^{1/2}]^2 \tag{3.50}$$

where γ^d is the dispersion component of the surface tension of the material and r_{11} and r_{33} are the separation distance between volume elements of the material. For water and systems with volume elements such as oxide ions, metal atoms, and CH_2 or CH groups which have nearly the same size, $6\pi r_{11}^2 = 1.44 \times 10^{-14}$. Values of γ^d for various substances have been determined and are reported in the literature [7].

The Hamaker theory is based on a summation of the separate contributions of each pair of molecules. Dzyaloshinski, Lifshitz, and Pitaevskii [8] developed a macroscopic theory based on continuum quantum electrodynamics without assuming the pairwise additivity of interatomic dispersion energies. Detailed descriptions are available in the literature [9-15]. The major difference between the Lifshitz theory and the Hamaker theory is that according to the Lifshitz theory, the Hamaker constant is not constant but depends on the distance between the particles and the temperature. This is why the term "Hamaker coefficient" is used. Another important feature is that the summation appearing in the Lifshitz theory is not confined to a single frequency but carried out over all frequencies of the electromagnetic spectrum and the retardation effect is included explicitly in the theory.

3.3 DLVO THEORY

The stability of colloid particles in aqueous media depends on the total potential energy of interaction V_t, that is,

$$V_t = V_R + V_A \tag{3.51}$$

Consider platelike particles. Figure 3.4 shows the total potential energy (V_t) for various values of Z ($ve\psi_0/kT$) as a function of the separation distance (h). The potential energies V_R and V_A are calculated using Eqs. (3.30) and (3.41), respectively, where $\kappa = 1 \times 10^7$, $v = 1$, and $A = 2 \times 10^{-12}$. Since V_R varies exponentially with the

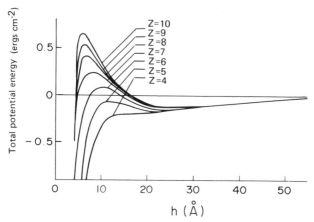

Figure 3.4 V_t as a function of h for various values of Z. $\kappa = 1 \times 10^7$ cm^{-1}, v = 1, and A = 2×10^{-12} erg. (From Ref. 2.)

distance and V_A varies inversely with the square of the distance, V_A surpasses V_R at short and long distances, thus producing attraction between the particles. At intermediate distances, the plot of V_t shows various shapes depending on Z. For large values of Z, there is a potential barrier and for small values of Z, there is an attractive potential. If the potential barrier is high compared to the kinetic energy of the particles, no coagulation of particles takes place.

The electrolyte concentration also has an effect on the potential barrier. Figure 3.5 illustrates the effect of the electrolyte concentra-

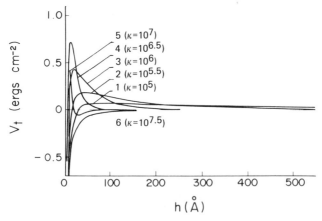

Figure 3.5 V_t as a function of h for various values of κ. Z = 4, v = 1, and A = 1×10^{-12} erg. (From Ref. 2.)

tion (as expressed by κ) on V_t versus h curves at $Z = 4$ and $v = 1$. When κ is small (low concentration), the potential barrier is broad and the barrier height is small (curve 1). The barrier becomes sharp and its height increases with increasing κ (curves 3 and 4). A further increase in κ causes the disappearance of the potential barrier and the plot of V_t shows that there is attraction between the particles (curve 6), and coagulation of the particles takes place. V_R increases with an increase in κ, but at smaller distances V_A dominates V_R and the potential barrier disappears. The electrolyte concentration at which coagulation commences is called the critical coagulation concentration. It is reasonable to assume that at the critical coagulation concentration, the potential barrier in the total potential energy curve disappears. Thus

$$V_t = 0 \tag{3.52}$$

and

$$\frac{dV_t}{dh} = 0 \tag{3.53}$$

The equations above mean that the maximum in the total potential energy curve touches the horizontal axis. Equation (3.54) is obtained from the equations above and Eqs. (3.30) and (3.41)

$$\kappa h = 1 \tag{3.54}$$

and the critical coagulation concentration,

$$n_c = \frac{107 \varepsilon^3 (kT)^5 \gamma^4}{A^2 (ve)^6} \tag{3.55}$$

For aqueous solutions at 25°C ($\varepsilon = 78.55$), the critical coagulation concentration C_c (mmol/liter) is given by

$$C_c = 8 \times 10^{-22} \frac{\gamma^4}{A^2 v^6} \tag{3.56}$$

If it is assumed that γ is constant, which is the case for large values of z [see Eq. (3.11)], the critical coagulation concentration is inversely proportional to v^6. The ratio of C_c to $v = 1$, $v = 2$, and $v = 3$ is expected to be $1:(1/2)^6:(1/3)^6$ or $100:1.6:0.13$. The Schulze-Hardy rule for the critical coagulation concentration is known. It states that for the coagulation of sols, the higher the counterion valency, the lower the critical coagulation concentration. Table 3.1 gives the critical coagulation concentrations for As_2S_3 (negatively charged) and Fe_2O_3 (positively charged) sols. It can be seen that the experimental values of C_c may be interpreted in terms of the v^{-6} law. However, this is not the complete explanation for the Schultze-

Table 3.1 Critical Coagulation Concentration (mmol/liter)

As_2S_3 (negatively charged)		Fe_2O_3 (positively charged)	
Electrolyte	Coagulation concentration	Electrolyte	Coagulation concentration
NaCl	51	NaCl	9.25
KCl	49.5	KCl	9.0
KNO_3	50	$(1/2)BaCl_2$	9.6
$(1/2)K_2SO_4$	65.5	KNO_3	12
LiCl	58		
$MgCl_2$	0.72	K_2SO_4	0.205
$MgSO_4$	0.81	$MgSO_4$	0.22
$CaCl_2$	0.65	$K_2Cr_2O_7$	0.195
$AlCl_3$	0.093		
$Al(NO_3)_3$	0.095		
$Ce(NO_3)_3$	0.080		

Source: Ref. 2.

Haryd rule, since the surface potential (z) is not very high under critical coagulation conditions, and accordingly γ cannot be regarded as constant. Overbeek [16] has given a more reasonable explanation for the Schulze-Hardy rule by taking into account the adsorption of counterions in the Stern layer, where the specific adsorption potential is a function of the valency of the counterions.

Experimental verification of the DLVO theory has been conducted using various systems: the stability of thin liquid films, the rate of coagulation of sols, and model experiments. Watanabe and Gotoh [17] developed a twin dropping mercury electrode technique by which the critical polarization potential for coalescence of mercury droplets was determined as a function of the electrolyte concentration. This demonstrated that the coalescence of mercury droplets may be interpreted in terms of the DLVO theory. This study was extended to the system containing surfactants [18] and to nonaqueous solutions [19] and mixed solvent [20] systems. Using a modified twin dropping mercury electrode, Usui, Yamasaki, and Shimoiizaka [21] also demonstrated that the coalescence of mercury in KF aqueous solutions (1×10^{-3} to 7.5×10^{-2} mol/liter) may be interpreted quantitatively in terms of the DLVO theory, where the double-layer interaction was analyzed in terms of ψ_δ. The Hamaker constant for mercury-mercury in water was determined to be 1.2×10^{-12} erg, in good agreement with the value obtained by Fowkes (1.3×10^{-12} erg) on the basis of Eq. (3.50). Recently, Israelachvili and Adams [22] have made direct measurements of forces acting between two molecularly smooth mica surfaces immersed in aqueous electrolyte solutions as a function of the separation distance. The DLVO theory accounts for the experimental results except for those of small separations where an additional repulsion was expected to operate. Similar results were indicated in a system containing cationic surfactant (CTAB) [23], although large deviations from the results predicted by the DLVO theory occurred in regions where the hydrophobic monolayer adsorption took place on mica surfaces.

The barrier height is proportional to the surface area of a particle, while the thermal energy of a particle is independent of the surface area, leading to the prediction that the smaller the particle, the smaller the potential energy barrier, and hence the easier the coagulation. This prediction is realized in practical systems. For spherical particles, the potential energy per particle is obtained and direct comparison with the thermal energy of Brownian movement is possible.

The total potential energy of interaction, V_t, has two potential minima: a deep primary minimum appearing at a small separation and a shallow secondary minimum appearing at a larger separation (Fig. 3.4). For platelike or rodlike particles, the coagulation into a secondary minimum is sometimes observed, because the interacting surface area becomes large and hence the depth of the secondary minimum becomes relatively deep. The secondary minimum coagulation accounts for the tactoid sol formation, the Schiller layer formation, and anisotropic phase separation of tobacco mosaic virus [24]. Rheological properties such as thixotropy and rheopexy are closely related to secondary minimum coagulation. Hachisu and Furusawa [25] studied the coagulation of tungstic acid sol in aqueous electrolyte solutions and interpreted the reversible coagulation observed in terms of the secondary minimum. Van de Ven and Mason [26] and Takamura, Goldsmith, and Mason [27] observed the rotation of polystyrene latex doublet particles captured in the primary and secondary minima in aqueous electrolyte solutions in shear flow and analyzed the results in terms of the DLVO theory.

3.4 COLLOID STABILITY—KINETICS OF COAGULATION

Coagulation in the absence of repulsive interaction between particles, where each collision between particles leads to coagulation, is called rapid coagulation. If the coagulation is conducted by the diffusion process, the following relation can be derived from the Smoluchowski theory [28]:

$$N_t = \frac{N_0}{1 + t/T} \quad (3.57)$$

with

$$T = \frac{1}{8\pi D a N_0} \quad (3.58)$$

where N_0 is the initial number of primary particles per cubic centimeter; N_t the total number of particles, including higher-order particles formed after coagulation at an elapsed time t; D (=$kT/6\pi\eta a$) is the diffusion coefficient; a the particle radius; and T the time of rapid coagulation (at $t = T$, $N_t = N_0/2$).

When there is a repulsive potential of interaction between particles, the rate of coagulation decreases to become

$$N_t = \frac{N_0}{1 + t/T_s} \quad (3.59)$$

where T_s is the time of slow coagulation and is larger than T. The ratio

$$W = \frac{T_s}{T} \quad (3.60)$$

gives a measure of the rate by which the coagulation is retarded; thus W is called the retardation factor or stability ratio.

Diffusion phenomena under the influence of the force field have not been completely solved, but a solution under the condition of a steady state is given by the Fuchs equation as follows:

$$J = \frac{8\pi D N_0}{\int_{2a}^{\infty} \exp(V/kT)(dR/R^2)} \quad (3.61)$$

where J is the number of particles colliding with a central particle per unit time due to Brownian movement, V the potential energy between particles, and R the radial distance. In the case of V = 0 the flux of rapid coagulation J becomes

$$J_r = 16\pi D N_0 a \quad (3.62)$$

The ratio of J_r to J corresponds to W; thus

$$W = 2a \int_{2a}^{\infty} \exp\left(\frac{V}{kT}\right) \frac{dR}{R^2} = 2 \int_{2}^{\infty} \exp\left(\frac{V}{kT}\right) \frac{ds}{s^2} \quad (3.63)$$

where $s = R/a$. It should be mentioned that Eq. (3.63) is valid only for the initial stage of coagulation, where the number of secondary

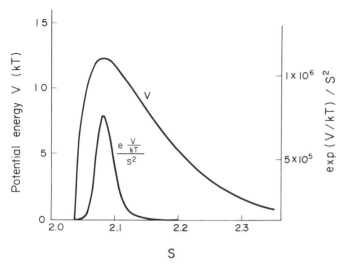

Figure 3.6 Potential energy curve (V) and corresponding curve of $\exp(V/kT)/S^2$ as a function of S. $a = 10^{-5}$ cm, $\kappa = 10^6$ cm^{-1}, $A = 10^{-12}$ erg, $\psi_0 = 28.2$ mV. (From Ref. 2.)

particles is small. In Fig. 3.6, V(s) and the corresponding $\exp(V/kT)/s^2$ are shown for $a = 1 \times 10^{-5}$ cm, $\kappa = 1 \times 10^6$ cm^{-1}, $A = 1 \times 10^{-12}$ erg, and $\psi_0 = 28.2$ mV, for which $W = 5.4 \times 10^4$ was obtained [2, p. 168]. The value of W should be obtained by numerical or graphical integration according to Eq. (3.63), which is rather tedious. In this case the following approximate equation (3.64) is useful, which follows from the fact that the maximum value of the total potential energy (V_{max}) contributes predominantly to the integration.

$$W = \frac{1}{2\kappa a} \exp\left(\frac{V_{max}}{kT}\right) \tag{3.64}$$

From Eq. (3.64), $W \cong 1 \times 10^5$ is obtained for $V_{max} = 15$ kT and $W \cong 3 \times 10^9$ for $V_{max} = 25$ kT. The colloid dispersion may be considered to be stable if V_{max} exceeds 15 kT.

Reerink and Overbeek [29], using Eq. (3.36) for V_R and Eq. (3.44) for V_A, derived the linear relationship between log W and log c as follows:

$$\log W = -K_1 \log c + \log K_2 \tag{3.65}$$

where K_1 and K_2 are constants. At 25°C

$$K_1 = -2.15 \times 10^7 \frac{a\gamma^2}{v^2} \tag{3.66}$$

where the units of c are m mol/liter. It can be seen that the slope (K_1) of the stability curve (log W versus log c) is proportional to the particle radius and permits us to calculate γ and hence ψ_δ. Reerink and Overbeek determined the rate of coagulation of negative AgI sol (pI = 4) having different particle radii in aqueous solutions of La(NO$_3$)$_3$, Ba(NO$_3$)$_2$, and KNO$_3$ from the initial slope of the turbidity versus time curve and constructed the stability curves shown in Fig. 3.7. A breakpoint in the figure corresponds to the critical coagulation concentration at which rapid coagulation began (i.e., W = 1). From Fig. 3.7, the linear relationship between log W and log c agreed with the conditions of the theory. The linear relationship was proved to hold also for other systems which had been used in the past. However, it should be mentioned that the theory insists on the proportionality of the slope to the particle radius. As can be seen in Fig. 3.7, there is no significant difference in the slope between fine sol B (a = 200-250 Å) and coarse sols A, C, D, and E (a = 500-2000 Å). The contradiction about the slope between the theory and the experiments was also indicated in some other sols using different particle radii [30]. Ottewill and Shaw [31] carried out stability experiments using well-defined monodispersed polystyrene latex particles of five different sizes (300, 515, 1213, 1840, and 2115 Å) in Ba(NO$_3$)$_2$ aqueous solutions. The linearity between log W and log c satisfied the theoretical prediction, but the slope did not agree with the theoretical requirement. In this study, zeta (ζ) potentials of latex particles were measured (interestingly, the ζ potential decreased with a decrease in the particle size) and strict analysis of d log W/d log c was made by taking ζ as equivalent to ψ_δ. Nevertheless, the theory still predicts an increase in slope with an increase in particle size, contradicting the experimental results. The Hamaker constant can be obtained from either the slope of the stability curve or the critical coagulation concentration. In the system of polystyrene-aqueous solutions

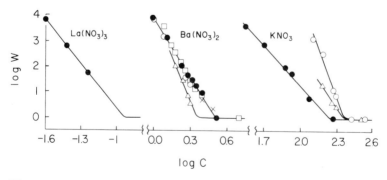

Figure 3.7 Stability curves for silver iodide sols. (From Ref. 29.) ● sol A; ○ sol B; □ sol C; × sol D; and △ sol E.

discussed above, the Hamaker constants were obtained as 1.3×10^{-14} to 1.1×10^{-13}, which are of predicted order of magnitude.

Several investigations have been made [32-35] to resolve the discrepancy in the slope between theory and experiment. Major considerations involved the correction in which the retardation of the intervening liquid flow at the final stage of particle attachment was taken into consideration. However, hydrodynamic corrections have not alleviated the discrepancy and this problem needs further study. Overbeek [36] suggested that the effect of electrolyte concentration on the surface potential and/or on the surface charge and the effect of the slow rate of surface charge adjustment during the coagulation process should be reexamined. Derjaguin, Dukhin, and Shilov [37] mentioned the role of the dynamic double-layer effect as significant in colloid stability problems.

Repeptization is of importance in both theoretical and practical aspects. There are some systems in which flocculated particles show repeptization with decreasing electrolyte concentration. It is necessary for the depth of the primary minimum of total potential energy curve to decrease in order to redisperse the flocculated system (see Fig. 3.8). The depth of the primary minimum and hence repeptization is closely related to the charge adjustment of the double layer as well as to ψ_δ and the Hamaker constant. Thus the repeptization phenomena will offer fruitful information in further development of the colloid stability theory.

3.5 HETEROCOAGULATION

Up to this point the coagulation of particles of identical nature (homocoagulation) has been discussed. There are many cases in which the

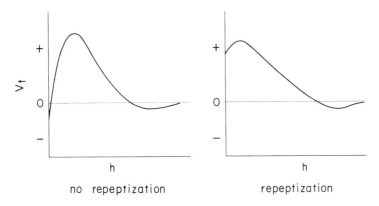

Figure 3.8 Schematic illustration of the primary minimum in total potential energy curve in relation to repeptization.

coagulation of particles of dissimilar nature (heterocoagulation) occurs. Heterocoagulation is also governed by the double-layer interaction and van der Waals interaction between particles, which is also the case for homocoagulation, but there are some differences that make heterocoagulation more complex than homocoagulation.

3.5.1 Dissimilar Double-Layer Interaction

The theory of dissimilar double-layer interaction was first presented by Derjaguin [38], who calculated the force of interaction between two plates immersed in an electrolyte solution by solving the Poisson-Boltzmann equation under the conditions that the surface potentials are ψ_1 and ψ_2, respectively, and the plate distance is h. The Poisson-Boltzmann equation [Eq. (2.32)] can be integrated to yield

$$\left(\frac{d\phi}{d\chi}\right)^2 = 2 \cosh \phi + C \tag{3.67}$$

where $\phi = ze\psi/kT$, $\chi = \kappa x$, and C is an integration constant. The force of interaction P (per square centimeter) between two plane-parallel double layers is given by

$$P = nkTW \tag{3.68}$$

$$W = -(C + 2) \tag{3.69}$$

The object of the calculation was to obtain P, and hence W, as a function of H ($=\kappa h$). This is quite tedious. Instead, Derjaguin obtained H as a function of W, by constructing a ϕ versus χ curve (isodynamic curves) for given values of C (and hence W) using Eq. (3.67). The major points obtained from the analysis of isodynamic curves were as follows (Fig. 3.9). For the case where ψ_1 and ψ_2 are of same sign but of different magnitudes ($|\psi_1| < |\psi_2|$), the force of interaction changes from a repulsion to an attraction after passing through a maximum. The separation distance at which force maximum occurs depends upon the ratio ψ_2/ψ_1, but the magnitude of the force barrier, W_m, is determined solely by the lower surface potential (ψ_1).

Devereux and de Bruyn [39] calculated the potential energy of interaction between plane-parallel dissimilar double layers using the Verwey-Overbeek method of calculation of the free-energy change associated with overlapping of the double layers. The calculation was carried out using a computor and the numerical results are tabulated in terms of $V' = \kappa V_{el}/2nkT$, where V_{el} is the potential energy of interaction per unit area. V' is a function of $\xi = \kappa D$, where D is the separation distance, for specific $Y_0 = ve\psi_1/kT$ and $Y_D = ve\psi_2/kT$ and the integration constant C.

Although the Devereux-de Bruyn calculation gives accurate results, it is not feasible for practical purposes. Hogg, Healy, and Fuerstanau [40] calculated the potential energy of interaction between dissimilar

Electrical Double Layers and Colloid Stability / 67

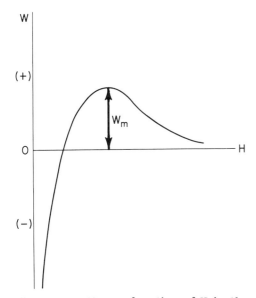

Figure 3.9 W as a function of H in the case where ψ_1 and ψ_2 are of the same sign and different magnitude.

flat double layers using the Debye-Hückel or linearized approximation. The general solution of the linearized Poisson-Boltzmann equation [Eq. (2.33)] is given by

$$\psi = A_1 \cosh \kappa x + A_2 \sinh \kappa x \tag{3.70}$$

where A_1 and A_2 are integration constants and are determined by the boundary conditions (i.e., $\psi = \psi_1$ at $x = 0$ and $\psi = \psi_2$ at $x = 2h$) (see Fig. 3.10). Thus the electric potential between plates is given as follows:

$$\psi = \psi_1 \cosh \kappa x + \frac{\psi_2 - \psi_1 \cosh 2\kappa h}{\sinh 2\kappa h} \sinh \kappa x \tag{3.71}$$

The electric potential between plates is illustrated schematically in Fig. 3.10 for the case where ψ_1 and ψ_2 are of the same sign. The dashed line represents the electric potential which respective double layers would have in the absence of interaction. The surface charge densities σ_1 and σ_2 are given by

$$\sigma_1 = - \frac{\varepsilon}{4\pi} \left(\frac{d\psi}{dx} \right)_{x=0} \tag{3.72}$$

$$\sigma_2 = + \frac{\varepsilon}{4\pi} \left(\frac{d\psi}{dx} \right)_{x=2h} \tag{3.73}$$

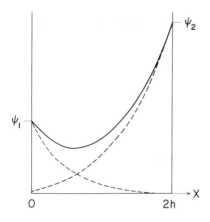

Figure 3.10 Schematic representation of the electric potential between unequally charged surfaces.

Thus σ_1 and σ_2 are obtained as a function of κh as follows:

$$\sigma_1 = -\frac{\varepsilon\kappa}{4\pi}(\psi_2 \operatorname{cosech} 2\kappa h - \psi_1 \coth 2\kappa h) \tag{3.74}$$

$$\sigma_2 = +\frac{\varepsilon\kappa}{4\pi}(\psi_2 \coth 2\kappa h - \psi_1 \operatorname{cosech} 2\kappa h) \tag{3.75}$$

In the linear approximation, the free energy of a double-layer system with layers $2h$ apart, G_{dl}^{2h}, is given by

$$G_{dl}^{2h} = -\frac{1}{2}(\sigma_1\psi_1 + \sigma_2\psi_2) \tag{3.76}$$

By referring to Eqs. (3.74) and (3.75), we obtain

$$G_{dl}^{2h} = \frac{\varepsilon\kappa}{8\pi}[2\psi_1\psi_2 \operatorname{cosech} 2\kappa h - (\psi_1^2 + \psi_2^2)\coth 2\kappa h] \tag{3.77}$$

At $2h = \infty$,

$$G_{dl}^{\infty} = -\frac{\varepsilon\kappa}{8\pi}(\psi_1^2 + \psi_2^2) \tag{3.78}$$

The potential energy of interaction V_{el} (per square centimeter) is given by

$$V_{el} = G_{dl}^{2h} - G_{dl}^{\infty} \tag{3.79}$$

Thus V_{el} is obtained:

$$V_{el} = \frac{\varepsilon\kappa}{8\pi}[2\psi_1\psi_2 \operatorname{cosech} 2\kappa h + (\psi_1^2 + \psi_2^2)(1 - \coth 2\kappa h)] \tag{3.80}$$

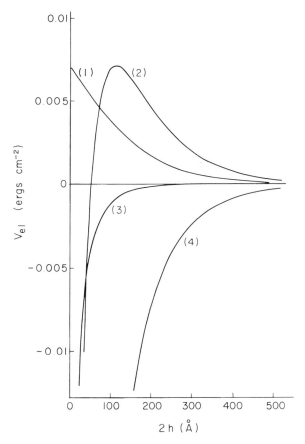

Figure 3.11 V_{el} as a function of 2h. c = 1 mmol/liter, 1-1 electrolyte ($\kappa = 10^6$ cm^{-1}). (1) $\psi_1 = \psi_2 = 10$ mV; (2) $\psi_1 = 10$ mV, $\psi_2 = 30$ mV; (3) $\psi_1 = 0$ mV, $\psi_2 = 10$ mV; (4) $\psi_1 = 10$ mV, $\psi_2 = -30$ mV. (From Ref. 43.)

Figure 3.11 shows V_{el} as a function of 2h for some selected pairs of ψ_1 and ψ_2 together with the symmetrical case for $\kappa = 1 \times 10^6$ (1-1 electrolyte). It is seen that V_{el} is always negative (attraction) when ψ_1 and ψ_2 are of opposite sign and when $\psi_1 = 0$. The maximum value of V_{el} is determined solely by the lower surface potential (ψ_1) and is given by

$$V_{el(max)} = \frac{\varepsilon \kappa}{4\pi} \psi_1^2 \tag{3.81}$$

Equation (3.80) becomes Eq. (3.29) for $\psi_1 = \psi_2 = \psi$.

The potential energy of interaction between two spherical particles, $V_{e\ell(sp)}$, is given by the Derjaguin method [3] as

$$V_{e\ell(sp)} = \frac{\varepsilon a_1 a_2 (\psi_1^2 + \psi_2^2)}{4(a_1 + a_2)} \left\{ \frac{2\psi_1 \psi_2}{(\psi_1^2 + \psi_2^2)} \ln \frac{1 + \exp(-\kappa H_0)}{1 - \exp(-\kappa H_0)} \right.$$

$$\left. + \ln [1 - \exp(-2\kappa H_0)] \right\} \qquad (3.82)$$

where H_0 is the shortest distance between sphere surfaces. Although Eq. (3.80) was derived with the assumption that ψ_1 and ψ_2 are small in magnitude, this equation serves as a good approximation when ψ_1, $\psi_2 < 50\text{-}60$ mV [40]. Barouch et al., [41] analyzed the Poisson-Boltzmann equation in its two-dimensional form to calculate the interaction energy of two unequal spheres and compared the results with those obtained by the Hogg-Healy-Fuerstenau (HHF) equation [Eq. (3.82)]. The Barouch-Matijević-Ring-Finlan (BMRF) equation gave lower values than the HHF equation, and Chan and White [42] commented on this.

It is seen in Fig. 3.11 that attraction occurs at close approach when ψ_1 and ψ_2 are of the same sign and of different magnitudes. This is because the sign of the surface charge of the plate with lower surface potential (σ_1) is reversed when the two plates approach closely. Figure 3.12 illustrates how the electric potential between two plates changes with the approach of the plates (as expressed by $2\kappa h$), where $\psi_1 = 10$ mV, $\psi_2 = 30$ mV, and $\kappa = 1 \times 10^6$ are assumed. Since the surface charge density is proportional to $(d\psi/dx)_{x=0}$ [see Eq. (2.38)], the charge reversal of σ_1 is easily seen when the two plates come close together. Strictly speaking, the sign of the force is determined by the relative magnitude of P_E and P_0 [see Eqs. (3.1) and (3.2)].

3.5.2 Constant Surface Charge Model

In Sec. 3.5.1 it was assumed that the surface potentials are held constant during the interaction of electrical double layers (constant surface potential model). This is reasonable if the electrochemical equilibrium by which the surface potential is determined is maintained at each stage of approach of the two interacting double layers. Frens, Engel, and Overbeek [44] have shown that a Brownian collision takes about 10^{-7} sec, while 1 sec is needed for a discharging current to restore the equilibrium in the distorted double layers. This indicates that it is not the surface potential but the surface charge which remains constant during the collision process of colloid particles. Thus Frens and Overbeek [45] determined the potential energy of interaction between identical double layers, V_R^σ, based on the constant charge model. For flat double layers V_R^σ is given by

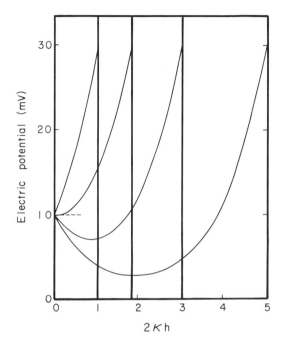

Figure 3.12 Electric potential between two unequally charged surfaces as a function of $2\kappa h$. $c = 1$ mmol/liter, 1-1 electrolyte; $\psi_1 = 10$ mV, $\psi_2 = 30$ mV. (From Ref. 43.)

$$V_R^\sigma = V_R^{Z_h} + \frac{8nkT}{\kappa}\left[(Z_h - Z_\infty)\sinh\frac{Z_\infty}{2} - 2\left(\cosh\frac{Z_h}{2} - \cosh\frac{Z_\infty}{2}\right)\right]$$
(3.83)

where $V_R^{Z_h}$ is the potential energy of interaction calculated by the constant potential model, Z_h and Z_∞ refer to Z ($ve\psi_0/kT$) at the separation distance of 2h and infinity, respectively.

The interaction of dissimilar double layers on the basis of the constant charge process was analyzed by Bell and Peterson [46]. They demonstrated that when ψ_1 and ψ_2 are of the same sign and of different magnitudes, the heterointeraction is always repulsion in contrast to the constant potential case where attraction appeared at small separations (see Fig. 3.11). Another significant difference is that if ψ_1 and ψ_2 are of opposite sign and of different magnitudes, the interaction became a repulsion when the plates drew close in contrast to the constant potential case where the interaction was always attraction. In the case of constant charge process, the Debye-Hückel approximation gives less reliable results because surface potentials increase in magnitude with decreasing separation distance. However, the analysis based on the linearized Poisson-Boltzmann equation will be helpful for understanding the difference in physical interpretation between the constant charge and the constant potential processes. The calculation of the potential energy of interaction in the constant charge process involves the calculation of changes only in G_{el}, because this process is not accompanied by the adsorption of potential determining ions. In the constant charge process, ψ_0 varies with a decrease in the separation distance, while in the constant potential process, σ varied. Thus the potential energy of interaction between dissimilar flat double layers in the constant charge process (V_{el}^σ) using the Debye-Hückel approximation is given by

$$V_{el}^\sigma = G_{el}^{2h} - G_{el}^\infty$$
$$= \frac{1}{2}(\sigma_1^{2h}\psi_1^{2h} - \sigma_1^\infty\psi_1^\infty) + \frac{1}{2}(\sigma_2^{2h}\psi_2^{2h} - \sigma_2^\infty\psi_2^\infty) \qquad (3.84)$$

where the superscripts 2h and ∞ refer to the separation distance 2h and infinity, respectively. Surface charge densities σ_1^{2h} and σ_2^{2h} are given by Eqs. (3.74) and (3.75), where ψ_1 and ψ_2 should be ψ_1^{2h} and ψ_2^{2h}, respectively. σ_1^∞ and σ_2^∞ are given by

$$\sigma_1^\infty = \frac{\varepsilon\kappa}{4\pi}\psi_1^\infty \qquad (3.85)$$

$$\sigma_2^\infty = \frac{\varepsilon\kappa}{4\pi}\psi_2^\infty \qquad (3.86)$$

From the conditions that $\sigma_1^{2h} = \sigma_1^\infty$ and $\sigma_2^{2h} = \sigma_2^\infty$, we obtain

$$\psi_1^{2h} = \psi_1^\infty \coth 2\kappa h + \psi_2^\infty \operatorname{cosech} 2\kappa h \qquad (3.87)$$

$$\psi_2^{2h} = \psi_1^\infty \operatorname{cosech} 2\kappa h + \psi_2^\infty \coth 2\kappa h \qquad (3.88)$$

Substituting Eqs. (3.85) to (3.88) for Eqs. (3.84), we obtain V_{el}^σ as follows [47]:

$$V_{el}^\sigma = \frac{\varepsilon\kappa}{8\pi}[2\psi_1\psi_2 \operatorname{cosech} \kappa h - (\psi_1^2 + \psi_2^2)(1 - \coth 2\kappa h)] \qquad (3.89)$$

Equation (3.89) can also be written as

$$V_{el}^\sigma = V_{el}^\psi + \frac{\varepsilon\kappa}{4\pi}(\psi_1^2 + \psi_2^2)(\coth 2\kappa h - 1) \qquad (3.90)$$

where V_{el}^ψ is the energy of interaction of the constant potential process given by Eq. (3.80) and the superscript ∞ of ψ_1 and ψ_2 is omitted for brevity. By comparison of Eq. (3.89) with Eq. (3.80) it is seen that V_{el}^σ differs from V_{el}^ψ only in the negative sign in front of the second term in brackets. Furthermore, since coth $2\kappa h > 1$ it is seen that V_{el}^σ is always more positive (repulsive) than V_{el}^ψ. Figure 3.13 shows V_{el}^σ as a function of $2\kappa h$ under the same conditions as those in Fig. 3.11 (i.e., $\psi_1 = \pm 10$ mV, $\psi_2 = 30$ mV, $v = 1$, and $\kappa = 1 \times 10^6$), together with V_{el}^ψ for comparison. It can be seen that V_{el}^σ is the mirror image of V_{el}^ψ. The minimum value of curve 3 is given by

$$V_{el(min)}^\sigma = -\frac{\varepsilon\kappa}{4\pi}\psi_1^2 \tag{3.91}$$

As seen in curve 3, the interaction changes from attraction to repulsion when ψ_1 and ψ_2 are of opposite sign (except the case of $\psi_1 = -\psi_2$). This is explained by considering that the sign of the surface potential of the smaller magnitude (ψ_1) reverses when two plates drawn close. In Fig. 3.14, the electric potential between two plates is illustrated as the plates approach (as expressed by $2\kappa h$). The solid lines refer to the case of $\psi_1 = 10$ mV and $\psi_2 = 30$ mV and the dashed lines to $\psi_1 = -10$ mV and $\psi_2 = 30$ mV. Note that ψ_1 and ψ_2 vary in such a way as the plates approach that $d\psi/dx$ at the wall remains unaltered.

As can be seen in Fig. 3.13, there are considerable differences between V_{el}^ψ and V_{el}^σ. However, it should be noted that the difference becomes significant in the region of $\kappa h < 1$. The coagulation of particles depends on the potential barrier appearing around $h = 1/\kappa$ [see Eq. (3.54)]. Therefore, as far as the coagulation experiments are concerned, it will not be important which model, V_{el}^ψ or V_{el}^σ, is used for double-layer interaction. The significant difference may be expected when repeptization or detachment of particles is considered, where a primary minimum appearing at a very short separation plays an important role.

The potential energy of interaction at constant surface charge for spherical particles is calculated by the Derjaguin method and was given by Wies and Healy [48] as follows:

$$\begin{aligned}V_{el(sp)}^\sigma &= \frac{\varepsilon a_1 a_2}{4(a_1 + a_2)}(\psi_1^2 + \psi_2^2)\left\{\frac{2\psi_1\psi_2}{(\psi_1^2 + \psi_2^2)}\ln\frac{1 + \exp(-\kappa H_0)}{1 - \exp(-\kappa H_0)}\right.\\ &\quad \left. - \ln[1 - \exp(-2\kappa H_0)]\right\}\\ &= V_{e\ell(sp)}^\psi - \frac{\varepsilon a_1 a_2}{2(a_1 + a_2)}(\psi_1^2 + \psi_2^2)\ln[1 - \exp(-2\kappa H_0)]\end{aligned}$$

(3.92)

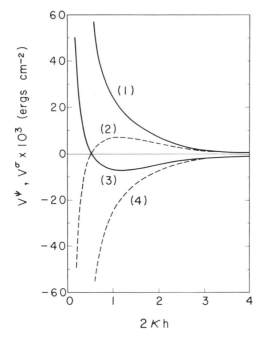

Figure 3.13 V_{el}^{σ} as a function of $2\kappa h$. (1) V^{σ} and (2) V^{ψ}, for both $\psi_1 = 10$ mV, $\psi_2 = 30$ mV; (3) V^{σ} and (4) V^{ψ}, for both $\psi_1 = -10$ mV, $\psi_2 = 30$ mV; $\kappa = 10^6$ cm^{-1}, v = 1. (From Ref. 47.)

In some cases (e.g., face-to-edge interaction of clay particles) a mixed model in which a constant potential for one plate (a) and a constant charge for the other plate (b) is appropriate to describe the system. Kar, Chander, and Mika [49] have given the following potential energy of interaction:

$$V_{el}^{\psi-\sigma} = \frac{\epsilon\kappa}{8\pi}[2\psi_1\psi_2 \text{ sech } 2\kappa h + (\psi_2^2 - \psi_1^2)(\tanh 2\kappa h - 1)] \quad (3.93)$$

$V_{el}^{\psi-\sigma}$ lies between V_{el}^{ψ} and V_{el}^{σ} under corresponding conditions.

As has been mentioned, for the case of constant charge process, the Debye-Hückel approximation must be accepted with reservation. Readers are referred to the literature [46,50-54] for exact calculations of the double-layer interaction at constant surface charge and for various surface conditions.

3.5.3 Regulation Model

The constant potential and constant charge models represent idealized and extreme situations. The actual interaction must lie somewhere between them. Chan et al. [55,56], as an extension of the Ninham and

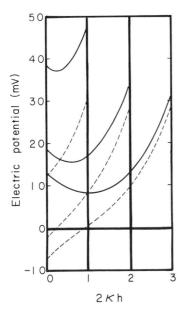

Figure 3.14 Electric potential between dissimilar plates as a function of $2\kappa h$ on the basis of constant charge model. Solid line, $\psi_1 = 10$ mV, $\psi_2 = 30$ mV; dashed line, $\psi_1 = -10$ mV, $\psi_2 = 30$ mV. (From Ref. 47.)

Parsegian model [57], presented the regulation model by which the surface charge and the surface potential can be obtained as functions of the separation distance assuming neither constant potential nor constant charge. The model assumes the surface reaction described by Eqs. (2.79) and (2.80), from which the surface charge densities σ_1 and σ_2 are determined by using Eq. (2.85). Based on the Gouy-Chapman model of the electrical double layer, σ_1 and σ_2 should also be required to satisfy the following conditions:

$$\left(\frac{d\psi}{dx}\right)_{x=0} = -\frac{4\pi\sigma_1}{\varepsilon} \tag{3.94}$$

$$\left(\frac{d\psi}{dx}\right)_{x=2h} = +\frac{4\pi\sigma_2}{\varepsilon} \tag{3.95}$$

The surface potentials, ψ_1 and ψ_2, are determined in such a way that σ_1 and σ_2 obtained by using Eq. (2.85) should be equal to σ_1 and σ_2 obtained from Eqs. (3.94) and (3.95). Since heterointeractions are rather complex, ψ_0 is shown in Fig. 3.15 as a function of the separation distance for homointeractions for $pK_+ = 3$, $pK_- = 9$, $pH_{PZC} = 6$,

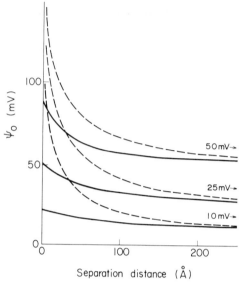

Figure 3.15 ψ_0 as a function of separation distance in the case of homointeraction. $pK_+ = 3$, $pK_- = 9$, $pH_{PZC} = 6$, $c = 1 \times 10^{-3}$ mol/liters 1-1 electrolyte. Solid line, regulation model; dashed line, constant charge model. (From Ref. 55.)

and $c = 1 \times 10^{-3}$ mol/liter (1-1 electrolyte). In the same figure, ψ_0 for a constant charge process is included for comparison. By knowing ψ_1, ψ_2, σ_1, and σ_2 as functions of the separation distance the potential energy of interaction can be calculated. For heterointeractions, there are cases where the potential energy curves show complex behavior which has never been predicted by previous theories. As an example, in Fig. 3.16 the pressure, surface potentials, and surface charge densities as functions of separation distance are shown for a system in which ψ_1 and ψ_2 are of the same sign and of different magnitudes. The regulation model is based on surface site equilibria and may be called the constant chemical potential model. Perfect surface regulation corresponds to the constant potential model, poor regulation to the constant charge model. James, Homola, and Healy [58] examined heterocoagulation between an amphoteric (negatively charged) latex and a conventional negative sulfate latex and found evidence of regulation-type behavior. Prieve and Ruckenstein [59] calculated the rate of particle deposition in view of the heterocoagulation, where surface dissociation equilibria as well as the constant charge case were discussed.

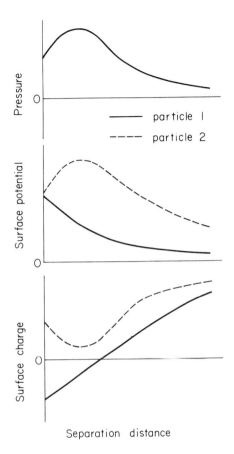

Figure 3.16 Pressure, surface potentials, and surface charge densities as functions of separation distance in the case where ψ_1 and ψ_2 are of the same sign and of different magnitude on the basis of regulation model. (From Ref. 56.)

3.5.4 London-van der Waals Interaction Between Dissimilar Substances

The London-van der Waals interaction energy between particles of dissimilar nature is also described by equations presented in Sec. 3.2. However, there are some points to be mentioned about the Hamaker constant which characterizes the heterointeraction in a liquid medium.

The Hamaker constant in the system in which particle 1 and particle 2 are embedded in a liquid medium, 3, $A_{12/3}$, is described as follows:

$$A_{12/3} = A_{12} + A_{33} - (A_{13} + A_{23}) \tag{3.96}$$

where A_{12}, A_{33}, A_{13}, and A_{23} are Hamaker constants for systems 1-2, 3-3, 1-3, and 2-3 in a vacuum. Applying the geometrical mean law, $A_{ij} = (A_{ii}A_{jj})^{1/2}$, we obtain

$$A_{12/3} = (A_{11}^{1/2} - A_{33}^{1/2})(A_{22}^{1/2} - A_{33}^{1/2}) \qquad (3.97)$$

Equation (3.97) indicates that if

$$A_{11} < A_{33} < A_{22} \quad \text{or} \quad A_{11} > A_{33} > A_{22} \qquad (3.98)$$

then $A_{12/3}$ becomes negative. This means that the London-van der Waals interaction between particles 1 and 2 in a liquid medium 3 becomes repulsive. Phenomena in relation to negative van der Waals forces were reviewed by van Oss, Absolom, and Neumann [60] and Visser [61].

Heterocoagulation is also governed by the total potential energy between particles, which consists of double-layer interaction energy and London-van der Waals interaction energy. Derjaguin pointed out that where the Hamaker constant is negative, a singular coagulation phenomenon is expected; coagulation takes place in low electrolyte concentration, while no coagulation occurs in high electrolyte concentration. Details are given in Refs. 38 and 43.

3.5.5 Examples of Heterocoagulation

Experimental studies on heterocoagulation were carried out in various fields, such as water treatment, detergency, and flotation. Model experiments using the rotating disk method were reported [62-64]. Usui, Yamasaki, and Shimoiizaka [21], using a twin dropping mercury electrode, demonstrated that the coalescence of unequally polarized mercury droplets in aqueous solutions (10^{-3}-10^{-2} mol/liter) of KF, which show no specific adsorption on mercury surfaces, was interpreted quantitatively in terms of heterocoagulation theory. Heterocoalescence of mercury droplets in aqueous solutions of KI, which exhibit strong specific adsorption on mercury surfaces, was explained semiquantitatively in terms of ψ_δ. Usui and Yamasaki [65] examined the attachment between mercury and glass in aqueous solutions of KF (10^{-3}-10^{-1} mol/liter) by varying the polarization potential of mercury and demonstrated that the heterocoagulation theory accounts for the experimental results up to the KF concentration of 1.5×10^{-1} mol/liter. Above this concentration, the attachment became weak. They also used the heterocoagulation theory to study the detachment between mercury and glass in the same system and estimated the additional repulsion, which was thought to be responsible for the weakening of the attachment at higher concentrations. A brief review up to 1970 is available in the literature [43].

Recently, Tokiwa and Imamura [66,67] studied the deposition of hematite dirt particles on various fabrics in detergent systems in view of heterocoagulation. They stated that the ratio of the depth of the primary minimum to the potential barrier height is of importance in washing processes. Suzawa et al. [68,69] studied the deposition of polystyrene latex onto various fabrics and analyzed the results in terms of the heterocoagulation theory. Watanabe and Tagawa [70] carried out an interesting experiment in which soil particles were removed from the capillary walls of cotton cloth by electroosmotic flow which was produced by an electric field applied to the cloth. Healy, Wiese, and co-workers [71,72] investigated the heterocoagulation of mixed oxide colloidal dispersions, SnO_2-TiO_2 and Al_2O_3-TiO_2, in terms of the stability ratio as a function of pH, where dissolution of oxides made the coagulation phenomena complex. James, Homola, and Healy [58] extended their study to a system in which an amphoteric latex was used to check the theory. Kitahara and Ushiyama [73] examined the homo- and heterocoagulation in a mixed dispersion of polystyrene latices having different zeta potentials and different sizes.

In a binary mixture of two kinds of particles (1 and 2) the overall stability ratio W_t is given by [40]

$$\frac{1}{W_t} = \frac{n_1^2}{W_{11}} + \frac{n_2^2}{W_{22}} + \frac{2n_1 n_2}{W_{12}} \tag{3.99}$$

where W_{11} and W_{22} are stability ratios for homocoagulation and W_{12} is that for heterocoagulation, for the system of particles 1 and 2, and n_1 and n_2 are the fraction of primary particles 1 and 2 ($n_1 + n_2 = 1$).

Bleier and Matijević [74] carried out a quantitative check of the heterocoagulation theory, including Eq. (3.99), using poly(vinyl chloride) latex and well-defined monodispersed spherical chromium(III) hydrous oxide. They found semiquantitative agreement between the experimental data and the theory and offered a few explanations of the discrepancies. They extended the heterocoagulation study to other binary (PVC latex-Ludox HS silica) [75] and ternary (PVC-Ludox silica-aluminum species) [76] systems. Hansen and Matijević [77] analyzed the adsorption of fine latex particles on much larger spherical hydrated aluminum oxide and hematite particles. Sasaki, Matijević, and Barouch [78] studied the coagulation behavior of polystyrene latex particles and aluminum oxide sol. In the two cases cited above, the experimental results were better explained by the BMRF equation based on the two-dimensional Poisson-Boltzmann equation [41]. Kolakowski and Matijević [79] and Kuo and Matijević [80, 81] carried out a series of experiments regarding adhesion and removal of particles on glass or steel as a function of pH using monodispersed chromium hydroxide and ferric oxide. The particle removal studies promise a better understanding of heterocoagulation.

The effect on heterocoagulation of particle size and the size ratio between particles was studied by Harding [82]. Samygin, Barskii, and Angelova [83] demonstrated that the coagulation rate between small and large particles is larger than that between small particles. In a mixed system containing large particles, the inertia contribution to the force should be taken into consideration. Selective flocculation as one of the applications of heterocoagulation is of technical importance, for which readers are referred to articles of Kitchener and co-workers [84,85].

3.6 THIN LIQUID FILMS

Surface tension is regarded as the reversible work of bringing molecules from the interior of the bulk phase to the surface in order to increase the surface by a unit area. A molecule b in the bulk interacts with its surrounding molecules and experiences no net force, while a molecule S in the surface interacts with fewer molecules and is pulled inward (see Fig. 3.17). In other words, surface molecules have an excess of free energy compared to molecules in the bulk. When the thickness of a free liquid film becomes small, the chemical potential of a molecule f in the film becomes larger than that of a molecule b in the bulk (i.e., $\mu_f > \mu_b$) (see Fig. 3.18). If such a film is connected to a bulk liquid, molecules in the thin film move spontaneously toward the bulk (drainage) to thin the film. Derjaguin defined the disjoining pressure π of a thin film as follows [86]:

$$\Delta \mu (h) = \mu_h - \mu_b = -\pi \bar{v} \tag{3.100}$$

where h is the thickness of the film and $\bar{v}$ is the molar volume of film material. π consists of various components [87]:

$$\pi = \pi_{el} + \pi_{VW} + \pi_a + \pi_{st} \tag{3.101}$$

where π_{el} refers to the overlapping of the electrical double layers, π_{VW} the London-van der Waals forces, π_a the overlapping of adsorbed

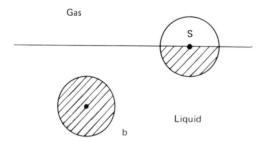

Figure 3.17 Schematic illustration of molecules in the surface and in the bulk. Circles represent the sphere of action.

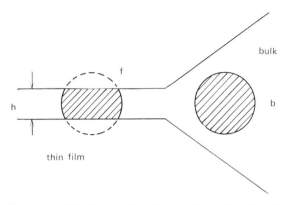

Figure 3.18 Schematic illustration of molecules in the thin film and in the bulk.

molecules, and π_{st} the deformation of the structure of the polar liquid from that of the bulk. The π_{st} (structural component) is of significance in both theoretical and experimental aspects and is a subject for future study. If π consists only of $\pi_{VW}(<0)$, then $\Delta\mu(h) > 0$ and thinning of the film proceeds.

For soap films in which double-layer interaction comes into play, π_{el} (>0) opposes π_{VW} and thickens the film. Approximate equations for π_{el} and π_{VW} are given by

$$\pi_{el} = 64 n k T \gamma^2 \exp(-\kappa h) \tag{3.102}$$

$$\pi_{VW} = -\frac{A}{6\pi h^3} \tag{3.103}$$

The equations above are identified with Eqs. (3.13) and (3.42), respectively, indicating that the stability of films is substantially the same as the colloid stability with respect to the acting forces. Derjaguin and Titijevskaya [88] measured the thickness of thin films formed between two bubbles in electrolyte solutions containing fatty acid soaps as a function of the bubble pressure with which the disjoining pressure of the film is equilibrated. The π versus h curve obtained was explained using the double-layer interaction theory with an appropriate value of ψ_0. Scheludko and Exerowa [89,90] made a microscopic circular thin liquid film in a narrow glass tube (R = 3 mm) by drawing a liquid from a branched capillary tube and measured the equilibrium film thickness (h) by varying the electrolyte (KCl) concentration (c) (Fig. 3.19). The film thickness was measured by automatic photoelectric recording of reflection intensity, whereby a film thickness as thin as some tens of angstrom units could be measured

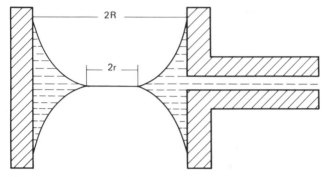

Figure 3.19 Formation of a microscopic circular thin liquid film. (From Ref. 90.)

precisely. Here the disjoining pressure π was identified with the capillary pressure and was approximated by

$$\pi = \frac{2\gamma}{R} \tag{3.104}$$

where R is the radius of the cylinder in which a microscopic thin liquid film is formed and γ is the surface tension of film liquid. In the case of low electrolyte concentrations, (large film thickness) the contribution of π_{VW} can be neglected and hence the equilibrium film thickness is established in such a way that $\pi = \pi_{el}$. Experimentally determined h versus c curves were reasonably explained by Eq. (3.102) when $\psi_0 = 90$ mV was assumed. In the case of high electrolyte concentrations (small film thickness), where π_{VW} can no longer be ignored, π_{VW} was estimated as the difference between π and π_{el} It was found that π_{VW} determined in this way obeyed the one-third power law [see Eq. (3.103)].

Lyklema and Mysels [91] measured the equilibrium thickness of soap (sodium dodecyl sulfate and sodium tetradecyl sulfate) films with various ionic strengths of inorganic supporting electrolyte (LiCl) to check the DLVO theory. Experimental results were explained qualitatively by the DLVO theory, but some discrepancies were indicated in the quantitative aspects. In the case of high ionic strength, it was necessary to assume an additional repulsion to explain the results, while in the case of low ionic strength, the London-van der Waals forces obeyed the one-third power law (nonretarded), in contrast to the theory, which predicted the one-fourth power law (retarded).

There are various systems of thin liquid films other than gas-liquid-gas: for example, gas-liquid-liquid and gas-liquid solid. The former is concerned with the spreading of oil on water and the latter with flotation. In the system consisting of gas (1), liquid (2), and liquid (solid) (3), the Hamaker constant A is given by [see Eq. (3.96)]

$$A = A_{13} + A_{22} - A_{12} - A_{23} \tag{3.105}$$

By neglecting the terms including the gas phase (1),

$$A = A_{22} - A_{23} \tag{3.106}$$

Using the geometric mean law for A_{23}, we obtain

$$A = A_{22}^{1/2}(A_{22}^{1/2} - A_{33}^{1/2}) \tag{3.107}$$

It is evident that if $A_{22} < A_{33}$, then $A < 0$ ($\pi_{VW} > 0$) and the liquid film (2) is stable. If $A_{22} > A_{33}$, then $A > 0$ ($\pi_{VW} < 0$) and the liquid film (2) is unstable. Schulze and Cichos [92] studied the stability of thin aqueous films formed between a quartz plate and a small air bubble, in which the negative Hamaker constant (-1×10^{-13} erg) explained the experimental results in solutions of high ionic strength. Nakamura and Uchida [93] analyzed the thinning phenomena of the dimple between a glass plate and an air bubble in terms of the heterocoagulation theory. For other studies on thin liquid films, readers are referred to the literature [94-96]. A useful review of recent studies of thin films has been given by Buscall and Ottewill [97].

3.7 CONCENTRATED DISPERSION

3.7.1 Introduction

There are three main subjects in the study of colloid science; one is on the attributes of individual particles, the size, shape, surface potential, and so on; the second is the stability of dispersion or the interaction between the particles; and the last is the statistical properties of stably dispersed colloid systems, which is based on the knowledge and theories of the first two.

Statistical studies thus far made have been only on very dilute systems that correspond to the ideal gas systems. They were observation of Brownian movement of individual particles [98], experiments on sedimentation equilibrium [98,99], measurements of osmotic pressure, and so on. In spite of the fact that systems which colloid scientists encounter in practice (in industry or biology) are almost always concentrated systems, the statistical study on concentrated systems has been scarce thus far [100].

This is probably because of the difficulty in preparing monodisperse systems. Until recently, colloid scientists could not prepare a significant amount of stable monodisperse colloid of high concentration. In this sense, the appearance of monodisperse latex in 1947 [101] can be said to be an epoch-making event in colloid science.

It is now possible to prepare a stable suspension of highly monodisperse microspheres of synthetic polymers with concentrations of up to 0.4 in volume fraction, and a number of studies have been done on the structure of iridescent latices [102-109], the transition behavior from the milky white state to the iridescent state, on liquidlike structures in the milky white latices [110-113] and so on.

Among them, the most interesting is the transition behavior from the milky white state to the iridescent state. Its peculiarity is that in the first impression it looks as if it contradicts the principle of diffusion. It was finally explained by the Kirkwood-Alder transition, and at the same time requested us to refine our naive understanding of some fundamental properties of matter.

Before entering the main discourse, it is necessary to define concentrated systems. By the word "concentrated" we mean a situation where interaction between the particles is so strong that the behavior of the system deviates greatly from that of the ideal gases. This is seen in colloids when the diffuse part of the double layers extends over the whole space of the system, so that there is no place in the system free of the influence of the particles; this statement, however, does not hold when the particles are hard spheres or the double layers are very thin compared with the particle diameter. In this case, the volume fraction of the particles is taken as the measure. More generally, the "concentratedness" is defined by the concept of short-range order [111-113]. However, we do not discuss this matter, but tentatively say "concentrated" when a hard-sphere system has a volume fraction larger than 0.3. Since our main object is the Kirkwood-Alder transition that occurs at a higher concentration, this arbitrary definition is enough for our purpose. Interested readers are referred to the References.

Furthermore, the shape of the particles is not restricted to spherical; concentrated dispersions of anisometric particles also exhibit interesting behavior [114,115]. However, those are not in the scope of the present chapter.

3.7.2 Ordered Structure in Monodisperse Latices

A monodisperse latex is a suspension of minute (<1 μm) spheres of a synthetic high polymer with uniform size. It is prepared by emulsion polymerization. An electron micrograph of a polystyrene latex is shown in Fig. 3.20. The suspending particles are in most cases negatively charged and sheathed by an atmosphere of positive charge. Because of these electrical double layers, the particles interact with each other by a repulsive potential to make the dispersion stable.

As a rule, for spherical particles of moderate size ($\leq$ 5000 Å in diameter), the van der Waals attractive force is not strong; the interaction potential curve for a stably dispersed system of such particles has only a negligibly shallow secondary minimum. This is especially the case with polystyrene latices, the Hamaker constant of which is relatively small. Therefore, a stably dispersed polystyrene latex is effectively a system of repulsive interaction.

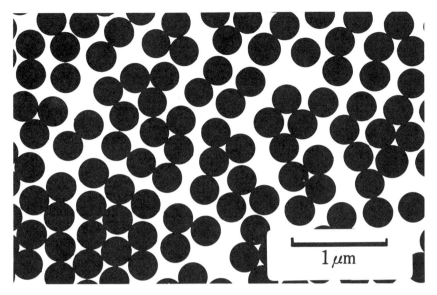

Figure 3.20 Electron micrograph of a polystyrene latex.

One of the most interesting characteristics of a monodisperse latex is the exhibition of iridescence upon desalting; a latex* that is milky white in appearance turns into a state of iridescence by dialysis, and comes back to the original milky white state when an electrolyte is added. This indicates that the appearance of the iridescence is the result of the expansion of electrical double layers by the decrease of electrolyte concentration of the medium, and vice versa.

As the iridescence must be due to some interference effect of scattered light by an ordered structure, and as the particles are spherical and the interaction between the particles is also of spherical symmetry, the structure would be that of a close-packing state of spheres, which must have been produced as a result of the expansion of the effective size of the particles.

There are two types of close packing, cubic type and hexagonal type. In the former, two-dimensional hexagonal arrays of the spheres superimpose one on another in the manner of ABCABC . . . , in the latter in the manner ABABAB . . . The former constitute the face-centered cubic lattice, whose (111) plane is the above-mentioned hexagonal plane. The actual structure in the latex was determined through

---

*For example, one 2000 Å in diameter, 0.2 in volume fraction, and where the electrolyte (KCl) concentration of the medium is 5×10^{-3} mol/liter.

spectroscopic study to be face-centered cubic by Luck et al. [102]. Of course, there are some irregularities in the manner of superposition of the (111) plane, and these are called stacking faults [105].

It is desirable to observe the structure directly with one's own eyes; and this is possible by the use of a metallurgical microscope [106]. Several examples are shown in Figs. 3.21-3.23. In these figures hexagonal arrays of the particles are seen. They must be (111) net plane of a face-centered cubic lattice. Lattice defects, dislocations (edge), and grain boundaries are seen. In Fig. 3.22, the latex is heavily desalted by ion-exchange resin, and the particle concentration is very low (0.01%); hexagonal and square lattices forming twin structures are seen. They must be (111) and (100) planes of face-centered cubic structure. The picture in Fig. 3.23 represents the state of coexistence of the two states, with the boundary actively fluctuating.

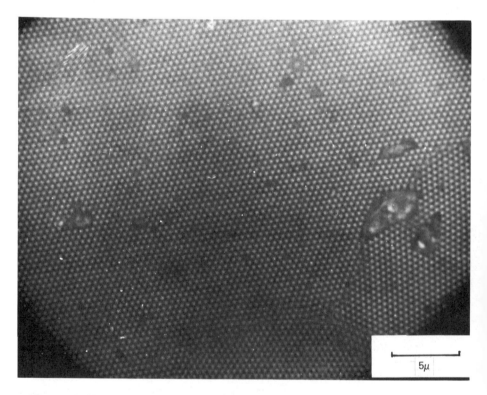

Figure 3.21 Hexagonal arrays of latex particles, representing the (111) net plane of face-centered cubic lattice. Edge dislocations and lattice defects are seen.

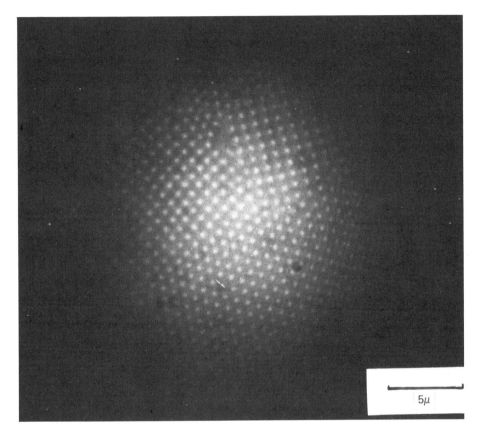

Figure 3.22 Heavily desalted latex of low-volume fraction. An enormous distance is seen between the particles. Hexagonal and square patterns coexist, forming a twin structure.

3.7.3 The Mechanism of the Ordered Formation

As already mentioned, the ordered structure in the latex which is responsible for the iridescence is the realization of the state of closest packing caused by the intensification of the repulsive interaction. This is agreeable only as far as we see the initial disordered state and the final ordered state. However, when we look into the process of the transition, a strange characteristic of the latex is revealed. In Fig. 3.24, a pictorial drawing is given which shows the process of transition in the course of dialysis. As seen, at a certain stage of the dialysis, a small amount of iridescent sediment deposits at the bottom of the dialysis tube. It is to be noted that the boundary is a sharp horizontal line. Thereafter, the sediment increases with the

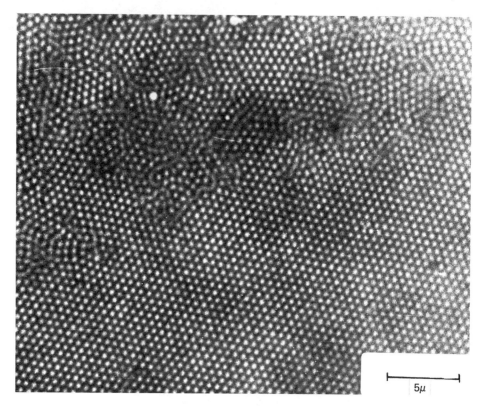

Figure 3.23 State of coexistence of ordered and disordered structures.

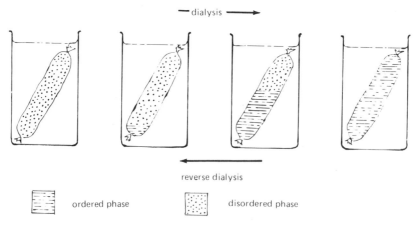

Figure 3.24 Diagrammatic picture of the process of dialysis.

process of dialysis, and finally the entire volume of the latex is occupied by the iridescence. The process can be reversed; the iridescence is made to disappear by the addition of electrolyte to the outer liquid of dialysis.

This behavior is surprising, since polystyrene has a higher density (ρ = 1.054) than water. The iridescent sediment must be more concentrated than the milky white supernatant. As already mentioned, in a stably dispersed latex the effect of the secondary minimum is negligible. Hence the latex should not separate into two phases, dense and less dense; this would be possible only when some attractive force is acting effectively.

How can these two phases coexist? Why does diffusion not take place to smear out the uneven distribution? The situation seems to contradict the fundamental principles of thermodynamics. Should we revise the theory of double-layer interaction in order to deduce some attractive force? The answer is no; the strange phenomenon can be explained by invoking the concept of the Kirkwood-Alder [116] transition.

3.7.4 Alder Transition

In the statistical physics of liquids, there has been the problem of whether any attractive force is necessary for a liquid to transform into a solid state. Kirkwood [117] first stated that when a hard-sphere system is gradually compressed, the system will transfer into the state of long-range order long before the state of the close packing is reached.

Later, Alder and Wainright [118] undertook the calculation of molecular dynamics for a two-dimensional system containing 870 elastic disks in which no attractive force is acting. They found that upon compression of the system volume, a transition from disordered state to ordered state takes place at a certain stage. The transition is of the first order and produces the state of coexistence of the two phases, the dense ordered state and the less dense disordered state.

In Fig. 3.25, the equation of state obtained by computation is shown. The abscissa is the area expressed by A/A_0, where A_0 is the area for close packing, and the ordinate is the pressure expressed by the ratio P/P_0, where P_0 is the ideal gas pressure at the close packing, NkT/A_0, where N is the number of particles. The equation of state states that when the system is compressed, the pressure increases smoothly in the earlier stage, and at a certain point the pressure starts to fluctuate heavily (vertical lines indicate the extent of the fluctuation). On further compression the fluctuation ceases, and thereafter the pressure increases smoothly again. In the fluctuating region, the two states, the ordered and the disordered, coexist.

By connecting the center of the fluctuating pressure, a curve similar to the van der Waals loop is obtained. If the number of particles

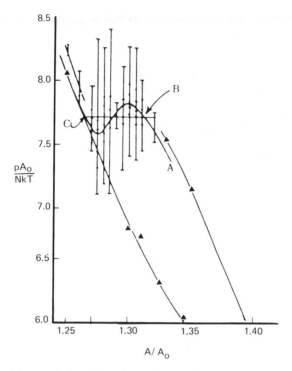

Figure 3.25 Equation of state for system of 870 hard disks.

is very large, this loop would be replaced by a horizontal line. The right end of the line represents the disordered state (the freezing point) and the left end represents the ordered state (the melting point). On the horizontal line, the two states coexist. In Fig. 3.26, the oscillographic display of the coexisting state is shown.

The discussion above describes a two-dimensional case. In a three-dimensional case, molecular dynamics require a great deal of calculation, and thus the transition points have been determined by other means [119]. This is shown in Fig. 3.27. The value of V/V_0 (V_0 is the system volume at the close packing) for the melting and freezing points are 1.31 and 1.51, respectively; in volume fractions they are 0.55 and 0.5, respectively. This transition in a hard-sphere system is called the Kirkwood-Alder transition [116].

Now the phase transition in a monodisperse latex is explained as follows. The expansion of the double-layer thickness (caused by the desalting of the system) increases the effective size of the particle or effective volume fraction of the particles. When the desalting reaches the level where the effective volume fraction is 0.5, a transition from the disordered to the ordered state occurs and a small amount of the

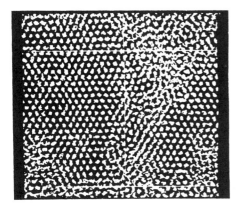

Figure 3.26 Oscillograph display of the coexisting state.

ordered phase deposits as an iridescent sediment. As the dialysis proceeds, the iridescent part increases and finally occupies the entire volume when the effective volume fraction reaches the value of 0.55. The coexisting state of the milky white and the iridescent phases corresponds to the horizontal part of the curve in Fig. 3.27. Thus the coexistence in the latex of the ordered and disordered phases of different density is explained at least qualitatively.

The transition behavior of a monodisperse latex has been studied extensively by Hachisu and co-workers [107-109] and the result is summarized in a phase diagram in Fig. 3.28. The ordinate is the volume fraction and the abscissa is the KCl concentration. The area is divided into three domains. The upper right is the ordered region, the lower left is the disordered region, and between them is the region of coexistence of the two phases. In the phase diagram, the coexisting behavior at high electrolyte concentration, where the effect of the electric double layer is small, is in good agreement with the calculated value of Alder transition.

3.7.5 Exposition of Alder Transition

Although the strange behavior of monodisperse latex has been explained by the Alder transition, we may not be fully satisfied with the explanation; we may feel a conflict between the concept of the Alder transition and our familiar belief that, without some attractive force, it is impossible for two phases of different concentrations to exist in equilibrium. We would contend as follows.

Let us imagine that there are two phases 1 and 2 of different concentrations (phase 1 being more concentrated) in contract. Free energies of the respective phases are

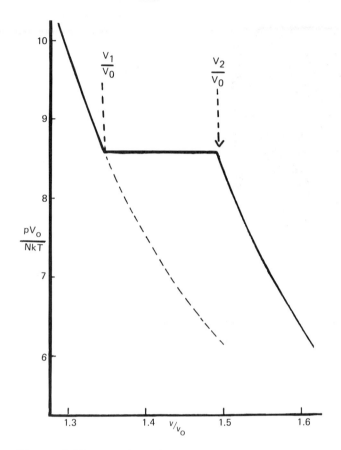

Figure 3.27 Equation of state for hard-sphere system.

$$F_1 = U_1 - S_1 T \quad \text{and} \quad F_2 = U_2 - S_2 T$$

Here, $S_1 < S_2$ (since phase 1 is more concentrated) and $U_1 = U_2$ (no interaction for a hard-sphere system). Therefore, the conclusion is that $F_1 > F_2$ and that phases 1 and 2 cannot coexist in equilibrium. However, this discussion is wrong, since the Helmholtz free energy has been used. In the present discussion Gibbs free energy is to be used, because the transition from phase 1 to phase 2 is made under constant T and P. The Gibbs free energies for the two phases are

$$G_1 = U_1 - S_1 T + PV_1 \quad \text{and} \quad G_2 = U_2 - S_2 T + PV_2$$

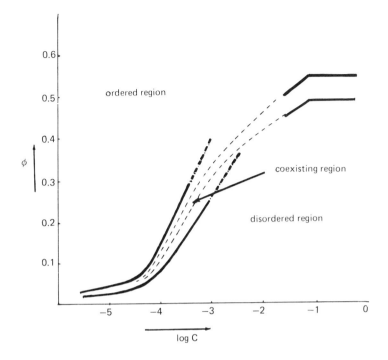

Figure 3.28 Phase diagram of polystyrene latex. Results of two independent studies are given by the solid line. The supposed true boundary line are indicated by the dashed line.

Since $U_1 = U_2$, the equilibrium condition $G_1 = G_2$ reduces to

$$T(S_2 - S_1) = P(V_2 - V_1)$$

This does not conflict with the condition that $V_2 > V_1$ and $S_2 > S_1$. Namely, thermodynamics never rejects the possibility that two phases of different densities coexist in equilibrium in a hard-sphere system. It is to be noted that the discussion above concerns the equation of state in Fig. 3.27, the two phases corresponding to the left and right ends of the horizontal part of the curve.

Thus it has been shown that the Alder transition never contradicts the principles of thermodynamics. The feeling of conflict would come from our naive understanding that on the equation of state there cannot be a horizontal region on the curve unless some attractive force is acting between the particles. This is the result of a careless extrapolation of our experience of dilute systems to the case of highly concentrated systems.

REFERENCES

1. B. V. Derjaguin and L. Landau, *Acta Physicochim., URSS, 14,* 633 (1941).
2. E. J. W. Verwey and J. Th. G. Overbeek, *Theory of the Stability of Lyophobic Colloids,* Elsevier, Amsterdam, 1948, pp. 96, 139, 168.
3. B. V. Derjaguin, *Kolloid-Z., 69,* 155 (1934).
4. G. M. Bell, S. Levine, and L. N. McCartney, *J. Colloid Interface Sci., 33,* 335 (1970).
5. H. C. Hamaker, *Physica, 4,* 1058 (1937).
6. J. Th. G. Overbeek, in *Colloid Science* (H. R. Kruyt, ed.), Vol. 1, Elsevier, Amsterdam, 1952, p. 268.
7. F. M. Fowkes, *Ind. Eng. Chem., 56,* 40 (1964).
8. I. E. Dzyaloshinski, E. M. Lifshitz, and L. P. Pitaevskii, *Adv. Phys., 10,* 165 (1961).
9. H. Krupp, *Adv. Colloid Interface Sci., 1,* 111 (1967).
10. J. Gregory, *Adv. Colloid Interface Sci., 2,* 396 (1969).
11. I. N. Israelachvili and D. Tabor, in *Progress in Surface and Membrane Sciences,* Vol. 7 (J. F. Danielli, M. D. Rosenberg, and D. A. Cadenhead, eds.), Academic Press, New York, 1973, p. 1.
12. V. A. Parsegian, in *Physical Chemistry: Enriching Topics from Colloid and Surface Science* (H. van Olphen and K. J. Mysels, eds.), Theorex, La Jolla, Calif., 1975, p. 27.
13. J. Mahanty and B. W. Ninham, *Dispersion Forces,* Academic Press, London, 1976.
14. J. Visser, *Adv. Colloid Interface Sci., 3,* 331 (1972); also in *Surface and Colloid Science,* Vol. 9 (E. Matijević, ed.), Wiley, 1976, p. 3.
15. D. B. Hough and L. R. White, *Adv. Colloid Interface Sci., 14,* 3 (1980).
16. J. Th. G. Overbeek, *Pure Appl. Chem., 52,* 1151 (1980).
17. A. Watanabe and R. Gotoh, *Kolloid-Z., 191,* 36 (1963).
18. A. Watanabe, M. Matsumoto, and R. Gotoh, *Kolloid-Z., 201,* 147 (1965).
19. M. Matsumoto, *Bull. Inst. Chem. Res. Kyoto Univ., 47,* 354 (1969).
20. M. Matsumoto, *Nippon Kagaku Zasshi, 91,* 708 (1970).
21. S. Usui, T. Yamasaki, and J. Shimoiizaka, *J. Phys. Chem., 71,* 3195 (1967).
22. J. N. Israelachvili and G. E. Adams, *J. Chem. Soc. Faraday Trans. I, 74,* 975 (1978).
23. R. M. Pashley and J. N. Israelachvili, *Colloids Surf., 2,* 169 (1981).
24. J. Th. G. Overbeek, in *Colloid Science* (H. R. Kruyt, ed.), Vol. 1, Elsevier, Amsterdam, 1952, p. 302.

25. S. Hachisu and K. Furusawa, *Sci. Light*, *12*, 157 (1963).
26. T. G. M. van de Ven and S. G. Mason, *J. Colloid Interface Sci.*, *57*, 517, 505 (1977).
27. K. Takamura, H. L. Goldsmith, and S. G. Mason, *J. Colloid Interface Sci.*, *72*, 385 (1979).
28. J. Th. G. Overbeek, in *Colloid Science* (H. R. Kruyt, ed.), Vol. 1, Elsevier, Amsterdam, 1952, p. 278.
29. H. Reerink and J. Th. G. Overbeek, *Discuss. Faraday Soc.*, *18*, 74 (1954).
30. J. Th. G. Overbeek, in *Colloid Science* (H. R. Kruyt, ed.), Vol. 1, Elsevier, Amsterdam, 1952, p. 320.
31. R. H. Ottewill and J. N. Shaw, *Discuss. Faraday Soc.*, *42*, 154 (1966).
32. L. A. Spielman, *J. Colloid Interface Sci.*, *33*, 562 (1970).
33. E. P. Honig, G. J. Roebersen, and P. H. Wiersema, *J. Colloid Interface Sci.*, *36*, 97 (1971).
34. J. M. Deutch and B. U. Felderhof, *J. Chem. Phys.*, *59*, 1669 (1973).
35. H. Brenner, *Chem. Eng. Sci.*, *16*, 242 (1961).
36. J. Th. G. Overbeek, *J. Colloid Interface Sci.*, *58*, 408 (1977).
37. B. V. Derjaguin, S. S. Dukhin, and V. N. Shilov, *Adv. Colloid Interface Sci.*, *13*, 141 (1980).
38. B. V. Derjaguin, *Discuss. Faraday Soc.*, *18*, 85 (1954).
39. O. F. Devereux and P. L. de Bruyn, *Interaction of Plane-Parallel Double Layers*, MIT Press, Cambridge, Mass., 1963.
40. R. Hogg, T. W. Healy, and D. W. Fuerstenau, *Trans. Faraday Soc.*, *62*, 1638 (1966).
41. E. Barouch, E. Matijević, T. A. Ring, and J. M. Finlan, *J. Colloid Interface Sci.*, *67*, 1 (1978); *70*, 400 (1979).
42. D. Y. C. Chan and L. R. White, *J. Colloid Interface Sci.*, *74*, 303 (1980).
43. S. Usui, in *Progress in Surface and Membrane Science*, Vol. 5 (J. F. Danielli, M. D. Rosenberg, and D. A. Cadenhead, eds.), Academic Press, New York, 1972, p. 223.
44. G. Frens, D. J. C. Engel, and J. Th. G. Overbeek, *Trans. Faraday Soc.*, *63*, 418 (1967).
45. G. Frens and J. Th. G. Overbeek, *J. Colloid Interface Sci.*, *38*, 376 (1972 .
46. G. M. Bell and G. C. Peterson, *J. Colloid Interface Sci.*, *41*, 542 (1972).
47. S. Usui, *J. Colloid Interface Sci.*, *44*, 107 (1973).
48. G. R. Wiese and T. W. Healy, *Trans Faraday Soc.*, *66*, 490 (1970).
49. G. Kar, S. Chander, and T. S. Mika, *J. Colloid Interface Sci.*, *44*, 347 (1973).

50. V. M. Muller, in *Research in Surface Forces* (B. V. Derjaguin, ed.), Vol. 3, Consultants Bureau, New York, 1971, p. 236.
51. J. E. Jones and S. Levine, *J. Colloid Interface Sci.*, *30*, 241 (1969).
52. H. Ohshima, *Colloid Polym. Sci.*, *252*, 158, 257 (1974); *253*, 150, 158 (1975); *254*, 484 (1976); *257*, 630 (1979).
53. J. Gregory, *J. Chem. Soc. Faraday Trans. II*, *69*, 1723 (1973); *J. Colloid Interface Sci.*, *51*, 44 (1975).
54. G. M. Bell and G. C. Peterson, *J. Colloid Interface Sci.*, *60*, 376 (1977); *Can. J. Chem.*, *59*, 1888 (1981).
55. D. Chan, J. W. Perram, L. R. White, and T. W. Healy, *J. Chem. Soc. Faraday Trans. I*, *71*, 1046 (1975).
56. D. Chan, T. W. Healy, and L. R. White, *J. Chem. Soc. Faraday Trans. I*, *72*, 2844 (1976).
57. B. W. Ninham and V. A. Parsegian, *J. Theor. Biol.*, *31*, 405 (1971).
58. R. O. James, A. Homola, and T. W. Healy, *J. Chem. Soc. Faraday Trans. I*, *73*, 1436 (1977).
59. D. C. Prieve and E. Ruckenstein, *J. Colloid Interface Sci.*, *60*, 337 (1977).
60. C. J. Van Oss, D. R. Absolom, and A. W. Neumann, *Colloids Surf.*, *1*, 45 (1980).
61. J. Visser, *Adv. Colloid Interface Sci.*, *15*, 157 (1977).
62. J. K. Marshall and J. A. Kitchener, *J. Colloid Interface Sci.*, *22*, 342 (1966).
63. M. Hull and J. A. Kitchener, *Trans. Faraday Soc.*, *65*, 3093 (1969).
64. G. E. Clint, J. H. Clint, J. M. Corkill, and T. Walker, *J. Colloid Interface Sci.*, *44*, 121 (1973).
65. S. Usui and T. Yamasaki, *J. Colloid Interface Sci.*, *29*, 629 (1969).
66. T. Imamura and F. Tokiwa, *Nippon Kagaku Kaishi*, *1972*, 2177 (1972); *1973*, 648, 2051 (1973); *1974*, 405 (1974); *1975*, 943 (1975); *1976*, 869 (1976).
67. F. Tokiwa and T. Imamura, *Yukagaku*, *26*, 143 (1977).
68. T. Suzawa, H. Tamai, H. Shirahama, and K. Yamamoto, *Nippon Kagaku Kaish* , *1979*, 16 (1979).
69. H. Tamai, T. Hakozaki, and T. Suzawa, *Colloid Polym. Sci.*, *258*, 870 (1980).
70. A. Watanabe and M. Tagawa, *J. Colloid Interface Sci.*, *75*, 218 (1980).
71. T. W. Healy, G. R. Wiese, D. E. Yates, and B. V. Kavanagh, *J. Colloid Interface Sci.*, *42*, 647 (1973).
72. G. R. Wiese and T. W. Healy, *J. Colloid Interface Sci.*, *52*, 458 (1975).
73. A. Kitahara and H. Ushiyama, *J. Colloid Interface Sci.*, *43*, 73 (1973).

74. A. Bleier and E. Matijević, J. Colloid Interface Sci., 55, 510 (1976).
75. A. Bleier and E. Matijević, J. Chem. Soc. Faraday Trans. I, 74, 1346 (1978).
76. E. Matijević and A. Bleier, Croat. Chem. Acta, 50, 93 (1977).
77. F. K. Hansen and E. Matijević, J. Chem. Soc. Faraday Trans. I, 76, 1240 (1980).
78. H. Sasaki, E. Matijević, and E. Barouch, J. Colloid Interface Sci., 76, 319 (1980).
79. J. E. Kolakowski and E. Matijević, J. Chem. Soc. Faraday Trans. I, 75, 65 (1979).
80. J. Kuo and E. Matijević, J. Chem. Soc. Faraday Trans. I, 75, 2014 (1979).
81. J. Kuo and E. Matijević, J. Colloid Interface Sci., 78, 407 (1980).
82. R. D. Harding, J. Colloid Interface Sci., 40, 164 (1972).
83. V. D. Samygin, A. A. Barskii, and S. M. Angelova, Colloid J. (English Trans.) 30, 435 (1969).
84. R. J. Pugh and J. A. Kitchener, J. Colloid Interface Sci., 35, 656 (1970).
85. B. Yarar and J. A. Kitchener, Trans. Inst. Min. Metall., 79, C23 (1970).
86. A. Scheludko, Adv. Colloid Interface Sci., 1, 391 (1967).
87. B. V. Derjaguin, Z. M. Zorin, N. V. Churaev, and V. A. Shishin, in Wetting, Spreading and Adhesion (J. F. Paddy, ed.), Academic Press, London, 1978, p. 201.
88. B. V. Derjaguin and A. S. Titijevskaya, Discuss. Faraday Soc., 18, 24 (1954).
89. A. Scheludko and D. Exerowa, Kolloid-Z., 165, 148 (1959); 168, 27 (1960).
90. A. Scheludko, Colloid Chemistry, Elsevier, Amsterdam, 1966.
91. J. Lyklema and K. J. Mysels, J. Am. Chem. Soc., 87, 2539 (1965).
92. H. J. Schulze and C. Cichos, Z. Phys. Chem. (Leipzig), 251, 252, 145 (1972).
93. M. Nakamura and K. Uchida, J. Colloid Interface Sci., 78, 479 (1980).
94. K. J. Mysels, K. Shinoda, and S. Frankel, Soap Films, Pergamon, London, 1959.
95. J. A. Kitchener, in Recent Progress in Surface Science, Vol. 1 (J. F. Danielli, K. G. A. Pankhurst, and A. C. Riddiford, eds.), Academic Press, New York, 1964, p. 51.
96. J. S. Clunie, J. F. Goodman, and B. T. Ingram, in Surface and Colloid Science, Vol. 3 (E. Matijević, ed.), Wiley-Interscience, New York, 1971, p. 167.
97. R. Buscall and R. H. Ottewill, in Colloid Science (D. H. Everett, ed.), Vol. 2, The Chemical Society, London, 1975, p. 191.

98. J. Perrin, *The Atom*, Constable, London, 1923.
99. M. McDowell and F. L. Usher, *Proc. R. Soc. A*, *138*, 133 (1932).
100. R. F. Steiner, Ph.D. thesis, Harvard University, 1950.
101. J. W. Vanderhoff, H. J. Van del Hul, J. M. Tausk, and J. Th. G. Overbeek, in *Clean Surfaces: Their Preparation and Characterization for Interface Studies* (G. Goldfinger, ed.), Marcel Dekker, New York, 1970.
102. W. Luck, M. Klier, and H. Weslau, *Bunsenges. Phys. Chem.*, *67*, 75 (1963).
103. I. M. Krieger and F. M. O'Neill, *J. Am. Chem. Soc.*, *90*, 3144 (1968).
104. P. A. Hiltner and I. M. Krieger, *J. Phys. Chem.*, *73*, 2386 (1969).
105. P. A. Hiltner, Y. S. Papier, and I. M. Krieger, *J. Phys. Chem.*, *75*, 1881 (1971).
106. A. Kose, K. Ozaki, K. Takano, Y. Kobayashi, and S. Hachisu, *J. Colloid Interface Sci.*, *44*, 330 (1973).
107. S. Hachisu, Y. Kobayashi, and A. Kose, *J. Colloid Interface Sci.*, *42*, 342 (1973).
108. A. Kose and S. Hachisu, *J. Colloid Interface Sci.*, *46*, 460 (1974).
109. Y. Kobayashi and S. Hachisu, *J. Colloid Interface Sci.*, *46*, 470 (1974).
110. J. C. Brown, P. N. Pusey, and R. Dietz, *J. Chem. Phys.*, *62*, 1936 (1975).
111. K. Takano and S. Hachisu, *J. Colloid Interface Sci.*, *66*, 124 (1978).
112. E. A. Nieuwenhius and A. Vrij, *J. Colloid Interface Sci.*, *72*, 321 (1979).
113. A. K. Van Helden and A. Vrij, *J. Colloid Interface Sci.*, *78*, 312 (1980).
114. J. Th. G. Overbeek, in *Colloid Science* (H. R. Kruyt, ed.), Vol. 1, Elsevier, Amsterdam, 1952, pp. 326, 332.
115. Y. Maeda and S. Hachisu, *Colloids Surf.* *6*, 1 (1983).
116. M. Wadachi and M. Toda, *J. Phys. Soc. Jpn.*, *32*, 1147 (1972).
117. I. G. Kirkwood, *J. Chem. Phys.*, *7*, 919 (1939).
118. B. J. Alder and T. E. Wainright, *Phys. Rev.*, *127*, 359 (1962).
119. B. J. Alder, H. G. Hoover, and D. A. Young, *J. Chem. Phys.*, *49*, 3688 (1968).

4
Electrokinetics

Sei Hachisu* The University of Tsukuba, Sakura, Ibaraki, Japan

4.1. Outline of Electrokinetic Phenomena 99
4.2. Theories of Electrokinetic Phenomena 102
    4.2.1. Introduction 102
    4.2.2. Electroosmosis 103
    4.2.3. Streaming potential and sedimentation potential 107
4.3. Electrophoresis 108
    4.3.1. Theoretical treatment ignoring relaxation 108
    4.3.2. Effect of relaxation 111
    4.3.3. Present status of theory of electrophoresis 114
4.4. Nonequilibrium Thermodynamics of Electrokinetic Phenomena 115
    References 117

4.1. OUTLINE OF ELECTROKINETIC PHENOMENA

In Chap. 1 the origin of electric charges at the interface has been described and the three important potentials relating to the interface, ϕ, χ, and ψ_0, have been discussed. It has also been stated that the surface potential ψ_0 is produced by deposition or adsorption at the interface of ions, and its direct measurement is, in principle, possible. The electrokinetic phenomena described in this chapter are dependent on the surface potential ψ_0, and its value can be approached by studying these phenomena in a quantitative way with great care. A brief outline of the phenomena is given below.

---

*Currently Professor Emeritus.

When electrical charges are generated at an interface, specifically a solid/liquid interface, a diffuse layer of electric charges is formed on the liquid-phase side of the interface. If an external electrid field is applied to this system, a relative movement of the two phases will take place with respect to each other since the sign of the charge is opposite on both sides of the interface (the electric force acts on the two phases in opposite directions). Conversely, if a relative motion is caused by an external force tangential to the interface, an electric field will be generated. These phenomena are called electrokinetic phenomena (Fig. 4.1). Electroosmosis and electrophoresis are examples of the first kind, while streaming potential and sedimentation potential are examples of the second kind.

Electroosmosis

Let P in Fig. 4.2 be a porous solid placed in a glass assembly filled with an electrolyte solution. As there is a potential difference between the solid surface and the solution, the latter will start to move through the porous solid when an electric potential difference, E, is applied to the gap between the electrodes, A and B. This movement of liquid can be measured as the motion of the meniscus formed in the capillary at one end of the assembly. The use of reversible electrodes such as

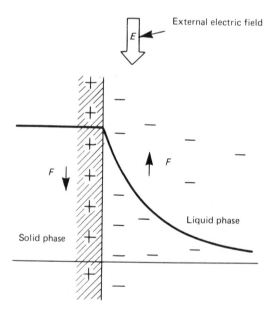

Figure 4.1 Occurrence of electrokinetic phenomena. Solid phase and liquid phase are subjected to forces acting on them in opposite directions.

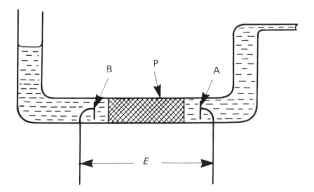

Figure 4.2 Apparatus for measuring electroosmosis or streaming potential. P, porous plug; A and B, electrodes; E, potential applied or generated.

Ag-AgCl electrodes is recommended when a large electric current flows in the solution. This is because the strong current causes electrolysis of the solution accompanying the generation of bubbles on the electrode surface, which may disturb the measurement. Zn-$ZnSO_4$ (saturated) electrodes and a salt bridge are necessary in case a much larger current passes through the solution.

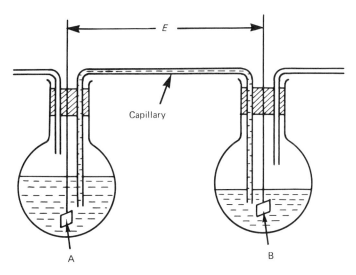

Figure 4.3 Apparatus for measuring electrokinetic phenomena using a single capillary. A and B, electrodes; E, potential applied or generated.

A capillary seems to be superior to a porous plug because it is easier to clean, gives better reproducibility, and provides data that can be analyzed more easily (Fig. 4.3). However, the use of a capillary makes measurement difficult because an extremely small volume of the liquid passes through it. Moreover, it is also disadvantageous to use a capillary because only glass or quartz can produce capillaries. On the other hand, a porous plug has the advantage that it can be made of any material.

Streaming Potential

A measurable potential difference is set up between the electrodes when the electrolyte solution in the apparatus in Fig. 4.2 is forced to flow through the porous plug by the application of pressure. As the potential difference is generally very small, it is desirable to use reversible electrodes such as Ag-AgCl electrodes.

Electrophoresis

The phenomenon in which colloid particles migrate under the influence of an electric field is called electrophoresis. This motion of colloid particles is very similar to that of ions in an electric field.

Sedimentation Potential

The motion of a large number of colloid particles at a constant velocity in one direction, sedimentation, for example, produces an electric potential in the direction of particle motion. The potential difference can be detected with reversible electrodes. This is sometimes called the Dorn effect.

4.2 THEORIES OF ELECTROKINETIC PHENOMENA

4.2.1 Introduction

All electrokinetic phenomena are caused by the flow of charge and liquid in an electrical double layer relative to the surface. In treating these phenomena theoretically it is assumed in many cases that the double layer takes the form of a plane condenser and the electric charges on the liquid side are concentrated in a plane parallel to the interface. This is called the Helmholtz layer, although Helmholtz himself made no mention of the structure of the electrical double layer. Perrin is said to have introduced the condenser model. With the exception of electrophoresis, those electrokinetic phenomena in which the solid phase is at rest can be described by Poisson's equation (4.1) alone,

$$\Delta \psi = - \frac{4\pi \rho}{\varepsilon} \tag{4.1}$$

without knowledge of the structure of the double layer. Here, $\Delta = \partial^2/\partial x^2 + \partial^2/\partial y^2 + \partial^2/\partial z^2$ and ε is the dielectric constant of the medium. Although the dielectric constant is not necessarily a constant throughout the system, it is usually assumed to be so. A constant viscosity and a laminar flow of the liquid are also assumed. It can be assumed that the movement of the liquid relative to the solid surface never begins immediately next to the surface, but that a stationary layer of the liquid with a thickness of one to several molecules is always present. The outside boundary of this layer is called the slipping plane. The electrical potential in the slipping plane plays a decisive role in theories of electrokinetic phenomena and is usually called the zeta (ζ) potential. In view of the presence of the Stern layer there still remains some uncertainty about whether the location of the slipping plane is inside or outside the layer or on the outer boundary of the layer. However, the plane is generally thought to be located on the outside.

As will be described later, not only the particle itself but also the surrounding liquid will be set in motion by an external electric field in electrophoresis and this causes complications. As a result, the theoretical treatment of electrophoresis has long been in a bewildering state. However, recent developments in the theory have, in principle, solved the problem.

4.2.2 Electroosmosis

Let us assume that the electrical double layer has the structure depicted in Fig. 4.4. That is, the Stern layer is in contact with the interface, whose potential is ψ_s, and the diffuse layer extends outward from the end of the Stern layer. There is an abrupt change in potential in the Stern layer, but this should be distinguished from that in χ potential. It should also be remembered that the dielectric constant of the liquid in this layer is much lower than that of the bulk water. The slipping plane is assumed to be located on the outside of the Stern layer.

When an electric field parallel to the interface is applied, a stationary state is soon reached. In this stationary state a layer dx at a distance x from the solid surface moves with a uniform velocity parallel to the wall. There are two different forces acting on the layer and these two forces should be balanced (Fig. 4.5).

One is the force due to the external electric field E and is given by $E_\rho \, dx$ (ρ is the space charge density) per unit area. This force acts separately on each of the ions in the layer and is transferred to the layer as a whole by the internal friction of the liquid. The other is the frictional force exerted on the layer by its neighboring layers moving with different relative velocities and is given by

$$\eta \left(\frac{dv}{dx}\right)_{x+dx} - \eta \left(\frac{dv}{dx}\right)_x$$

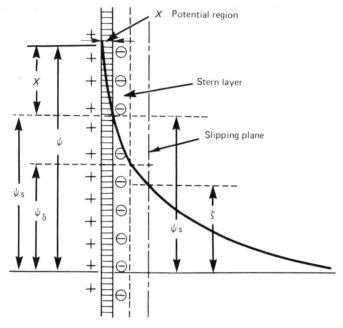

Figure 4.4 Scheme showing χ-potential region, Stern layer, and slipping plane. The location of slipping plane is uncertain. ψ_s, instead of ψ, is used. In general, ψ is customarily used to represent the potential in the diffuse layer, $\psi(x)$.

per unit area. As a consequence, in the stationary state

$$E\rho \, dx = \eta \left(\frac{dv}{dx}\right)_{x+dx} - \eta \left(\frac{dv}{dx}\right)_x = \frac{d^2v}{dx^2} dx \qquad (4.2)$$

where v is the velocity of the layer, dx. Inserting Poisson's equation (4.1)

$$\frac{d^2\psi}{dx^2} = -\frac{4\pi\rho}{\varepsilon}$$

we obtain

$$-\frac{\varepsilon E}{4\pi} \frac{d^2\psi}{dx^2} = -\eta \frac{d^2v}{dx^2} \qquad (4.3)$$

The integration of Eq. (4.3) is carried out over the whole region in which there is movement of the liquid. The first integration gives

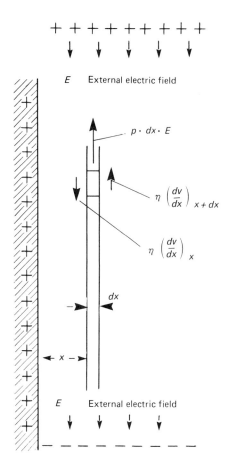

Figure 4.5 Two sorts of force acting on the electric charges in the diffuse layer.

$$-\frac{\varepsilon E}{4\pi}\frac{d\psi}{dx} + C_1 = \eta\frac{dv}{dx}$$

where the integration constant C_1 can be determined to be zero using the boundary conditions that $d\psi/dx = 0$ and $dv/dx = 0$ when $x \to \infty$. Recalling that $v = 0$ and $\psi = \zeta$ on the slipping plane, a second integration results in

$$-\frac{E\varepsilon}{4\pi}\psi + C_2 = \eta v$$

with integration constant $C_2 = E\varepsilon\zeta/4$. Hence we get

$$\eta v = \frac{E\varepsilon}{4\pi}(\zeta - \psi) \tag{4.4}$$

which is reduced to

$$\eta v_\infty = \frac{E\varepsilon}{4\pi}\zeta \tag{4.5}$$

when x approached infinity. This equation can also be easily derived on the basis of the plane-condenser model.

Equation (4.5) gives the osmotic velocity of liquid, or, in a practical sense, the velocity of a liquid column situated a small distance from the capillary wall. As mentioned earlier, it is noted that no special condition is attached to the double-layer structure except the presence of the slipping plane in the derivation of the equation.

Now let us calculate the volume of liquid transported by electroosmosis. Although the above-mentioned osmotic velocity of liquid v_∞ is a measurable quantity, it is frequently more convenient to measure the volume of liquid transported by electroosmosis. If a capillary in which there is an osmotic flow has a constant cross-sectional areas S, the volume of liquid transported per second is given by

$$\text{volume} = sv_\infty = \frac{S\zeta\varepsilon E}{4\pi} \tag{4.6}$$

In this expression it is assumed that the whole mass of liquid in the capillary moves with a constant velocity of v_∞. Since the velocity of the liquid in the diffuse double layer is lower than v_∞, as seen from Eq. (4.4), the assumption above is correct only when the double-layer thickness is very small compared to the diameter of the capillary. This is usually the condition found, because the double-layer thickness is generally smaller than 10^{-5} cm and the diameter of the capillary is in most cases larger than 10^{-3} cm.

Elimination of S from Eq. (4.6) by the use of Ohm's law in the following form gives

$$E = \Omega i = \frac{i}{s\lambda} \tag{4.7}$$

where Ω is the resistance and λ the specific conductance of the liquid. Therefore, the volume of liquid transported per unit time is given by

$$\text{volume} = \frac{\varepsilon\zeta i}{4\pi\eta\lambda} \tag{4.8}$$

which is sometimes called the Helmholtz-Smoluchowski equation. However, the application of Ohm's law in the form of Eq. (4.7) may cause a certain discrepancy because the equation ignores the effect of surface conductance, a phenomenon in which excess conductance occurs along the capillary surface as a consequence of the accumulation of ions in the double layer. This excess conductance may be of the same order of magnitude as the conductance through the bulk liquid in the capillary in the case of dilute solutions. For this reason, there is no longer any proportionality between i and λ and Eq. (4.7) should be written in the form

$$sE\lambda + LE\lambda_s = i \tag{4.9}$$

where λ_s is the specific surface conductance and L the circumference of the capillary. Then Eq. (4.8) becomes

$$\text{volume} = \frac{\varepsilon i}{4\pi\eta(\lambda + L\lambda_s/s)} \tag{4.10}$$

Necessary corrections can be made by measuring separately the resistance of the capillary when filled with concentrated and dilute electrolyte solutions or by applying Eq. (4.6) to the measurement in which the conductance through the capillary plays no significant part.

Equation (4.8) can be applied to the osmotic flow in a porous plug whenever the effect of surface conductance is negligible and hence reliable results are obtained. Those necessary corrections that are practicable for a single capillary cannot be made in the case of a porous plug.

Electroosmosis can also be determined by measuring the pressure produced by the electroosmotic flow of the liquid. This method is also applicable to the electroosmosis through a porous plug if the effect of surface conductance is negligibly small.

4.2.3 Streaming Potential and Sedimentation Potential

When a pressure difference is applied between the ends of a capillary (or porous plug), a flow of liquid passes through the capillary. As this flow carries the charge of the double layer with it, a potential difference arises between the ends of the capillary. The potential difference produces a conduction current directed opposite to the liquid flow and the two are soon balanced. In the stationary state, the convection current, which is proportional to the pressure difference P, and the conduction current, which is proportional to the potential difference E, are equal in magnitude, and consequently the streaming potential E is proportional to the pressure P. The proportionality constant is given by the Helmholtz-Smoluchowski equation as

$$\frac{E}{P} = \frac{\varepsilon\zeta}{4\pi\eta\lambda} \tag{4.11}$$

This relation is independent of the dimensions of the capillary or plug under the following conditions:

1. The flow of liquid through the capillary is a laminar one, which is satisfied in most cases.
2. The radius of the capillary is much larger than the thickness of the double layer.
3. The specific conductance determines the magnitude of the conduction current and is equal to that of bulk solution. In a strict sense, the surface conductance should be negligibly small.

The ratio E/P is constant except when a turbulent flow occurs at high pressures. However, even at low pressures, experimentally determined E/P values are sometimes lower than those predicted from Eq. (4.11) if dilute electrolyte solutions are used. This was once ascribed to a low ζ potential but it is now known to be due to surface conductance. In fact, the effect of surface conductance is marked at low bulk conductance, reducing the streaming potential E.

Necessary corrections can be made in a way quite similar to that used in electroosmosis measurements. Namely, the corrected formula for a capillary of radius r becomes

$$\frac{E}{P} = \frac{\varepsilon \zeta}{4\pi \eta (\lambda + 2\lambda_s / v)} \tag{4.12}$$

Although many corrected formulas have been proposed for porous plugs, all of them are oversimplified and hence only give values of the ζ potential lower than the true values.

Interestingly, electroosmosis and streaming potential are related to each other through a constant as revealed by comparison of Eq. (4.8) with Eq. (4.11) in the following way:

$$\frac{\text{volume}}{i} = \frac{E}{P} = \frac{\varepsilon \lambda}{4\pi \eta \lambda} \tag{4.13}$$

This relation holds even if the effect of surface conductance is taken into consideration, which is a special case of Onsager's reciprocal theorem in nonequilibrium thermodynamics. This means that electroosmosis and the streaming potential give the same information on the properties of the double layer.

The sedimentation potential arises when solid particles move through liquid, in contrast to the motion of liquid through the pores in solid in the case of streaming potential. Although there exists a strict theory on this phenomena [1], only a simple equation is given here:

$$E = \frac{\varepsilon \zeta g \Delta n a^3}{3 \lambda \eta} \tag{4.14}$$

where λ is the specific conductance of liquid, g the gravitational constant, Δ the difference in density of the solid and liquid, n the number of solid particles in a unit volume, and a the radius of the particle when it is assumed to be spherical.

4.3. ELECTROPHORESIS

4.3.1 Theoretical Treatment Ignoring Relaxation

In the phenomenon of electrophoresis, not only does the particle, which is solid, migrate in one direction but part of the liquid moves in the opposite direction while the solid remains at rest in electroosmosis and the other related phenomena. This creates an extremely complicated situation and causes problems in the development of the theory. Recent developments in the theory have, however, brought us very close to the solution, at least theoretically. Nevertheless, there still remains many discrepancies between theory and experiment and we cannot yet satisfactorily analyze and interpret the data obtained for various systems. An outline of the theoretical treatment is given below.

Electrophoresis can be analyzed in two fundamentally different ways, both of which eventually lead to the same result. The first approach starts from the idea that the external electric field acts on the *charge of the particle,* and the electrophoretic velocity of the particle is given by Stokes' law,

$$v = \frac{QE}{6\pi\eta a} \tag{4.15}$$

where Q and a are the charge and radius of the particle, respectively. This equation should, of course, be corrected for the movement of the double layer. The counterions surrounding the particle are forced by the external field to migrate in a direction opposite to that of the particle, and this movement of the ions is transferred to the liquid around them, causing a flow of liquid in the opposite direction. As a consequence, the particle does not move in a stationary liquid but moves in a liquid flowing in the opposite direction to reduce its electrophoretic velocity. The exact calculation of this effect gives

$$v = \frac{\varepsilon \zeta E}{6\pi\eta} \tag{4.16}$$

which is Hückel's equation. This equation was obtained by correcting Eq. (4.15) for the electrophoretic retardation in the theory of strong electrolytes.

An alternative approach is the one used by Smoluchowski a long time before Hückel. Here the phenomena of electrophoresis is regarded as the *inverse of electroosmosis,* accounting for the double layer. In electroosmosis the liquid moves whereas the solid is at rest. In contrast, the solid is assumed to move in the stationary liquid in elecrrophoresis. In both cases the movement of liquid relative to the solid surface is governed by the *forces acting on the double layer.* Consequently, electrophoretic velocity is given by Eq. (4.5):

$$v = \frac{\varepsilon \zeta E}{4\pi\eta} \tag{4.17}$$

This is called Smoluchowski's equation. In the derivation of this equation the particle was assumed to migrate in a uniform electric field E. The shape of the particle is not involved in the equation, just as the shape of the capillary pore plays no role in electroosmosis. It should be noted, however, that the equation holds only under the following conditions:

1. The thickness of the double layer is much smaller than the particle size.
2. The particle is an insulator and the surface conductance is low enough to cause no disturbance in the external electric field.

These conditions were not given very serious consideration for a long time, and hence Eq. (4.17) was regarded as applicable to practically all cases.

There seems to be a significant discrepancy between these two results. Hückel's equation gives an electrophoretic velocity lower by a factor 2/3 than that obtained from Eq. (4.17). Accordingly, it appears that the two approaches are not equivalent. Henry, however, showed that the discrepancy is not of a fundamental nature, as will be shown below.

In the derivation of Eq. (4.17) by Smoluchowski, the deformation of the electric field in the liquid due to the presence of the insulating particle was implicitly taken into account. Hückel, on the other hand, made his calculations on the assumption that the field in the double layer and in the bulk of the liquid is not disturbed by the presence of the particle. This assumption by Hückel is only permissible either if the conductivity of the particle is exactly the same as that of the liquid, or if the particle size is so small compared with the double-layer thickness that the deformation of the field is localized to the innermost part of the double layer.

Although Henry's method [2] is similar to that adopted by Hückel, it is an accurate mathematical treatment, taking full account of the deformation of the electric field by the presence of the particle. This method generates the following equation, valid in all cases:

$$v = f(\kappa a) \frac{\varepsilon \zeta E}{\pi \eta} \qquad (4.18)$$

where $f(\kappa a)$ is a function of κa, κ being the reciprocal of the double-layer thickness. Namely, κa is the particle radius expressed in the double-layer thickness as the unit. Henry gave the values of $f(\kappa a)$ only for those cases which are analytically solved. Thus the value of f is equal to 1/4 for a cylindrical particle with its axis parallel to the direction of the electric field, and is 1/8 when the particle lies in a direction vertical to the field. In the case of a spherical particle, the value of $f(\kappa a)$ is shown to vary from 1/6 to 1/4 as the values of κa increase. These are illustrated in Fig. 4.6.

In this way, Henry's equation is in complete accord with Smoluchoski's equation, giving f = 1/4 when $\kappa a \gg 1$, and is also in accord with Hückel's equation, predicting that f = 1/6 if $\kappa a \ll 1$. This equation also takes into consideration that the conductivity of metallic particles affects the distribution of the electric field in their neighborhood. It is expected that this has no influence on the electrophoretic velocity if the double-layer thickness is much larger than the particle

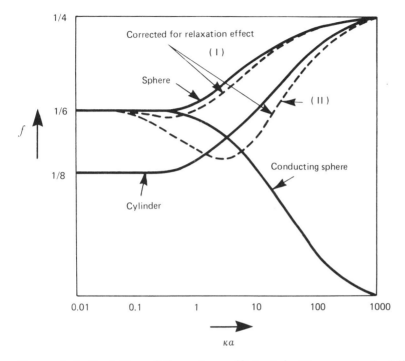

Figure 4.6 Variation of Henry's coefficient f with κa. Dashed lines: (I) $\zeta \leqslant 25$ mV, (II) $\zeta = 100$ mV. (From Ref. 3.)

radius ($\kappa a \ll 1$). However, when the double layer is thin ($\kappa a \gg 1$) the velocity approaches zero. In most practical cases of metallic particles, however, the effect of the conductance is removed by the polarization at the particle surface. In electrophoresis experiments the applied field is at most of the order of 10 V/cm. Then the potential difference produced between the upper and lower sides (front and back sides) of the particle is only several micro volts if the particle diameter is of the order of 10^{-6} cm, thus the polarization makes metals behave as insulators.

4.3.2 Effect of Relaxation

There still remains a problem to be solved. In the derivation of Eq. (4.17) the effect of surface conductance was assumed to be negligible. However, what will be the case if this condition is not fulfilled? The problem can be tackled in two different ways.

In the framework of Smoluchowski's theory, the electric current is concentrated in the neighborhood of the moving particle, owing to surface conductance, and this causes the accumulation of electrical

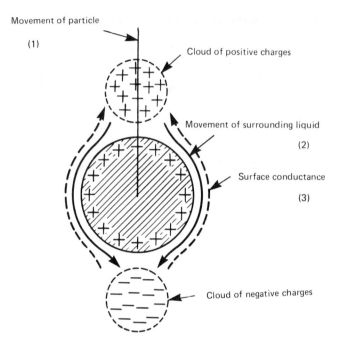

Figure 4.7 Three elements in electrophoresis. 1, electric force acting on the particle; 2, frictional force due to movement of the surrounding liquid; 3, electric force due to polarization caused by surface conductance.

charges opposite in sign to each other on the front and back sides of the particle, creating a counter electric field around it (Fig. 4.7). When a stationary state is reached, the counter field is balanced at a certain point by a counter current (very similar to the counter current in the case of streaming potential). The electrophoretic velocity is reduced, as the counter field tends to pull the particle backward.

In Hückel's and Henry's treatments, on the other hand, the relaxation-time effect, which is familiar to us from the theory of the conductance of strong electrolytes, is not considered. This effect is due to the deformation of the double layer. The movement of the counterions in the double layer in a direction opposite to that of the particle decreases the electrophoretic velocity of the particle since it causes the liquid surrounding the particle to flow in the opposite direction, as mentioned earlier. In addition, asymmetry in the double layer produced by the migration of the counterions generates an electric force acting on the particle in the opposite direction, the degree of the asymmetry being determined by the balance between the migration and diffusion of the counterions. This allows us to treat the effect as a relaxation phenomenon.

The two methods outlined above should, of course, fundamentally attack the same problem, although one method considers the effect of surface conductance and the other treats the deformation of the double layer.

Although various methods are available to evaluate the influence of the relaxation-time effect on the electrophoretic velocity, none has yet produced a satisfactory result due to the mathematical difficulties involved. Nevertheless, it can be safely concluded that the electrophoretic velocity is greatly reduced by a high surface conductance, especially when the particle size is small.

According to Overbeek [3], the counter electric field of relaxation is proportional to the ζ potential for asymmetric electrolytes, and to the square of it for symmetric electrolytes. The electrophoretic velocity is generally expressed by

$$v = \frac{\varepsilon E \zeta}{6\pi\eta} f(\kappa a, \zeta) \qquad (4.19)$$

In contrast to Henry's equation, f is a function of κa and ζ. In Fig. 4.6 the dashed lines represent $f(\kappa a, \zeta)$ for a spherical particle as a function of κa at ζ potentials of 25 and about 100 mV, respectively. If an explicit form of $f(\kappa a, \zeta)$ is given, we can set up a table for calculating ζ potentials from electrophoretic velocities. However, the use of Eq. (4.19) is limited because it was obtained on the condition that the ζ potential is lower than 25 mV. Nevertheless, several important conclusions may be drawn from Overbeek's theory.

1. The relaxation correction is smaller than a few percent for ζ potentials lower than 25 mV.
2. For very small values of κa (less than 0.1) the correction may be neglected. However, such small values will never occur in a colloidal system. For instance, the lowest value of electrolyte concentration in the dispersion medium is of the order of $1\text{-}10^{-5}$ mol/liter, which means that the smallest value of κ is of the order of 10^5 and the smallest value of a is of the order of 10^{-6} cm. Thus the smallest value of κa is 0.1.
3. The correction may be significant for intermediate values of κa (0.1-100) unless ζ is very low. This often makes the value of ζ for colloid particles evaluated by electrophoresis measurements doubtful.
4. For large values of κa the relaxation effect becomes small, irrespective of the shape of the particle. Hence Smoluchowski's equation remains valid even if the relaxation effect is taken into consideration. This condition is fulfilled in coarse colloid systems containing high concentrations of electrolytes.

Of the foregoing conclusions, the third is especially important. For a long time it has been the custom in colloid chemistry to evaluate the ζ potential by electrophoresis measurements, and it is now clear that many of these values cannot be trusted. The fourth conclusion was experimentally verified by Abramson, who found that the electrophoretic velocity of particles of quartz and other substances coated with egg albumin is independent of the size and shape of the particles. As the particles used in his experiments were larger than 10^{-4} cm and the experiments were carried out in buffer solutions with κ larger than 10^6, the condition of κa was fully satisfied.

4.3.3 Present Status of Theory of Electrophoresis

In recent years, important progress has been made in the theory of electrophoresis by Wiersema, Loeb, and Overbeek [4], who proposed a set of general equations as an extension of the hydrodynamic calculations made by Henry, taking account of both retardation and relaxation phenomena. This set consists of seven simultaneous differential equations and cannot be solved analytically. These authors solved the equations numerically using an electronic computer.

The solutions they obtained are qualitatively in line with Overbeek's conclusions described earlier, but some remarkable conclusions can be drawn from them. One of the conclusions is that the electrophoretic velocity never exceeds a value of 48×10^{-4} (cm/sec)(V/cm) at 25°C for values of κa around 5, however high the ζ potential may be (calculation was made up to 150 mV). This is due entirely to the relaxation-time effect and raises a serious problem in the calculation of ζ from electrophoretic velocity. For example, Smoluchowski's equation gives a very low ζ potential value of 50 mV or so when it is calculated on the basis of the electrophoretic velocity determined at $\kappa a = 5$, even though the "true" ζ potential is 150 mV. However, if the true ζ potential is lower than 50 mV, the discrepancy becomes much smaller and we can obtain the approximate "true" value by applying Henry's coefficient. The reason the word "true" is used here is that we can gain no reliable value for ζ in the true sense directly from electrophoresis measurements. We have the concept of a true ζ potential only on the assumption that the theory of Wiersema et al. accurately describes the electrokinetic phenomena.

Thus it is extremely difficult to evaluate ζ by electrophoresis, and although electroosmosis is theoretically more suitable for this purpose, it is not applicable to disperse systems. Although it is, of course, possible to determine ζ by the use of the table given by Wiersema et al., we have to measure the electrophoretic velocity of the particle at various values of κa. For a given colloidal system, this can be done by measuring the electrophoretic velocities of colloidal particles of various sizes assuming that the ζ potential is independent of the particle size.

At present, if we put aside our purely scientific interest, the safest way to evaluate ζ would probably be by application of Smoluchowski's equation to the electrophoretic velocity measured at $\kappa a \gg 1$.

Electrokinetics / 115

In practical problems, there are many cases that do not necessarily require the value of ζ. It would then be wise to correlate the electrophoretic velocity with the various properties of the system.

4.4 NONEQUILIBRIUM THERMODYNAMICS OF ELECTROKINETIC PHENOMENA

As stated previously, when a pressure difference (force) is applied to both ends of a capillary containing an electrolyte solution, not only does flow of the liquid (conjugate flow) take place through the capillary, but a flow of electric charges (nonconjugate flow) is produced simultaneously. There are many other phenomena in nature in which a force causes a seemingly unrelated flow in addition to the specific flow produced by the force.

Onsager [5] proposed in 1931 that the flow of a component in a system is linearly dependent on all thermodynamic forces operating in the system. The resulting set of equations, known as the phenomenological equations, may be written in the form

$$
\begin{aligned}
J_1 &= L_{11}X_1 + L_{12}X_2 + \cdots + L_{1n}X_n \\
&= L_{21}X_1 + L_{22}X_2 + \cdots + L_{2n}X_n \\
&\rule{6cm}{0.4pt} \\
J_n &= L_{n1}X + L_{n2}X_2 + \cdots + L_{nn}X_n
\end{aligned}
\tag{4.20}
$$

or

$$J_i = \sum_{K=1}^{n} L_{iK} X_K \quad (i = 1, 2, \ldots, n) \tag{4.21}$$

where J_i's and X_i's are flows and forces which conjugate with each other, and L_{ik}'s are phenomenological coefficients. It was also shown by Onsager that the matrix of these phenomenological coefficients is symmetric, or that

$$L_{iK} = L_{Ki} \quad (i \neq k)$$

Equation (4.22) is valid as long as the flows and forces operating in the phenomenological equations are taken in such a way that

$$\sigma = \sum_{i=1}^{n} J_i X_i \tag{4.23}$$

where σ is the dissipation function for a system (rate of entropy production in the system) in which the irreversible processes are taking place, and X_i is the force which conjugates with the flow J_i. This set of conjugate force and flow, X_i and J_i, can be replaced by another set of conjugate force and flow, X_α and J_α, if the latter quantities are measured more easily than the former and this replacement does not alter the value of the product of flow and force in Eq. (4.23).

Now, the thermodynamic analysis of electrokinetic phenomena will be considered. The condition of uniform electrolyte concentration in the system leads to a considerable simplification since the dissipation function takes the form

$$\sigma = J_V P + iE \qquad (4.24)$$

in which J_V is the flow of volume, i the electric current, P the pressure difference, and E the electric potential difference. The system can be described in terms of the two flows and two forces appearing in this expression as follows:

$$J_V = L_{11} P + L_{12} E \qquad (4.25a)$$

$$i = L_{21} P + L_{22} E \qquad (4.25b)$$

The relation $i = L_{22} E$, obtained by putting $P = 0$ in Eq. (4.25b), is a form of Ohm's law.

From Eq. (4.25a) we find that

$$\left(\frac{J_V}{E}\right)_{P=0} = L_{12}$$

which is the flow of volume induced by an electromotive force under the condition of a uniform pressure, or the electroosmotic volume flow. On the other hand, the application of a pressure difference in the absence of an external electric field results in an electric current, the streaming current; namely, from Eq. (4.25b),

$$\left(\frac{i}{P}\right)_{E=0} = L_{21}$$

By applying Onsager's law, we obtain the relation

$$\left(\frac{J_V}{E}\right)_{P=0} = \left(\frac{i}{P}\right)_{E=0} \qquad (4.26)$$

The electroosmotic volume flow may also be expressed in a different manner by referring the flow not to the potential difference but to the associated current. Putting the condition $P = 0$ into Eq. (4.25b), we get

$$\left(\frac{J_V}{i}\right)_{P=0} = \frac{L_{12}}{L_{22}} \qquad (4.27)$$

which is called the coefficient of electroosmosis.

When an electric potential difference is given at zero current, a pressure difference will develop. The ratio of E to P is given by

$$\left(\frac{E}{P}\right)_{i=0} = -\frac{L_{12}}{L_{22}} \tag{4.28}$$

which is known as the coefficient of streaming potential.

Comparison of Eq. (4.27) with Eq. (4.28) shows that

$$\left(\frac{J_v}{i}\right)_{P=0} = -\left(\frac{E}{P}\right)_{i=0} \tag{4.29}$$

if the Onsager law is valid. This relationship is identical with Eq. (4.13). Similar relationships can be derived based on Eqs. (4.25.a) and (4.25b) and the Onsager law.

REFERENCES

For general reference, see H. R. Kruyt, ed., *Colloid Science*, Vol. 1, Elsevier, Amsterdam, 1952.

1. F. Booth, *J. Chem. Phys.*, 22, 1956 (1954).
2. D. C. Henry, *Proc. R. Soc. Lond.*, 133, 106 (1931).
3. J. Th. Overbeek, *Philips Res. Rep.*, 1, 135 (1946); *Kolloid Beih.*, 54, 287 (1943).
4. P. H. Wiersema, A. L. Loeb, and J. Th. Overbeek, *J. Colloid Sci.*, 22, 78 (1966).
5. L. Onsager, *Phys. Rev.*, 37, 405 (1931).

5
Nonaqueous Systems

Ayao Kitahara Science University of Tokyo, Tokyo, Japan

5.1. Electrochemistry of Nonaqueous Systems 120
    5.1.1. Conductivity in nonaqueous solutions 120
    5.1.2. The origin of surface charge 124
5.2. Double Layers and the DLVO Theory in Nonaqueous Media 128
    5.2.1. Characteristics of the double layer in low-permittivity media 128
    5.2.2. DLVO theory in nonaqueous systems 130
5.3. Electrokinetics 133
    5.3.1. Zeta potential by electrophoresis and the streaming potential 133
    5.3.2. Change in the zeta potential with concentration of dispersants 134
5.4. Effect of Water on the Zeta Potential and Stability 138
5.5. Stabilization of Nonaqueous Dispersions by Adsorbed Layers 140
    References 141

Primarily nonpolar solvents such as hydrocarbons or perhalogenated hydrocarbons as solvents in nonaqueous systems are described here, although polar organic solvents are also used occasionally. Nonaqueous systems are more complex than aqueous systems because there exist no well-defined ionic species, and the small amount of water in nonaqueous systems that cannot easily be controlled often gives puzzling effects. However, nonaqueous systems furnish a wide range of applications, for example, paints, printing inks, developing liquids in electroreproduction, cosmetics, dry cleaning, engine oils, and others.

Many studies have been done for nonaqueous systems from fundamental as well as applied viewpoints, inspired by the development of aqueous systems. Some reviews of surface and colloid chemical studies of nonaqueous dispersions have been published, two typical reviews written in English being Refs. 1 and 2.

5.1. ELECTROCHEMISTRY OF NONAQUEOUS SYSTEMS

Two main topics are described in this section: conductivity of nonaqueous solutions and the origin of surface charges in particles dispersed in nonpolar liquids.

5.1.1 Conductivity in Nonaqueous Solutions

Carefully purified hydrocarbons can have stable excess electronic charge carriers formed by any kind of irradiation, such as γ-rays. They show a conductivity of 0.5-0.6 $cm^2/V \cdot sec$ for benzene and toluene and 0.07-0.1 $cm^2/V \cdot sec$ for pentane and hexane under high electric strength (1.6-3.3 kV/cm) at room temperature [3]. However, hydrocarbons used as the media of dispersion contain some dispersants as surfactants or polymers and small amounts of water resulting from the dispersants or adsorbed from the environment. A dispersant or water scavenges existing electrons in hydrocarbons and introduces some ions there.

According to the classical work by Fuoss and Kraus, solutions of tetraisoamylammonium nitrate in a mixture of dioxane and water have the equivalent conductivity between 10^{-2} and 10^{-5} Ω^{-1} cm^2 mol^{-1} at lower regions of dielectric constant [4]. Bjerrum has shown theoretically that an electrolyte composed of large ions is easy to dissociate in low-dielectric-constant liquids [5]. The experiments by Fuoss and Kraus have also verified the Bjerrum theory [4]. Their results of the conductivity could be illustrated by the stabilized large ions or the formation of triplet ions.

We consider here another source of stabilized ionic species contributing to the conductivity. Water molecules act as the stabilizer due to the hydration to ions in an aqueous solution. Reversed micelles formed by oil-soluble surfactants in nonaqueous media are able to solubilize some polar species, such as water and water-soluble organic or inorganic salts and ions. By this action, ions dissociated from ionic oil-soluble surfactants can eventually be solubilized in a reversed micelle formed by the parent surfactant molecules, as shown in Fig. 5.1. This prediction was first made by Nelson and Pink to verify their experimental results of the conductivity of metal soap solutions in toluene in the following way [6]:

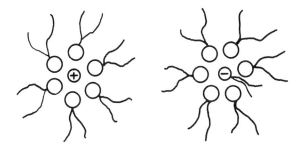

Figure 5.1 Scheme of the solubilization of dissociated surfactant ions in reversed micelles formed by surfactant molecules.

$$MA_2 \rightleftarrows MA^+ + A^-$$
$$nMA_2 \rightleftarrows (MA_2)_n \tag{5.1}$$
$$(MA_2)_n + MA^+ \rightleftarrows (MA_2)_n MA^+$$
$$(MA_2)_n + A^- \rightleftarrows (MA_2)_n A^-$$

where MA_2 denotes a molecule of metal soap. The data of the molar conductivity (the relation between Λ versus $\sqrt{c}$) are complex and vary among different kinds of soaps, and cannot be simply elucidated. However, the molar conductivity ranges over the order 10^{-4}-10^{-6} Ω^{-1} cm^2 mol^{-1} at a concentration range of 0.01-10 g/100 ml. A similar behavior was seen in other works [7]. The complexity of the conductivity curves seems to indicate the occurrence of a variety of complex ions.

The author and his collaborators calculated the number of ions produced in nonaqueous solutions with the use of Walden's law. The specific conductivity (λ) of the solution can be expressed as follows, irrespective of whether it is aqueous or nonaqueous:

$$\lambda = ne(u_+ + u_-) \tag{5.2}$$

where n is the ionic concentration (number/cm^3), e the elementary charge, and u_+ or u_- the ionic mobility. Because

$$u_+ + u_- = \frac{\Lambda_0}{F} \tag{5.3}$$

Eq. (5.2) reduces to the following:

$$\lambda = \frac{ne\Lambda_0}{F} \tag{5.4}$$

Table 5.1 Ion Concentration and Debye Length Obtained from Conductivity Data in Cyclohexane Solutions of Aerosol OT

Concentration of Aerosol OT (mM)	Concentration of water (ppm)	Specific conductivity (Ω^{-1} cm^{-1} × 10^{11})	Ion concentration (ions cm^{-3} × 10^{-11})	Debye length (μm)	κa
1.17	10	0.003	0.0015	95	0.0055
9.79	10	1.14	0.59	4.9	0.10
39.0	10	6.05	3.03	2.1	0.24
58.8	10	11.6	5.80	1.5	0.34
99.2	10	22.4	11.2	1.1	0.45
14.0	26	3.4	1.8	2.8	0.18
14.0	106	4.2	2.1	2.5	0.20
14.0	190	4.2	2.1	2.5	0.20
14.0	420	6.7	3.4	2.0	0.25

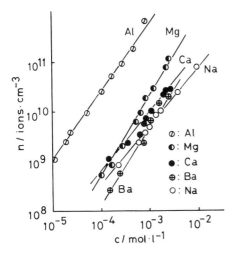

Figure 5.2 Relation between the concentration of ions and the total surfactant concentration. Surfactants: derivatives of Aerosol OT; ⓛ, Al-salt; ◐, Mg-salt; ●, Ca-salt; ⊕, Ba-salt; ○, Na-salt (Aerosol OT). (From Ref. 9.)

where Λ_0 is the limiting molar conductivity and F the Faraday constant.

In nonaqueous solutions of electrolytes, all electrolytes behave as weak ones. Hence Λ_0 cannot be obtained from extrapolation of the relation Λ and concentration. Furthermore, ionic mobility of ions dissociated from a surfactant is unknown in nonaqueous solutions. We then used Walden's law* to calculate Λ_0, although the validity of the law may be open to discussion. Λ_0 and η in aqueous solution were adopted in order to use the law. The result of Aerosol OT[†] in cyclohexane is listed in Table 5.1 together with the calculated thickness of the double layer around particles dispersed in the solution [8]. The relation between the ionic concentration and the surfactant concentra-

---

*$\Lambda_0 \eta$ is constant in all solutions, where η is the viscosity of the solution.
[†]Sodium 1,2-bis(2-ethylhexyl oxycarbonyl)-1-ethanesulfonate,

$$\begin{array}{l} CH_2 \cdot COOC_8H_{17} \\ | \\ CH \cdot COOC_8H_{17} \\ | \\ SO_3Na \end{array}$$

tion is shown in Fig. 5.2 for the cyclohexane solution of the derivatives from Aerosol OT [9].

5.1.2 The Origin of Surface Charge

The existence of surface charge on particles dispersed in nonaqueous media can be verified by electrophoresis or electrodeposition in the dispersions. Two possibilities are generally accepted to explain the origin of surface charge in nonaqueous dispersions: preferential adsorption of dissociated cation or anion, and the dissociation of any dissociative group on the surface.

In the absence of adsorbable ions in nonaqueous media, the dissociation of any surface group controls surface charge. Lyklema [1] proposed the following scheme of dissociation based on the idea of a proton donor and a proton acceptor:

$$PH_2^+ + S^- \rightleftarrows PH + SH \rightleftarrows P^- + SH_2^+ \qquad (5.5)$$

where PH and SH are the surface group of the particle and the solvent molecule, respectively. The scheme is applicable to a system composed of any particle and solvent of proton-donating or proton-accepting capacity (Brönsted acid or base). Lyklema collected many experimental facts studied by other authors and showed the applicability of the mechanism. Among them, the work of Verwey [10] is worth noting. Verwey determined the sign of the charge by electroosmosis measurement. According to his results, the order of acidity is $SiO_2 > TiO_2 > ZrO_2$ in any of the solvents, showing that the oxide on the left side in the series above has a more negative charge. Also, the order of acidity of the solvents is $CH_3OH > C_2H_5OH > (CH_3)_2CO > H_2O$ for any of the oxides, showing that the oxide on the left side is more positive in the solvent on the left side.

Acid-base theory can be applied to systems of aprotic substances because nonaqueous dispersions contain a little water unless a highly elaborate drying operation is carried out. The dissociation of water as a weak electrolyte furnishes protons and hydroxyl ions. The experimental results of Tamaribuchi and Smith [11] on the determination of the sign of surface charge by the electrodeposition method are shown in Table 5.2. The result can be illustrated by acid-base theory. The illustration may be described schematically as follows:

$$P + S + H^+ + OH^- \begin{array}{c} \nearrow PH^+ + S + OH^- \qquad (5.6a) \\ \searrow POH^- + S + H^+ \qquad (5.6b) \end{array}$$

When particle surface P is more basic than solvent molecule S, the charging process follows Eq. (5.6a) and the sign of the particles becomes positive, and vice versa. In Table 5.2, the order of basicity is

Table 5.2 Sign of Pigment Charge in Several Solvents by Electrodeposition

Particle	Solvent[a]	Particle charge
Carbon black	2,2,4-Trimethylpentane	+
	Benzene	−
	Methyl ethyl ketone	−
	Ethyl acetate	−
Toluidine red	2,2,4-Trimethylpentane	+
	Benzene	−
	Methyl ethyl ketone	−
	Ethyl acetate	−
Titanium dioxide	2,2,4-Trimethylpentane	+
	Benzene	+
	Methyl ethyl ketone	+
	Ethyl acetate	−

[a]Water content of solvents used (%): 2,2,4-trimethylpentane, 0.0040; benzene, 0.0270; methyl ethyl ketone, 0.0650; ethyl acetate, 0.0700.

titanium oxide > toluidine red > carbon black and ethyl acetate > methyl ethyl ketone > benzene > 2,2,4-trimethylpentane. Furthermore, the applicability of acid-base theory to the systems of Table 5.2 was verified by the sign of the charge on toluidine red or carbon black suspended in 2,2,4-trimethylpentane, which was reversed by the addition of 0.5% w/v basic copolymer of methyl vinyl pyridine and alkyl methacrylate to the solvent. Acid-base theory can be utilized to illustrate the sign of the charge in other cases: for example, TiO_2 is negative in n-butylamine and positive in 1-butanol [12], and α-Fe_2O_3 is negative in cyclohexanone and positive in isopropyl alcohol [13].

Ionic surfactants or similar dispersants are often used to improve the wettability or dispersibility of powders in nonaqueous media. Those substances dissociate to a minor extent and are stabilized because of large ions according to Bjerrum's theory, as seen in Sec. 5.1.1, or by the solubilization of ions by reversed micelles formed from ionic surfactants, as seen in Sec. 5.1.1. The surface charge of particles is then produced by the preferential adsorption of either of the ions thus stabilized. The factor determining preferential adsorption is the physicochemical affinity of either of the ions to particle surface. One of the affinities is acid-base interaction and an example of adsorption of a proton or hydroxyl ion derived from water was shown in the illustration of acid-base theory. The other affinity is the hydrophilic or oleophilic property.

Table 5.3 Sign of Particle Charge in Dispersant Solutions of Nonaqueous Media

Particle	Solvent	Dispersant	Sign of charge	References
Cu phthalocyanine	Heptane	Aerosol OT	−	28
Carbon black	Xylene	Aerosol OT	−	28
	Benzene	Tetraisoamylammonium picrate	−	18
	Benzene	Ca diisopropyl salicylate	+	18
	Cyclohexane	Aerosol OT	−	8
	n-Heptane	Aerosol OT	−	8
	Benzene	Aerosol OT	−	8
	Cyclohexane	Tetraosyethylene nonyl phenyl ether	0	8
	n-Heptane	Tetraoxyethylene nonyl phenyl ether	0	8
	Benzene	Mg(AOT)$_2$[a]	+	14
	Benzene	Ca(AOT)$_2$[a]	+	14
	Benzene	Ba(AOT)$_2$[a]	+	14
BaSO$_4$	Cyclohexane	Aerosol OT	+	8
	n-Heptane	Aerosol OT	+	8

Nonaqueous Systems / 127

	Cyclohexane	Tetraoxyethylene nonyl phenyl ether	0	8
	n-Heptane	Tetraoxyethylene nonyl phenyl ether	0	8
	Xylene	Span 80	+	17
Fe_2O_3	Xylene	Tetraisoamylammonium picrate	0	17
	Xylene	Cu oleate	+	17
TiO_2	Xylene	Melamine resin	−	12, 44
	Xylene	Alkyd resin	+	44
	Xylene	Aerosol OT	+	28
	Cyclohexane	Aerosol OT	+	45
	Cyclohexane	$Mg(AOT)_2$[a]	+	45
	Cyclohexane	$Ca(AOT)_2$[a]	+	45
	Cyclohexane	$Ba(AOT)_2$[a]	+	45
	Cyclohexane	$Al(AOT)_3$[a]	+	45
Al_2O_3	Xylene	Span 40	0	17
	Xylene	Aerosol OT	+	17, 28

[a]Metal salts of di-2-ethylhexyl sulfosuccinate.

Let us consider one example. Aerosol OT or AOT, a typical oil-soluble surfactant, is a good dispersant for carbon black or TiO_2 in nonaqueous media. AOT dissociates to produce Na^+ and RSO_3^-, where RSO_3^- shows an oleophilic group, 1,2-bis(2-ethylhexyl oxycarbonyl)-1-ethanesulfonate ion. Inorganic particles such as oxides and $BaSO_4$ have an affinity to Na^+, and organic particles such as carbon black and organic pigments have an affinity to RSO_3^-.

The author rearranged the results of many electrokinetic measurements to show the sign of surface charge in Table 5.3. The data of the table are illustrated through the idea of preferential adsorption, except for the system of carbon black and polyvalent surfactants such as calciumdiisopropyl salycilate and $M(RSO_3)_n$, where M is a di- or trivalent metal.

The sign of the latter systems cannot simply be elucidated by the affinity idea. Parreira [14] gave the following explanation for the sign. Divalent salts dissociate only in the first stage in nonaqueous media, as shown in the following equation:

$$M(RSO_3)_2 \rightleftarrows (RSO_3M)^+ + RSO_3^- \qquad (5.7)$$

The difference in oleophilicity between the cation and the anion is minor and the surface of carbon black has the polar groups C=O. The ion-dipolar force C=O---$(RSO_3M)^+$ prevails on the surface and furnishes a positive charge.

It is reasonably clear from Table 5.3 that nonionic surfactants have zero charge and that in the presence of resin dispersants, the sign can be explained by the acid-base mechanism. Differences in hydrocarbon solvents do not change the sign, although they produce quantitative differences in the value of the zeta (ζ).

5.2 DOUBLE LAYERS AND THE DLVO THEORY IN NONAQUEOUS MEDIA

We learned of the occurrence of surface charges on particles and of the ionic character in media of low dielectric constant in Sec. 5.1.2. Hence there occurs an electric double layer composed of the surface charge and the gegen ions around particles similar to the aqueous media. Let us examine next the characteristic of the double layer and the interaction between double layers in nonaqueous media.

5.2.1 Characteristics of the Double Layer in Low-Permittivity Media

The double-layer potential around a spherical particle is expressed with polar coordinates by Gouy-Chapman as follows:

$$\psi(r) = \frac{\psi_0 a}{r} \exp[-\kappa(r - a)] \qquad (5.8)$$

where ψ_0 is the surface potential of a particle of radius a, κ the reciprocal of the Debye length, and r the distance from the center of the particle. κ is described as follows:

$$\kappa = \left(\frac{8\pi e^2 nZ^2}{\varepsilon kT}\right)^{1/2} \tag{5.9}$$

The leading characteristic of nonaqueous media is a low dissociation degree of electrolytes and the resulting low ion concentration (n), as seen in Table 5.2 and in the conductivity data in Sec. 5.1.1. The order of magnitude of the lowering of n in nonaqueous media compared with aqueous media surpasses that of ε. As a result, the overall effect of n and ε in Eq. (5.9) makes κ much lower. Very low κ is an important characteristic of low-permittivity media.

If we assume as a limiting case that $\kappa = 0$, Eq. (5.8) reduces to the following equation:

$$\psi(r) = \frac{\psi_0 a}{r} \tag{5.10}$$

On the other hand, we have a general formula for total surface charge of a spherical particle:

$$Q = \psi_0 \varepsilon a(\kappa a + 1) \tag{5.11}$$

This equation reduces to the following one as the limiting case when $\kappa = 0$:

$$Q = \psi_0 \varepsilon a \tag{5.12}$$

Combining Eqs. (5.10) and (5.12) gives the following:

$$\psi(r) = \frac{Q}{\varepsilon r} \tag{5.13}$$

This equation expresses the simple coulombic law for the potential, that is, the assumption that $\kappa = 0$ corresponds to the distribution of gegen ions over an infinitely wide range (i.e., n = 0) and to the neglect of the double layer. In other words, the existence of gegen ions or the double layer which is typically seen in aqueous media disturbs or corrects the application of the simple coulombic law.

Equation (5.8) for very low κ or Eq. (5.10) gives a very slow decay of the double-layer potential. This is shown schematically in Fig. 5.3 compared with the sharp decay in aqueous media. This feature gives to nonaqueous systems the advantage that ψ_0 or ψ_δ is safely substituted by ζ, although the difference between ψ_0 or ψ_δ and ζ is often considerable and difficult to estimate in aqueous systems.

We can often observe a ζ potential of magnitude comparable to that of aqueous systems in nonaqueous systems. Why is there such a ζ potential for systems of very low ion concentration or surface charge?

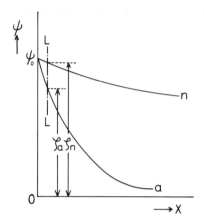

Figure 5.3 Double-layer potentials in aqueous and nonaqueous media. a, Aqueous; n, nonaqueous; LL', slipping plane.

This question is answered by Eq. (5.12); that is, the low values of ε and a make ψ_0 or ζ quite high, although Q is very low.

A problem exists in low-permittivity media, due to a very low surface charge. Here let us calculate the surface charge with the use of typical data: $\psi_0 = \zeta = 50$ mV $= 1.74 \times 10^{-4}$ esu, $\varepsilon = 2$, and a $= 10^{-5}$ cm. We can obtain from Eq. (5.12) a value for Q of 3.4×10^{-9} esu = 8 elementary charges per particle. Thus the area occupied by a charge is 1.6×10^4 Å^2. We observe from this calculation a very dilutely distributed charge of particle in nonaqueous media. Osmond [15] expressed doubts concerning the use of double-layer interaction in nonaqueous systems based on a smeared-out charge model developed in aqueous systems. This has not been proved. The discrete charge theory of Bell and Levine [16] may be a possible reason for it.

5.2.2 DLVO Theory in Nonaqueous Systems

The Derjaguin-Landau-Verwey-Overbeek (DLVO) theory, developed and verified in aqueous systems, should hold in nonaqueous systems because the latter also have an ionic double layer similar to that in the former. An existing important difference is the ionic concentration and therefore the magnitude of κ.

The double-layer interaction energy can generally be described as follows for the case $\kappa a \ll 1$:

$$V_R = \frac{\varepsilon a^2 \psi_0^2}{R} \exp(-\kappa h) \qquad (5.14)$$

where R = 2a + h and h is the shortest distance between the walls of two particles. If we assume $\kappa = 0$ as the limiting case, similar to the discussion in Sec. 5.2.1, Eq. (5.14) reduces to

$$V_R = \frac{\varepsilon a^2 \psi_0^2}{R} \tag{5.15}$$

This equation combined with Eq. (5.12) gives us

$$V_R = \frac{Q^2}{\varepsilon R} \tag{5.16}$$

This equation shows a simple coulombic interaction between two particles which have charge Q and are separated by R in centers. This conclusion is reasonable, because the gegen ions have been neglected in the limiting case. Equation (5.16) is often used as the approximate electric interaction in nonaqueous systems since the original work of Koelmans and Overbeek [17]. The exponential term in Eq. (5.14) should, however, be estimated for a more accurate calculation of the interaction, especially in a medium of increased permittivity or higher ion concentrations.

The electric potential energy described by Eq. (5.14) for low κ or Eq. (5.15) decays slowly with increase in h similar to the ψ curve of Eq. (5.10). The slow decay interaction is a characteristic of nonaqueous media. The resulting insensitiveness of the total potential energy, especially V_{max}, for the change of the Hamaker constant has been exemplified by Lyklema [1] as depicted in Fig. 5.4. On the other hand, the potential energy curve differs considerably with change in the Hamaker constant and its accurate calculation or measurement is very important in aqueous systems.

The behavior of nonaqueous media described above indicates that the ζ potential controls the stability of nonaqueous dispersions. This is true in dilute solutions and several experimental results have been published. The original work by Koelmans and Overbeek [17] theoretically showed the existence of the critical ζ potential for coagulation or flocculation. Van der Minne and Hermanie [18] and Briant and Bernelin [19] claimed the existence of critical values experimentally. The study by McGown and Parfitt [26] on the rate of flocculation of rutile (TiO_2) in xylene solution of Aerosol OT is a typical example of the relation between the ζ potential and stability. Good consistency between experimental and theoretical values is shown in Fig. 5.5, which depicts the relation between the stability ratio (W) and the ζ potential. It is difficult to find a sharp critical ζ potential under which flocculation occurs suddenly, although a vague critical range may be noted between 35 and 45 mV in Fig. 5.5. A similar relation between the stability ratio and the ζ potential was obtained by Cooper and Wright [20] for the dispersion of copper phthalocyanine in Aerosol OT-heptane solutions.

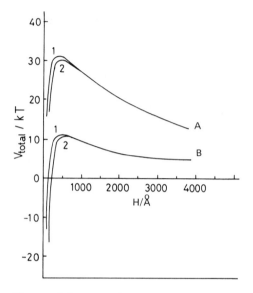

Figure 5.4 Total interaction potential energy (V_{tot}) between two spherical particles in nonaqueous media. Hamaker constants: 1.5×10^{-13} erg, 2×10^{-12} erg. Zeta potential: A, 75 mV; B, 45 mV. a = 1000 Å, $\varepsilon = 2$. (From Ref. 1.)

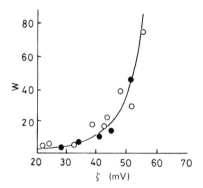

Figure 5.5 Stability ratio versus ζ potential for TiO_2 dispersed in xylene solutions of Aerosol OT. ○, Positive charges; ●, negative charges; solid line, theoretical. (From Ref. 26.)

In concentrated dispersions, the effect of surface potential on stability is different from that in dilute dispersions. Feat and Levine theoretically concluded that the double-layer force in a concentrated dispersion of two charged colloidal conducting particles is attractive and concluded that double-layer effects alone would not stabilize a highly concentrated suspension in a hydrocarbon medium [21]. Albers and Overbeek [22] showed theoretically that a decrease in the effect of the double layer on the stability of dispersions results from the overlapping of the potential energy curve. They also calculated a decrease in V_{max} in concentrated dispersions to illustrate the stability of water/oil emulsions which are independent of the ζ potential. Relating to this conclusion, Parfitt [23] has stated that there are hopeful signs that the electrical repulsion theory extended to dispersions having high particle concentration will clarify the interplay between electric and steric effects in systems of practical interest.

5.3 ELECTROKINETICS

5.3.1 Zeta Potential by Electrophoresis and the Streaming Potential

We saw in Sec. 5.2 that the ζ potential is nearly equal to surface potential in nonaqueous dispersions and that it is an important controlling factor in the stability of dilute dispersions. Therefore, measurement

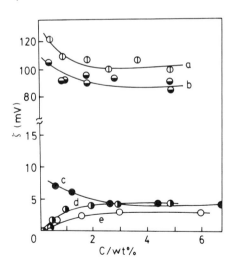

Figure 5.6 Change in ζ potential of $BaSO_4$ dispersed in Aerosol OT-heptane or cyclohexane solutions with concentration of Aerosol OT(C). a, Electrophoresis in cyclohexane; b, electrophoresis in heptane; c, electroosmosis in heptane; d, streaming potential in heptane; e, electroosmosis in cyclohexane. (From Ref. 24.)

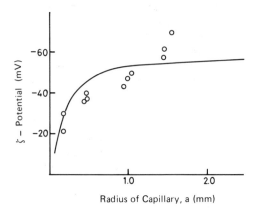

Figure 5.7 Change in ζ potential by streaming potential method with capillary radius (a) in glass capillary. Solid line: drawn by $\zeta = \zeta_0 - A/a$. $\zeta_0 = -62$ mV, $A = 8.8$ mm·mV. (From Ref. 25.)

of the ζ potential is very important. It was found by the author and his collaborators [24] that the values of the ζ potential depend on the methods used. Experimental results on barium sulfate particles dispersed in Aerosol OT- nonaqueous media are shown in Fig. 5.6.

It is shown in Fig. 5.6 that the values by electrophoresis are much higher than those by electroosmosis and the streaming potential. An important origin of the difference results from different particle concentration in the cells; that is, particles are very dilute in the former and very concentrated in the latter. It is considered that overlapping of double layers occurs in the latter and as a result, the effective ζ potential decreases. The author and other collaborators [25] have studied the effect of overlapping of double layers of glass particles or capillaries on the streaming potential method. Benzene solutions of tetrabutylammonium iodide were used as nonaqueous media. The ζ potential increased up to -15 mV and -62 mV with increase in glass particle size and glass capillary radius, respectively. However, it did not reach the -120 mV value obtained from electrophoresis of fine glass particles. The ζ potential of a glass capillary versus capillary radius is shown in Fig. 5.7.

5.3.2 Change in the Zeta Potential with Concentration of Dispersants

Many quantitative measurements of the ζ potential have been carried out with a change in concentration of dispersants such as surfactants, including metal soaps [7] or resins and polymers in nonaqueous media. Profiles of the ζ potential versus concentration of dispersant can be

Table 5.4 Classification of Concentration Effect of Surfactants on the ζ Potential

Surfactant	Particles	Solvent	Reference
Group A			
Tetraisoamylammonium picrate	Carbon black	Benzene	18
Aerosol OT	Copper phthalocyanine	Xylene	28
Aerosol OT	Carbon black (Elftex 5)	Benzene, cyclohexane	8
	Barium sulfate	Heptane	
Aerosol OT	Ferric oxide	Xylene	17
Aerosol OT	Barium sulfate		
Ca-AOT[a]			
Ba-AOT[a]	Titanium dioxide	Cyclohexane	45
Al-AOT[a]			
Melamine resin	Titanium dioxide	Xylene	44
Aerosol OT	Chlorinated copper phthalocyanine	Benzene, heptane	31
Group B			
Mg-AOT[a]			
Ca-AOT[a]	Carbon black (Stirling TB)	Benzene	14
Ba-AOT[a]			
Alkyd resin	Titanium dioxide	Xylene	44

[a]Metal salts of di-2-ethylhexyl sulfosuccinate.

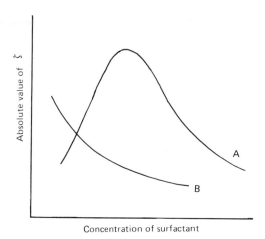

Figure 5.8 Scheme of change in absolute values of ζ potential with concentration of surfactants. (From Ref. 9.)

classified into two types, as shown schematically in Fig. 5.8. The curve of type A has a maximum, while that of B decreases monotonously. B is considered to correspond to the right side of the maximum of A. The published works are classified in Table 5.4 according to types A and B.

An analysis of type A done by the author and his collaborators [9] is introduced below. One kind of ion species produced from the dissociation of ionic dispersant (the positive ion is taken as an example) is tentatively assumed to be preferentially adsorbed on the particle surface according to the principle illustrated in Sec. 5.1.2. The adsorption equilibrium can be expressed by Eq. (5.17) at a lower concentration range of the dispersant:

$$S + P^+ \rightleftharpoons (SP)^+ \qquad (5.17)$$

where S, P^+, and $(SP)^+$ show the adsorption sites on particles, the positive ions in the solution, and the occupied surface sites adsorbed by some of the positive ions, respectively. The negative ions in the solution interact electrostatically with the increased positively charged sites with increase concentration of dispersant. Hence the following equilibrium is considered to hold:

$$(SP)^+ + N^- \rightleftharpoons (SPN)^\circ \qquad (5.18)$$

where N^- and $(SPN)^\circ$ are the negative ions in the solution and ion pairs formed on the particle surface, respectively.

If we write the equilibrium concentrations of S, P^+, N^-, $(SP)^+$, and $(SPN)°$ as $[S]$, C_+, C_-, σ_+, and $\sigma_\pm$, repectively, Eqs. (5.17) and (5.18) can be described as follows:

$$\frac{\sigma_+}{[S]C_+} = K_1 \qquad (5.19)$$

$$\frac{\sigma_\pm}{\sigma_+ C_-} = K_2 \qquad (5.20)$$

where K_1 and K_2 are equilibrium constants. If $[S]_0$ is the initial concentration of S, the following relation holds:

$$[S]_0 = [S] + \sigma_+ + \sigma_\pm \qquad (5.21)$$

Now we can obtain Eq. (5.22) from the preceding three equations:

$$\sigma_+ = \frac{[S]_0}{K_2 C_- + 1 + (1/K_1 C_+)} \qquad (5.22)$$

We can reasonably assume that C_+ or C_- increases with an increase in dispersant concentration (C), as shown in Fig. 5.2. Hence, when the concentration C, and therefore the ionic concentration (C_+ or C_-), is low, Eq. (5.22) reduces to

$$\sigma_+ = [S]_0 K_1 C_+ = [S]_0 K_1' C \qquad (5.23)$$

When C, and therefore C_+ (or C_-), is high, Eq. (5.22) reduces to

$$\sigma_+ = \frac{[S]_0}{K_2 C_-} = \frac{[S]_0}{K_2' C} \qquad (5.24)$$

On the other hand, surface potential, and therefore the ζ potential, is approximately proportional to surface charge according to Eq. (5.12) in nonaqueous systems. Hence Eqs. (5.23) and (5.24) can be rewritten as follows:

$$\zeta = K_1''[S]_0 C \qquad (5.25)$$

and

$$\zeta = \frac{K_2''[S]_0}{C} \qquad (5.26)$$

Equations (5.25) and (5.26) illustrate the left side and the right side from the maximum of type A in Fig. 5.8, respectively.

Another analysis of the maximum existing in the relation between the ζ potential and the concentration of a dispersant has been done by

Bell and Levine [16]. They ascribed the existence of the maximum to the discreteness of the surface charge.

5.4 EFFECT OF WATER ON THE ZETA POTENTIAL AND STABILITY

It is very difficult to remove a trace of water dissolved or adsorbed in nonaqueous systems. However, this may affect the surface potential and, in turn, the stability of dispersion, because dissociative protons and hydroxyl ions from water may be adsorbed selectively (see Sec. 5.1.2) and adsorbed water molecules make the particle surface basic. As for traces of water, the following work by McGown and Parfitt [26] is very interesting. A dispersion that was prepared by breaking an ampoule containing dry rutile powder (outgassed at 450°C, 10^{-5} torr for at least 3 hr) in dry Aerosol OT solution showed a negative ζ potential, which changed to positive after 15 hr. Dispersions of rutile in Aerosol OT solutions are usually positive. A similar reversal of the ζ-potential sign of α-Al_2O_3 was observed in alcohols by Romo [27]. Lyklema [1] indicated that a trace of water tends to be adsorbed at the surface of polar particles in nonpolar media, rendering it more basic. Polar particles therefore tend to have a positive ζ potential in nonpolar media due to the adsorption of a Lewis acid or a metal cation such as Na^+. This is another interpretation for the electric charge on particles which was discussed in Sec. 5.1.2.

An increase in water content increases the amount of cations adsorbed or dissolved in the water layer formed on particles. This increased amount of cations increases the positive value of the ζ potential. On the other hand, the positively charged water layer formed has an affinity for negative charges in the media. The resulting ζ-potential profiles reach a maximum and decrease with increase in water content. This behavior has typically been shown in the system rutile-Aerosol OT-xylene [28]. Micale, Lui, and Zettlemoyer [29] showed the presence of a maximum in the relation between electrophoretic velocity and water coverage on rutile dispersions in heptanol.

Carbon black has polar sites on the surface, although the degree varies widely according to the type of carbon black used. Further, dispersant molecules adsorbed on nonpolar sites make the surface polar. The polar parts can manifest the effect of water seen in the dispersion of polar particles as TiO_2. Thus the change in dispersibility of carbon black with the addition of a trace of water has been shown in a system with or without a surfactant [8,30]. The author and his collaborators reported the existence of maxima in the relation between the ζ potential and water content for carbon black dispersions in Aerosol OT (AOT) solutions, as shown in Fig. 5.9, although the maxima were not as pronounced as those in TiO_2.

The presence of polar sites developed by the adsorption of a dispersant on nonpolar particles is also possible for organic pigments.

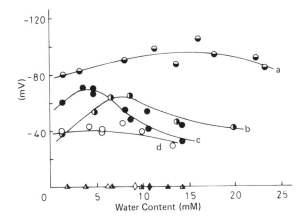

Figure 5.9 Effect of water on ζ-potential of carbon black dispersed in Aerosol OT solutions in nonpolar solvents. (From Ref. 8.) (a) in benzene (AOT:14.5 mM); (b) in cyclohekane (AOT:13.0 mM); (c) in heptane (AOT:11.3 mM); (d) in heptane (AOT:78.8 mM); ▵, ▴, ◊, ♦ in nonionic surfactant solutions.

The work of Cooper and Wright [31] is a good example. They showed the existence of maxima in profiles between the ζ potential and water concentration in β-phthalocyanine and chlorinated phthalocyanine dispersed in Aerosol OT or other oil-soluble surfactant solutions.

A change in the ζ potential with water content influences the stability of dispersions. An increase in the ζ potential before the maximum value increased the stability, and after the maximum value the stability did not change in the system rutile-Aerosol OT-xylene [32]. However, it was observed that the dispersion stability decreases with a decrease in the ζ potential after the maximum in carbon black systems [8].

The effect of water on the stability of carbon black dispersions in Aerosol OT or polyoxyethylene nonyl phenyl ether-cyclohexane solutions was studied over a wide range of water concentration from 2 mM to 2mM [33]. Water has been solubilized in the reversed micelle throughout this range. The dispersion of Sterling MT in Aerosol OT-cyclohexane solution (20 mM) was stable below a water concentration of about 10 mM. It began to flocculate increasingly above this concentration and flocculated perfectly over the range 100-800 mM of water. Furthermore, it restabilized above this range. The sediment height of carbon black in the dispersion after 6 hr is shown in Fig. 5.10 to illustrate the behavior of dispersibility described above. Other kinds of carbon black, such as Sterling R and MA-11 of the Mitsubishi Kasei Co., showed similar dispersibility behavior with an increase in water.

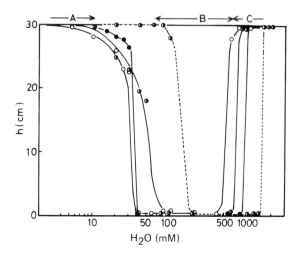

Figure 5.10 Change in sediment height (h) of carbon black dispersed in Aerosol OT-cyclohexane solutions with water contant. Aerosol OT concentration: solid lines, 20 mM; dashed lines, 40 mM. ◑, Sterling MT; ○, Sterling R; ●, MA-11. (From Ref. 33.)

The initial stabilizing range below 10 mM of water is considered to correspond to the stabilizing range before the maximum ζ-potential value which was described in the preceding paragraph [32]. The flocculation in Fig. 5.10 seems to result from a decrease in the ζ potential which was experimentally observed. The recovered stabilization at higher water concentrations (over 800 mM in Fig. 5.10) may be confusing. However, this can be explained by the formation of a weak network structure in the dispersion through carbon black and swollen water-solubilizing micelles (called water/oil microemulsion). The formation of the swollen micelles is ascertained from the occurrence of a faint blue opalescence with the addition of considerable amounts of water to the micellar solution. This network structure is suggested from the observation of a narrow hysteresis loop in the curve of shear stress and shear rate. The water concentration of the occurrence of faint blue opalescence in the solubilization experiment corresponded with that of the recovered stabilization in Fig. 5.10.

5.5 STABILIZATION OF NONAQUEOUS DISPERSIONS BY ADSORBED LAYERS

The effect of adsorbed layers formed by polymers or surfactants on the stabilization of nonaqueous dispersions is very important, as is the effect of electrical double layers. The former effect is more predominant in concentrated dispersion [21] and small particles [17]. Systems of polymer resin and pigment such as paints and printed inks are examples of concentrated dispersions.

A great deal can be written on the many theoretical and experimental studies [13,33-38] done on the effect of adsorbed layers on stabilization in nonaqueous media. However, the present book focuses on electrical phenomena at the interface. Hence the description of stabilization and related phenomena of nonaqueous dispersions has been confined to electrical phenomena. Fortunately, a comprehensive book on the stabilization of colloidal dispersions by polymer adsorption has recently been published [39]. Readers may also find excellent reviews for nonaqueous systems as well as aqueous ones in this book and can refer to other references cited here [1,2,13,33-38].

In closing this chapter, it is appropriate to add a few words regarding dispersions in polar organic solvents. Some work done on these systems was discussed in Sec. 5.1.2. The stabilization of ferric oxide ($\alpha\text{-Fe}_2\text{O}_3$) in cyclohexanone and isopropyl alcohol was studied with a dispersant, a polyamide of fatty acid [13]. It was shown in this work that the adsorbed layer was more effective than the ζ potential for stabilization. The result may be reasonable, because the experiment for stabilization was done in a concentrated dispersion of 10%. Recently, a pioneering work compared dispersions in polar solvents with those in aqueous systems [40]. The ζ potential and stabilization of dispersions of magnetic ferric oxides in 2-butanone (methyl ethyl ketone) were also studied with polymer dispersants, comparing them with nonmagnetic ferric oxide [41]. In this work the ζ potential did not relate to stabilization in either magnetic or nonmagnetic systems.

Fowkes and co-workers [42,43] proposed a proton transfer mechanism in an attempt to explain the nature of the surface charge of dispersions in nonaqueous media. This mechanism appears to explain the electrophoretic behavior of particles suspended in hydrocarbon liquids in the presence of polymeric or micellar substances capable of undergoing a proton transfer with the particle surface.

REFERENCES

1. J. Lyklema, *Adv. Colloid Interface Sci.*, 2, 65 (1968).
2. G. D. Parfitt and J. Peacock, in *Surface and Colloid Science*, Vol. 10 (E. Matijević, ed.), Plenum Press, New York, 1978, p. 163.
3. R. M. Minday, L. D. Schmidt, and H. T. Davis, *J. Phys. Chem.*, 76, 442 (1972).
4. C. A. Kraus and F. M. Fuoss, *J. Am. Chem. Soc.*, 55, 21 (1933).

5. D. A. MacInnes, *The Principles of Electrochemistry*, Reinhold, 1939, p. 369.
6. S. M. Nelson and R. C. Pink, *J. Chem. Soc.*, *1954*, 4412 (1954).
7. B. R. Vijayendran, in *Colloidal Dispersions and Micellar Behavior* (K. L. Mittal, ed.), ACS Symposium Series No. 9, American Chemical Society, Washington, D.C., 1975, p. 211.
8. A. Kitahara, S. Karasawa, and H. Yamada, *J. Colloid Interface Sci.*, *25*, 490 (1967).
9. A. Kitahara, M. Amano, S. Kawasaki, and K. Kon-no, *Colloid Polym. Sci.*, *255*, 1118 (1977).
10. E. J. W. Verwey, *Recl. Trav. Chim. Pays-Bas*, *60*, 625 (1941).
11. K. Tamaribuchi and M. L. Smith, *J. Colloid Interface Sci.*, *22*, 404 (1966).
12. L. A. Romo, *J. Phys. Chem.*, *67*, 386 (1963).
13. T. Sato, *J. Appl. Polym. Sci.*, *15*, 1053 (1971).
14. H. C. Parreira, *J. Electroanal. Chem.*, *25*, 69 (1970).
15. D. W. J. Osmond, *Discuss. Faraday Soc.*, *42*, 247 (1966).
16. G. M. Bell and S. Levine, *Discuss. Faraday Soc.*, *42*, 97 (1966).
17. H. Koelmans and J. Th. G. Overbeek, *Discuss. Faraday Soc.*, *18*, 52 (1954).
18. J. L. van der Minne and P. H. Hermanie, *J. Colloid Sci.*, *7*, 600 (1952); *8*, 38 (1953).
19. J. Briant and B. Bernelin, *Rev. Inst. Fr. Petrole*, *14*, 1767 (1959).
20. W. D. Cooper and P. Wright, *J. Colloid Interface Sci.*, *54*, 28 (1976).
21. G. R. Feat and S. Levine, *J. Colloid Interface Sci.*, *54*, 34 (1976).
22. W. Albers and J. Th. G. Overbeek, *J. Colloid Sci.*, *14*, 510 (1959).
23. G. D. Parfitt, *J. Colloid Interface Sci.*, *54*, 4 (1976).
24. A. Kitahara, H. Yamada, Y. Kobayashi, H. Ikeda, and Y. Koshiyama, *Kogyo Kagaku Zasshi*, *70*, 2222 (1967).
25. A. Kitahara, T. Fujii, and S. Katano, *Bull. Chem. Soc. Jpn.*, *44*, 3242 (1971).
26. D. N. L. McGown and G. D. Parfitt, *Discuss. Faraday Soc.*, *42*, 225 (1966).
27. L. A. Romo, *Discuss. Faraday Soc.*, *42*, 232 (1966).
28. D. N. L. McGown, G. D. Parfitt, and E. Willis, *J. Colloid Sci.*, *20*, 650 (1965).
29. F. J. Micale, Y. K. Lui, and A. C. Zettlemoyer, *Discuss. Faraday Soc.*, *42*, 238 (1966).
30. F. M. Meadus, I. E. Puddington, A. F. Sirianni, and B. D. Sparks, *J. Colloid Interface Sci.*, *30*, 46 (1969).

31. W. D. Cooper and P. Wright, *J. Chem. Soc. Faraday Trans. I*, *70*, 858 (1974).
32. D. N. L. McGown and G. D. Parfitt, *Kolloid Z. Z. Polym.*, *220*, 56 (1967).
33. A. Kitahara, T. Tamura, and S. Matsumura, *Chem. Lett.*, 1127 (1979); A. Kitahara, T. Tamura, and K. Kon-no, *Sep. Sci. Technol.*, *15*, 249 (1980).
34. F. Th. Hesselink, A. Vrij, and J. Th. G. Overbeek, *J. Phys. Chem.*, *75*, 2094 (1971).
35. R. H. Ottewill and T. Walker, *Kolloid Z. Z. Polym.*, *227*, 108 (1968).
36. D. H. Napper, *Trans. Faraday Soc.*, *64*, 1701 (1968); R. Evans and D. H. Napper, *Kolloid Z. Z. Polym.*, *251*, 329, 409 (1973).
37. V. K. Dunn and R. D. Vold, *J. Colloid Interface Sci.*, *54*, 22 (1976).
38. C. Thies, *J. Colloid Interface Sci.*, *54*, 13 (1976).
39. T. Sato and R. Ruch, *Stabilization of Colloidal Dispersions by Polymer Adsorption*, Marcel Dekker, New York, 1980.
40. N. D. Rooy, P. L. deBruyn, and J. Th. G. Overbeek, *J. Colloid Interface Sci.*, *75*, 542 (1980).
41. K. Kandori, K. Kon-no, and A. Kitahara, *Bull. Chem. Soc. Jpn.* *56*, 1581 (1983).
42. F. M. Fowkes, F. W. Anderson, R. J. Moore, H. Jinnai, and M. A. Mostafa, Electrical Charging of Particles, in *Colloids and Surfaces in Reprographic Technologies* (M. L. Hair and M. D. Croucher, eds.), ACS Symposium Series, American Chemical Society, Washington, D.C., 1982.
43. R. J. Pugh, T. Matsunaga, and F. M. Fowkes, *Colloids Surf.* (in press).
44. Y. Oyabu, H. Kawai, and Y. Nakanishi, *Shikizai*, *35*, 98 (1962).
45. A. Kitahara, T. Komatsuzawa, and K. Kon-no, *Proc. Int. Congr. Surf. Active Substances*, Vol. II, Ediciones Unidas Barcelona, 1968, p. 135.

PART II: MEASUREMENTS

6
Electrocapillary Measurements

Katsuo Takahashi The Institute of Physical and Chemical Research, Wako-Shi, Saitama, Japan

Hisako Tamai Tagaya Shiga University, Shiga, Japan

Ken Higashitsuji Marubishi Oil Chemical Co., Ltd., Osaka, Japan

Shigeharu Kittaka Okayama University of Science, Okayama, Japan

6.1. Thermodynamic Double-Layer Parameters 148
    6.1.1. Experimental techniques 149
6.2. Double-Layer Capacity 153
    6.2.1. Techniques for double-layer capacitance measurements 155
6.3. Oil/Water Interface 158
    6.3.1. Electrocapillary curves at oil/water interfaces 159
    6.3.2. Electrocapillary measurements at oil/water interfaces 160
    6.3.3. Adsorption at oil/water interfaces 160
    6.3.4. Binding at oil/water interfaces 162
    6.3.5. Tensametry at oil/water interfaces 165
    6.3.6. Applications 167
6.4. Point of Zero Charge of Suspension 169
    6.4.1. Introduction 169
    6.4.2. Determination of PZC 169
    6.4.3. Definition of the PZC 171
    6.4.4. Comparison of PZC values determined by different methods 172
    6.4.5. PZC as a measure of electrification of solids 174
    6.4.6. Dependence of the PZC on the temperature of measurement 175
    6.4.7. Capacitance of suspensions 175
    References 176

6.1 THERMODYNAMIC DOUBLE-LAYER PARAMETERS

The interface between a metal and an electrolyte solution is a unique boundary at which an electronic conductor is in contact with an ionic conductor. The charge separation across the double layer results from the electric potential applied to the interface from an outside source, the chemical potential difference of substances between electrode and solution phases, or the adsorption of charged particles on the electrode. The electrical double layer is formed at this interface. The nature of the electrical double layer has been investigated intensively for several reasons, such as that an electromotive force appears across the double layer, electrode reactions proceed by the electric field in the double layer, and electrostatic interactions between charged particles are dominated by the double layer around the particles.

The fundamental knowledge about the structure and behavior of the double layer, such as distribution of the potential and charges, orientation and dielectric properties of solvent molecules at the interface, and adsorption phenomena of ions and neutral molecules, has been obtained from studies of the ideally polarized interface. The ideally polarized interface is defined as an interface between electrode and solution at which charges accumulate on both sides of the interface by the imposed potential from an outside source without any charge transfer across the interface.

The typical ideally polarized electrode available for practical use is a mercury electrode (a dropping mercury electrode, DME, is commonly used). In this century, the mercury electrodes have been used by many workers: Gouy, Frumkin, Grahame, and others. The bulk of basic knowledge of double-layer phenomena has been obtained from their studies using the mercury electrode. Detailed theoretical and experimental descriptions about the double layer will be found in books and reviews listed in Refs. 1-11.

In this brief treatment, the fundamental relations between thermodynamic parameters of the double layer are graphically described by using Bockris et al.'s [12] experimental results (Fig. 6.1) on mercury electrode/aqueous hydrochloric acid solution/calomel electrode system.

The thermodynamic parameters can be related to each other by the well-known Gibbs adsorption isotherm:

$$-d\gamma = \sigma_M \, dE + \sum \Gamma_i \, d\mu_i \qquad (6.1)$$

The meaning of symbols and the equations relating to each parameter in Fig. 6.1 are listed in Table 6.1.

For example, one of the double-layer parameters, the surface excess Γ, can be obtained from observed C-E curves. The two methods of determination of Γ, (I) and (II), correspond to [c] → [b] → [a] → [d] → [g] → [j] and [c] → [f] → [i] → [h] → [g] → [j] in Fig. 6.1, respectively. Γ_+ values obtained by methods (I) and (II) are indicated in [g] of Fig. 6.1 by [●,▼,■,▲] and [○,▽,□,△], respectively.

6.1.1 Experimental Techniques

Electrocapillary (γ-E) Curve Measurement

The γ-E curve measured by a Lippman electrometer is the most fundamental quantity for thermodynamic study. A well-designed instrument of a Lippman electrometer system and measuring technqiue have been described by Hansen, Kelsh, and Grantham [13]. Similar techniques have also been described by Payne [9] and Perkins and Andersen [14]. By this method, accurate and reliable γ values can be obtained, but the solutions and mercury used have to be very carefully purified, and the measurements consum much labor and time.

The dropping time or ordinary DMEs is often measured for the γ-E measurements of the determination of the potential of zero charge as a convenient method. The method is not suitable for strict analysis of thermodynamic parameters. However, the dropping-time method has been used predominantly for accurate γ-E measurements [15-17]. An automated system for highly precise γ-E measurements based on the maximum bubble pressure principle has been developed by Mohilner's group [18-20].

In usual electrocapillary measurements, the interfacial tension with a sample solution is obtained relative to a standard value. Smolder's value [21] of 42.62 ± 0.02 μJ cm^{-2} for the interfacial tension at the electrocapillary maximum potential of the Hg/0.05 mol dm^{-3} Na$_2$SO$_4$ aqueous solution system at 25.0 ± 0.3°C is often used as a standard.*

Differential Capacity Measurements

The differential capacity-potential (C-E) curves render important information on the electrical double layer. Measuring techniques are described in the following section.

Determination of the Potential of Zero Charge, E^{zc}

E^{zc} is defined as a potential at the maximum interfacial tension on a γ-E curve. Consequently, the γ-E curves measured are useful to determine E^{zc}.

If there is no free surface charge on the electrode, no current is expected with expansion of the electrode surface area. Consequently, E^{zc} can be determined as the potential of zero charging

---

*Recent absolute determination of the interfacial tension measured by Vos and Los [22] shows the value of 42.553 ± 0.026 μJ cm^{-2} for a Hg/0.116 mol dm^{-3} KCl aqueous system at $E^{zc} = -0.5586 \pm 0.0012$ V versus RCE (or -0.5084 ± 0.0012 V versus NCE) at 25.00 ± 0.05°C.

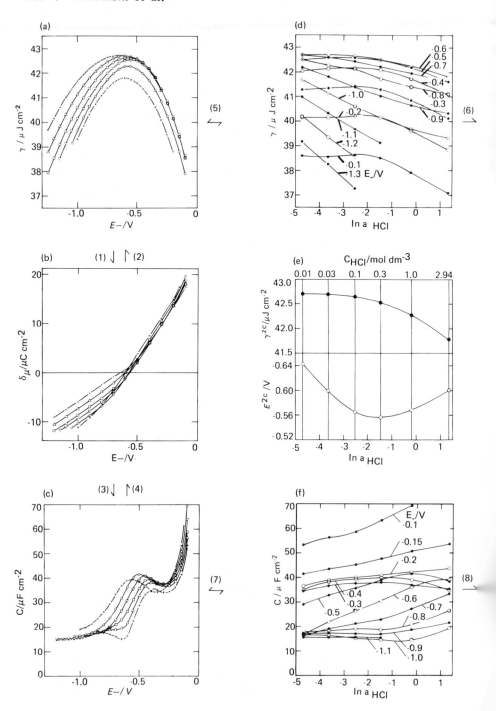

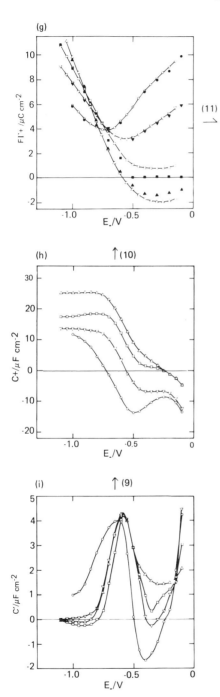
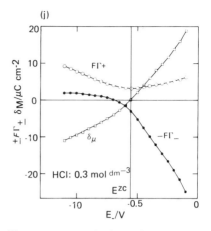

Figure 6.1 Relationships between electrical double-layer parameters. C-E curves: E_-^{zc} and γ^{zc} used are measured values for various concentrations of HCl. Concentrations of HCl in (a), (b), (c), (g), (h), and (i): ×, 2.94; ○, 1.0; ▽, 0.3; □, 0.1; △, 0.03; +, 0.01 mol dm^{-3}. (From Ref. 12.)

Table 6.1 Formulas in Fig. 6.1

(1) $\sigma_M = -\left(\dfrac{\partial \gamma}{\partial E_-}\right)_{c_{HCl}}$

(2) $\gamma = -\displaystyle\int \sigma_M \, dE_-$, where $\gamma = \gamma^{zc}$ at $E_- = E_-^{zc}$

(3) $C = \left(\dfrac{\partial \sigma_M}{\partial E_-}\right)_{c_{HCl}}$

(4) $\sigma_M = \displaystyle\int C \, dE_-$, where $\sigma_M = 0$ at $E_- = E_-^{zc}$

(5) γ-E plot $\leftrightarrow$ $\gamma - \ln a_{HCl}$ plot

(6) $\Gamma_+ = -(2RT)^{-1}(\partial \gamma / \partial \ln a_{HCl})_{E_-}$

(7) C-E plot $\leftrightarrow$ $C - \ln a_{HCl}$ plot

(8) $C' \equiv \dfrac{1}{2RT}\left(\dfrac{\partial C}{\partial \ln a_{HCl}}\right)_{E_-}$

(9) $C_+ \equiv \displaystyle\int C' \, dE_-$, where $C_+^{zc} = \dfrac{F C^{zc}}{2RT}\left(\dfrac{\partial E_-^{zc}}{\partial \ln a_{HCl}}\right)$

(10) $z_+ F\Gamma_+ = \displaystyle\int C_+ \, dE_-$, where $\Gamma_+^{zc} = \dfrac{-1}{2RT}\left(\dfrac{\partial \gamma^{zc}}{\partial \ln a_{HCl}}\right)_{E_-^{zc}}$

(11) $z_- F\Gamma_- = -(z_+ F\Gamma_+ + \sigma_M)$

Table 6.1 (continued)

E	electrode potential
E_-	electrode potential versus anion reversible reference electrode (electrode potential versus calomel electrodes contained in the sample solutions in Fig. 6.1)
μ	chemical potential
a_{HCl}	activity of HCl
c_{HCl}	concentration of HCl
γ	interfacial tension
σ_M	electrode surface charge density
C	differential double-layer capacity
C', C_+	defined by Eqs. (8) and (9) respectively
$\Gamma_\pm$	relative surface excess of the cation (+) and anion (−)
zc	quantity at the potential of zero charge

current from measurements of the charging current with expansion of the electrode surface. However, faradaic current (current with electrode reactions) interferes with observation of the charging current. A streaming electrode is useful for E^{zc} determination in this case [14, 23, 24].

The potential at the minimum of the diffuse double-layer capacitance on C-E curves corresponds to E^{zc} for dilute electrolyte solution systems without specific adsorption of charged species. This method gives high accuracy for (DME) systems, being the most useful method for solid electrode systems [14].

The E^{zc} value is a very important parameter for understanding the behavior of electrified interfaces and also one of the characteristic parameters of metals in contact with electrolyte solutions. However, the E_{zc} determination for solid electrodes is one of the most difficult techniques in electrochemistry, because the surface tension cannot be successfully measured for solid electrodes. Various methods of E^{zc} determination have been developed and E^{zc} values of metals have also been obtained by many workers, which are summarized in Refs. 8, 14, and 23-26.

6.2 Double-Layer Capacity

The measurement of double-layer capacity, meaning the measurement of interfacial impedance or admittance, has been used increasingly in studies of interfacial electrochemistry. The objectives of these studies are divided into the following areas:

1. *Thermodynamic properties of the double layer.* Differential capacity-potential curves are utilized for thermodynamic studies as described in Sec. 6.1. The impedance measurements are often used

jointly with electrocapillary measurements for thermodynamic studies and further discussion of the double-layer structure [12,13,27,28].

2. *Electrical properties of the double layer.* The electrical double layer can be considered by means of a parallel-plate condenser model, such as a simple Helmholtz-type model or a Bockris-type microscopic model [1,29]. In analogy with microscopic and macroscopic capacitors, behavior of solvent molecules, adsorbed substances, and solvated ions in the interfacial layer can be discussed from the observed double-layer capacities [1,4,5,30,31]. For example, the capacity is decreased by the adsorption of organic substance, because the thickness of the double layer is increased and the dielectric constant of the internal material of the capacitor is lowered by the adsorption. The surface coverage of the adsorbate can be estimated from the lowering of the capacity [32].

3. *Diffuse double layer.* The diffuse double layer formed by ions in distributed three-dimensional space in the interfacial layer can be also regarded as an electrical capacitor. The capacity is related to the distribution of ions by the well-known Gouy-Chapman theory [4,10,33]. The diffuse double-layer capacity can be clearly observed in polarized electrode/dilute electrolyte solution systems in the potential region of low surface charge density [4,10,34].

4. *Kinetic effects of the double-layer capacitance.* If the kinetic effect or the relaxation process is involved in the double-layer processes, the electrical properties of the double layer deviates from pure capacitive behavior. Both the capacitive and resistive components show frequency dispersion.

The kinetics of double-layer formation can be analyzed from the interfacial impedance. The rate of double-layer formation is so fast that frequency dispersion does not appear up to about 100 kHz. However, in the case of an extremely low concentration of electrolytes, frequency dispersion has been observed, which is regarded as the relaxation of mass-transport process in the double-layer formation [35,36].

The relaxation of adsorption-desorption processes can also be observed from the frequency dispersion of the interfacial impedance [36-39]. The rate-determining process can be classified by complex impedance plane analysis [40-42].

5. *Double layer with electrode reactions.* Studies on mechanisms and kinetics of electrochemical reactions are very common in many areas in electrochemistry [43]. Descriptions of interfacial impedance measurements are provided in Refs. 44-48.

Faradaic impedances attributed to electrode reactions of known or unknown substances appear generally in actual electrode/solution systems. For double-layer studies, faradaic impedances affect the double-layer capacitance as an nonideality of the ideally polarized interface. Usually, the residual faradaic impedances are not negligible for solid electrodes. The interfacial admittance Y_I, involving the

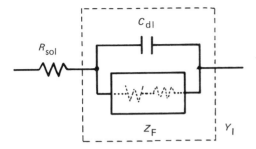

Figure 6.2 Equivalent circuit of the electrochemical cell. Y_I, Interfacial admittance; Z_F, faradaic impedance; C_{dl}, double-layer capacitance; R_{sol}, solution resistance.

double-layer capacitance C_{dl} and the faradaic impedance Z_F, can be shown by an equivalent circuit, as in Fig. 6.2. Since Z_F involves the resistive component, C_{dl} can be determined from the capacitive component of Y_I at high frequencies or by extrapolation to infinitely high frequency.

6. *Double layer on solid electrodes.* In recent years, the double-layer properties on single-crystal metal electrodes have been studied in view of the increasing interest on the nature of metal/solution interfaces [49-51]. Data for double-layer capacity are the most important source of our experimental knowledge, and sometimes spectroscopic measurements [52] or other techniques are combined for detailed discussion of the double-layer structure.

7. *Surface layer of electrodes.* The effect of the electrode surface layer on interfacial impedance has been studied as an aid to understanding the electrical and electrochemical properties of the surface layer [47,53,54]. Solvent-insoluble layers, such as those in metal oxides, semiconductor films, chemically modified electrode surfaces, and electrode surfaces coated with biological substances, are interesting subjects in the study of interfacial phenomena which may be related to problems of energy conversion, bioscience, and so on.

6.2.1 Techiques for Double-Layer Capacitance Measurements

Impedance Bridge Methods

A symmetrical Wien bridge system such as that shown in Fig. 6.3 was used by Grahme for DME/electrolyte solution systems in an ideally polarized potential region [55]. In the case of ideally polarized electrodes without kinetic effects, the capacitance C and resistance R read on the balanced bridge correspond to the double-layer capacitance C_{dl} and the solution resistance (involving resistances of the electrode and the wiring) R_{sol}, respectively. This method has been modified by many workers to improve the accuracy [40,41,56-58].

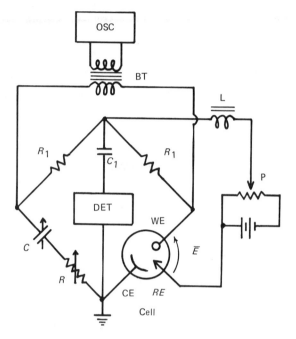

Figure 6.3 Schematic diagram of an impedance bridge for the cell impedance measurement. OSC, Sine-wave generator; DET, bridge balance detector (frequency-selective amplifier with oscilloscope); BT, bridge transformer; L, choke coil (~100 H); P, potentiometer for adjustment of the electrode potential $\bar{E}$; C, variable capacitor; R, variable resistor; WE, working electrode (DME); RE, reference electrode (calomel); CE, counter electrode (platinized Pt).

A transformer bridge system is a useful tool for reliable measurement of double-layer capacitances [12,58]. For measurement in the high-frequency region, (above about ~10 kHz) or in high-solution-resistance systems, the effects of stray capacitance lower the accuracy and reliability of measurement. Transformer bridges are effective in eliminating the effect of stray capacitance and noise, because the internal impedance of transformer bridges is markedly lower than that of Wien bridges.

Phase-Sensitive AC Voltammetry

Although the bridge method is suitable for accurate measurement of double-layer capacitance, it cannot be used as an automated recording method for impedance measurement.

Electrocapillary Measurements / 157

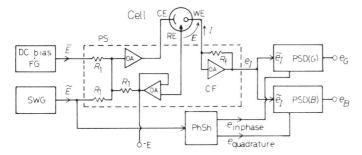

Figure 6.4 Schematic diagram of a phase-sensitive AC polarograph. WE, Working electrode; RE, reference electrode; CE, counter electrode; PS, potentiostat; CF, current follower; OA, operational amplifiers; FG, function generator; SWG, sine-wave generator; PhSh, phase shifter; PSD, phase-sensitive detectors or lock-in amplifiers.

Phase-sensitive detection technique for an AC current through the working electrode can be applied to measure both the conductive and susceptive components of cell admittance [44,45,59-61]. Figure 6.4 shows a schematic diagram of a cell admittance measurement system, called a phase-sensitive AC polarograph. In this system, the electrode potential E ($= \bar{E} + \tilde{E}$) is controlled against the reference electrode RE by the potentiostat PS. The electrode current I ($= \bar{I} + \tilde{I}$) is converted to the output voltage signal e_I by the current follower CF in the PS. $\bar{E}$ and $\bar{I}$ are the DC components of the electrode potential and current, respectively, and $\tilde{I}$ is the AC component of the current followed by the small-amplitude AC potential $\tilde{E}$. The AC component of e_I, $\tilde{e}_I$, being proportional to the cell admittance Y, is supplied to the input of phase-sensitive detectors (or lock-in amplifiers) PSD(G) and PSD(B), which are controlled by the reference AC signals $e_{inphase}$ and $e_{quadrature}$, respectively. The output voltage signal of PSD(G), e_G, and that of PSD(B), e_B, are proportional to the conductive component G and susceptive component B of the cell admittance Y, respectively, where Y is defined as a complex formula,

$$Y = G + jB$$

If we record e_G and e_B against the electrode potential $\bar{E}$, swept by the function generator, we can obtain the potential dependence of G and B automatically.

In the case of ideally polarized electrode system, the double-layer capacitance C_{dl} and solution resistance R_{sol} can be shown by a series combination of them in an equivalent circuit of the cell system, so that the C_{dl} and R_{sol} are calculated from observed G and B by the equations

$$R_{sol} = \frac{G}{G^2 + B^2} \quad \text{and} \quad C_{dl} = \frac{G^2 + B^2}{\omega B}$$

where ω is the angular frequency.

Transient Techniques

Some transient methods have been employed for double-layer studies. For example, the time dependence of the charging current I(t) with the potential jump ΔE can be measured by potential-step chronoamperometry [16,56]. In the simplest case of double-layer charging, the time constant $\tau = C_{dl}R_{sol}$ can be determined by the following relationship between I and t:

$$I = \frac{\Delta E}{R_{sol}} \exp\left(\frac{-t}{\tau}\right) \quad \text{or} \quad \log I = -\frac{t}{\tau} + \log \frac{\Delta E}{R_{sol}}$$

The charge-step chronopotentiometry is also used for double-layer studies [62-64]. In this method, a known quantity of charge, Δq, is supplied on the double layer and the capacitance C_{dl} can be determined by $C_{dl} = \Delta q/\Delta E$, where the ΔE observed is the electrode potential change with Δq.

In these transient methods, separation of the charging current from the faradaic current is very important. The observed current is analyzed for this purpose based on the difference between both the current-time relations of the charging process and the electrode reaction process.

For the study of the adsorption of electrochemically active substances, the transient methods are useful, because the amount of adsorption can be determined as the amount of charge consumed by the electrode reactions of adsorbed species [64-66]. Voltammetry and pulse polarography have both been used for similar purposes because a voltammogram gives unique patterns for electrode reactions of adsorbed materials [67,68].

Further techniques of automated impedance measurements and other digital signal acquisition techniques have been increasingly used for over a decade. Typical descriptions are given in Refs. 45 and 69-71.

6.3 OIL/WATER INTERFACE

Electrocapillarity is, in a wide sense, defined as a phenomenon of the change in interfacial tension resulting from the application of a potential difference between two phases in contact. Therefore, this phenomenon is, in principle, not restricted to metal/solution interfaces (Secs. 6.1 and 6.2) and is expected to take place at interfaces between any two immiscible phases (e.g., oil/water) [72-93].

Electrocapillary Measurements / 159

As in the case of mercury/solution interfaces, it can be expected that much information concerning the adsorption behavior and structure of adsorbed layers at oil/water interfaces would be obtained from a study of electrocapillarity. The introduction of this method means an increase in an independent variable at the interface (i.e., the interfacial potential difference), in addition to other variables, such as concentrations of surface-active materials, temperature, and nature of substrate.

6.3.1 Electrocapillary Curves at Oil/Water Interfaces

Figure 6.5 shows typical electrocapillary curves at oil/water interface. It is necessary to reduce the ohmic drops in the two bulk phases by adding electrolytes. When the oil phase [e.g., methyl isobutyl ketone (MIBK)] contains a surface-inactive electrolyte [e.g., tetramethylammonium iodide (TMAI)] and the aqueous phase is 1 mol/liter KCl solution, γ does not change with the polarizing potential E over the

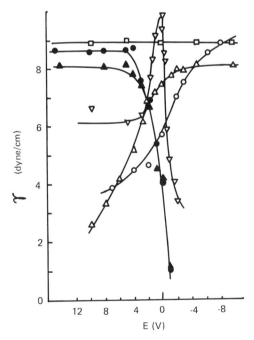

Figure 6.5 Typical electrocapillary curves at oil/water interface. Aqueous phase, 1 mol/liter KCl. Oil phase, MIBK containing: □, 1×10^{-4} mol/liter TMAI; ▲, 2.5×10^{-4} mol/liter HTAC; ●, 2.5×10^{-4} mol/liter HPC; ○, 2.5×10^{-4} mol/liter SHS; △, 1×10^{-4} mol/liter SDBSO; ▽, 5×10^{-5} mol SHS + 5×10^{-5} mol/liter HTAC. (From Ref. 80.)

range 0 to ±20 V. However, if an ionic surfactant is added to the oil phase, γ changes with E in a characteristic manner as a function of ionic type. In the case of anionic surfactants [e.g., sodium hexadecyl sulfate (SHS), sodium dodecylbenzenesulfonate (SDBSO), etc.], γ decreases over the positive polarization range (E > 0). In contrast, γ suppression takes place over the negative range (E < 0) in the case of cationic surfactants [e.g., hexadecyltrimethylammonium chloride (HTAC), hexadecylpyridinium chloride (HPE), etc.]. The coexistence of SHS and HTAC in the oil phase gives an additive effect, at least in a qualitative manner, and γ decreases over both positive and negative ranges with a maximum at E = 0 [80].

The characteristic effect of the ionic type of surfactant on the γ versus E curves mentioned above is also found for other organic solvents [80].

6.3.2 Electrocapillary Measurements at Oil/Water Interfaces

Theoretically, all methods of determining interfacial tension are available for this purpose as long as a polarizing potential is applied to the interface. Results using such methods as the dipping plate method [75], the modified ring method [78], and the drop volume method [80], have been reported. The drop volume method will be discussed as an example of these.

A drop of aqueous electrolyte solution (aqueous phase) is formed in the oil phase from the tip of a glass capillary by using the micrometer syringe (thin Teflon tubing can be used in place of the glass capillary). At the same time, a polarizing potential E is applied to the platinum electrode in the aqueous phase, with another platinum electrode in the oil phase as a counter electrode. The sign of E is conventionally taken as that of the aqueous phase with reference to that of the oil phase. It is thus possible to obtain the interfacial tension γ as a function of E by measuring the maximum drop volume V at various applied potentials, with the help of the Harkins-Brown formula [94, 95]:

$$\gamma = \frac{V(d_w - d_o)g\phi}{2r} \quad (6.2)$$

Here d_w and d_o are the densities of the aqueous and oil phases, respectively; r is the outer radius of the tip of the glass capillary (the inner radius of the tip for the Teflon capillary); g the acceleration of gravity; and φ the Harkins-Brown correction term, which is a function or $r/V^{1/3}$.

6.3.3 Adsorption at Oil/Water Interfaces

The mechanism of decreasing γ by the application of a polarizing potential has been discussed by various authors under the concept of electrocapillary adsorption and electroadsorption [72-79]. Blank and Feig [74] and Dupeyrat and Nakache [78] concluded that electroadsorption takes place when the solute is accumulated at the interface due to the difference in transport numbers of the solute in the two phases. However, if the suppression of γ is due to the difference in transport numbers of ionic species in the two phases, γ suppression must occur over the polarization range of the opposite sign, when a surfactant is added to another phase. According to the experiments of Watanabe and Tamai, however, the γ decreases over the same polarization range independent of whether the surfactant of the same ionic type is added to the oil or aqueous phase [85]. This experimental evidence leads to the conclusion that the present systems are similar to the mercury/inert aqueous solution interface, the ideally polarized interface.

If the interface can be treated as an ideally polarized interface, we can estimate the adsorbed amount of long-chain ions Γ_i from the slope of γ versus log c or γ versus log E curves [86,87]. Γ_i can be expressed by the equation [86,87]

$$\Gamma_i = K_1 c \exp\left(\frac{mw - zF\psi_0}{RT}\right) \tag{6.3}$$

where K_1 is a constant, m the number of methylene groups, w the desorption energy per methylene group, and ψ_0 the potential at the point where adsorption is taking place.

By applying the Gouy-Chapman theory, we obtain [96]

$$\psi_0 = \frac{RT}{F} \ln\left(\frac{139^2 \times 4}{A^{-2} z_e^2 c_e}\right) \tag{6.4}$$

for $\psi_0 > 100$ mV, T = 293°K, $|z| = 1$, and $\overline{A} = 10^{16}/\Gamma_i$, the available area per long-chain molecule in Å^2. Here c_e and z_e are the counterion concentration and valency, respectively. From Eqs. (6.3) and (6.4) we obtain [96]

$$\Gamma_i = K_2 c^{1/3} (z_e^2 c_e)^{1/3} \exp\left(\frac{mw}{3RT}\right) \tag{6.5}$$

where K_2 is a constant.

This equation shows that Γ_i is proportional to $c^{1/3}$ and $(z_e^2 c_e)^{1/3}$ for constant ionic strength $z_e^2 c_e$ and for constant surfactant concentration, respectively. The latter relation is a quantitative expression for the builder effect of electrolyte, that is, the increase in adsorption of surfactant by the addition of indifferent electrolyte.

As shown in Fig. 6.6, plots of Γ_i versus $c^{1/3}$ for sodium alkyl sulfates of different chain lengths exhibit a linear relationship predicted by Eq. (6.5) at the constant ionic strength of the aqueous

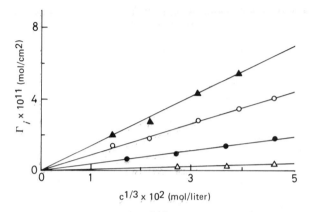

Figure 6.6 Γ_i versus $c^{1/3}$ for the adsorption of alkyl sulfate ions at the oil/water interface (E = +15 V). Aqueous phase, 1×10^{-2} mol/liter KNO_3. Oil phase, MIBK containing: ▲, sodium tetradecyl sulfate; ○, sodium dodecyl sulfate; ●, sodium decyl sulfate; △, sodium octyl sulfate. (From Ref. 86.)

phase [86,87]. From the relation between slope and m, the value of w was estimated to be 750 cal per mol of CH_2. This value of w is in good agreement with that obtained by Subba Rao et al. for n-heptane or benzene in contact with the aqueous solution of α-sulfo fatty acid esters (760-790 cal/mol CH_2) [97], as well as that obtained by Davies for paraffin in contact with 10^{-2} N HCl containing hexadecyltrimethylammonium chloride (810 cal/mol CH_2) [98].

6.3.4 Binding at Oil/Water Interfaces

Counterion Binding

At the oil/water interface, the hydrophilic ionic group of the adsorbed surfactant ion is believed to be immersed in the aqueous phase, whereas the hydrocarbon tail remains in the oil phase. Counterion binding will take place when the aqueous phase contains ion with strong affinity for the ionic head group, thus neutralizing the charge at the interface and decreasing the γ suppression.

For example, such counterion binding has been found to take place between hexadecylpyridinium chloride (HPC) and halide ions over the range E < 0. It was observed that γ suppression, due to the adsorption of HPC cation, is smaller for those counterions that bind strongly with the pyridinium group: the effect of halide ions on the γ suppression increased in the order $I^- < Br^- < Cl^- < F^-$ [81].

In the case of univalent cationic counterions for SHS, on the other hand, such a specificity was not found. In spite of the large differences between the cationic species, all data fell on the same curve [81].

Interfacial Isoelectric Point

One of the interesting phenomena related to counterion binding is that of the hydrogen ion. If an amphoteric surfactant, 2-heptadecyl-1-(2-hydroxyethyl)sodium carboxymethyl imidazolinium hydroxide (IM), is dissolved in MIBK at 10^{-4} mol/liter, the γ versus E curve changes its form when the pH of the aqueous phase, containing 10^{-2} mol/liter KCl, is changed. It is evident from Fig. 6.7 that IM behaves as an anionic surfactant at pH >8.84 and as a cationic surfactant at pH <7.32, respectively. Thus the interfacial isoelectric point (IEP) of IM is about 8 at this interface [83].

When the aqueous phase contains bovine serum albumin (BSA) in addition to 10^{-2} mol/liter NaCl, with the oil phase tetrabutylammonium chloride (TBAC) as the supporting electrolyte, the shape of γ versus E curves is extremely sensitive to the pH of the aqueous phase, being anionic for pH >5.281 and cationic for pH <5.273. Thus the IEP of BSA is 5.279 ± 0.003 at this interface [93]. As far as we know, such a high sensitivity in IEP measurement has never been found for proteins. The value of the IEP of BSA lies well between the value 5.5 obtained by Tanford, Swanson, and Shore measured by potentiometric titration [99] and 4.89 obtained by Longsworth and Jacobson measured by electrophoresis [100].

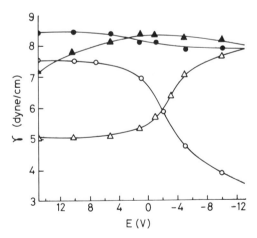

Figure 6.7 Influence of pH of aqueous phase on electrocapillary curves for amphoteric surfactant. Aqueous phase: 1×10^{-2} mol/liter KCl, pH; ○, 5.80; ●, 7.32; ▲, 8.84; △, 10.50. Oil phase: 1×10^{-4} mol/liter IM in MIBK. (From Ref. 93.)

When the aqueous phase contained 10^{-4} mol/liter KNO_3, the IEP of phosphatidyl choline (PC) was 3.2 and in good agreement with the isoelectric point measured by the electrophoresis of oil/water emulsions of the same composition [95]. This agreement indicates the absence of specific adsorption or counterion binding in these systems. However, this is not always the case and in fact the IEP was found to depend on the nature and concentration of coexisting supporting electrolytes [95].

The change in IEP with increasing electrolyte concentration is characteristic of the combination of phospholipid and cationic species. The behavior of PC and sphingomyeline (SM), especially, is markedly different from that of phosphatidyl ethanolamine (PE). IEP decreases in the case of the former two, whereas in the case of PE it increases with increasing concentration of $Th(NO_3)_4$. For PE, the increase in IEP is larger for counterions, with larger ionic radii increasing in the order $Th^{4+} > La^{3+} > Zn^{2+} > Ca^{2+} > K^+$. However, such a sequence was not found for PC or SM [89].

Competing Binding. The observation that the IEP of PE increases with increasing Th^{4+} ion concentration is explained by the competing binding between H^+ and Th^{4+} ions in the aqueous phase for the negative phosphate groups of PE adsorbed at the interface [90]. When head groups of PE are partially bound by Th^{4+} ions, the hydrogen ion concentration necessary to neutralize the charge at the interface decreases by an equivalent amount, giving rise to the increase in IEP.

Stern Effect. A tendency opposite to the competition discussed above—a decrease in the IEP of SM with increasing Th^{4+} concentration—can be explained by the local decrease in hydrogen ion concentration at the interface [89]. That is, when the Stern potential ψ_δ increases by the Langmuir-type adsorption of a counterion (i.e., Th^{4+}), the interfacial hydrogen ion concentration $[H^+]_0$ decreases compared with the bulk value $[H^+]$ due to electrostatic repulsion. Hence $[H^+]$ must be increased; that is, the pH must be decreased by hydrogen ions in order to neutralize the negative charge of lipid molecules adsorbed at the interface.

Mixed Adsorption. It is feasible to examine whether the binding is taking place at the interface between phospholipids and various counterions such as antibiotics, narcotics, and acetylcholine. For example, competing binding was found between dihydrostreptomycine sulfate (ST) and PC, whereas the Stern effect was found between colistin M, cyclic polypeptide, and PC. Both effects are influenced by the addition of cholesterol (CH). For instance, the IEP of PC shows a minimum at the mole ratio PC/CH = 1/3 when the aqueous phase contains ST or kanamicine in addition to 10^{-1} mol/liter NaCl [90]. It is suggested that PC and CH molecules make a sort of interfacial complex at this mole ratio.

6.3.5 Tensametry at Oil/Water Interfaces

To obtain a complete understanding of the process taking place at the oil/water interface, we must know the detailed structure of the electrical double layers at both sides of the interface, as well as the mechanism of the adsorption processes. Thus differential capacity and/or impedance measurements as a function of applied potential or current are expected to render some information. However, due to the high resistivity of the system, very few experiments have so far been made by this method (see Ref. 101). In this connection, "tensametry" has been introduced by Watanabe and Tagaya to investigate the adsorption mechanism of surface-active materials at the oil/water interface [102, 103].

Figure 6.8 is a schematic diagram of the experimental device [102]. It consists essentially of a dropping electrode [104-106] with a Teflon capillary tube F vertically placed and the aqueous phase A added dropwise into the oil phase B. The diameter of the glass capillary J, sealed in the connecting tube, and the height of the reservoir L are adjusted so that the flow rate of aqueous phase is controlled. A polarizing potential E applied to the platinum electrode D in the aqueous

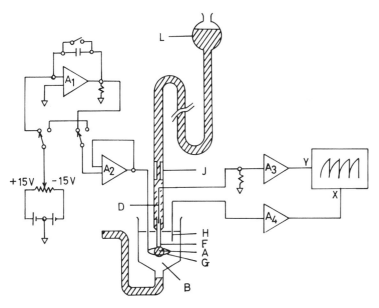

Figure 6.8 Schematic diagram of the device for tensametry. A, aqueous phase; B, oil phase; F, Teflon capillary tube; J, glass capillary tube; D, G, H, Pt electrodes; L, reservoir; A_1, integrator (time sweep); A_2, impedance converter; A_3, amplifier for i measurement; A_4, amplifier for E measurement.

phase, with platinum electrode G in the oil phase, makes it possible to polarize the oil/water interface by a constant potential or by a series of potential sweeps. The measurement of polarizing potential is carried out between D and another platinum electrode H in the oil phase. Due to the periodic rise and fall in the size of the aqueous drops, the current-time (I-t) curve shows a jagged shape, and γ can easily be obtained as a function of E from the measurement of the dropping period at the given flow rate of the aqueous phase.

Figure 6.9 shows a typical I-E curve (polarographic curve) for the aqueous phase of 10^{-2} mol/liter KCl, in contact with the oil phase of MIBK, containing octadecyltrimethylammonium chloride (OTAC). It is evident that γ decreases over the negative polarizing range (E < 0) (see Sec. 6.3.1). Moreover, it is evident that the maximum current during the life of individual drops I_{max} is higher in magnitude over the positive polarization range than over the negative range; the opposite tendency is found for the minimum current I_{min}.

The current densities at the maximum drop size i_{max} for the same system as above are plotted in Fig. 6.10 as a function of E [102]. The slopes (di/dE) in the negative polarization range are steeper than those in the positive range. For anionic surfactants, the opposite tendency can be observed. These results lead to the conclusion that

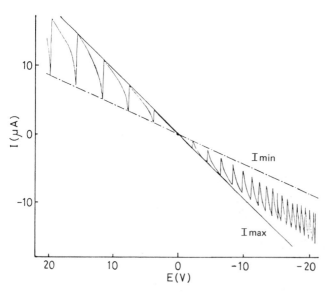

Figure 6.9 Typical I-E curve (polarographic curve). Rate of potential sweep, 1.0 V/sec; flow rate of aqueous phase, 1.5331 ml/sec. Aqueous phase: 1×10^{-2} mol/liter KCl. Oil phase: 5×10^{-5} mol/liter OTAC in MIBK.

Electrocapillary Measurements / 167

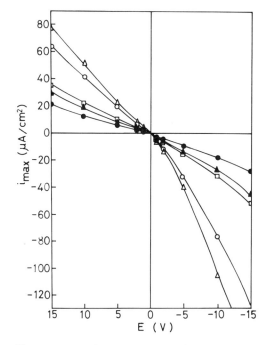

Figure 6.10 i_{max} versus E for the adsorption of OTAC at the oil/water interface. Aqueous phase: 1×10^{-2} mol/liter KCl. Oil phase: MIBK containing OTAC at concentration: ●, 5×10^{-6}; ▲, 7×10^{-6}; □, 1×10^{-5}; ○, 3×10^{-5}; △, 5×10^{-5} mol/liter.

the rectification takes place over the negative or positive polarization range depending on whether the surfactant adsorbed at the interface is cationic or anionic. No such effect was found without an ionic surfactant in the system.

The time dependence of the current densities i_t during the life of a drop is plotted in Fig. 6.11 as a function of $t^{-1/2}$, where the oil phase of MIBK contains SHS, in contact with the aqueous phase of 10^{-2} mol/liter KCl. Here we used an i value at E = +12.5 V. The linear relationship between i_t and $t^{-1/2}$ suggests that the current is determined by the rate of diffusion of surfactant ions from the bulk to the interface [103].

6.3.6 Applications

The electrocapillarity provides a new method of approach to various phenomena connected with adsorption at oil/water interfaces, as described below.

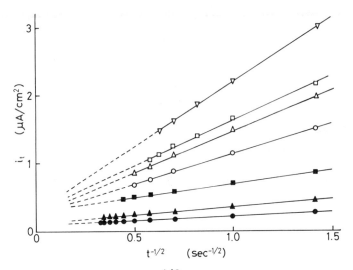

Figure 6.11 i_t versus $t^{-1/2}$ for the adsorption of SHS at oil/water interface (E = +12.5 V). Aqueous phase: 1×10^{-2} mol/liter KCl. Oil phase: MIBK containing SHS at concentration: ●, 1×10^{-6}; ▲, 4×10^{-6}; ■, 1×10^{-5}; ○, 3×10^{-5}; △, 4×10^{-5}; □, 5×10^{-5}; ▽, 1×10^{-4} mol/liter.

1. *Adsorption of phospholipids.* Phospholipids are typical surface-active materials found in biological processes. It is possible to obtain their isoelectric points at oil/water interfaces and to examine whether binding is taking place between phospholipid molecules and various counterions of biological importance, such as antibiotics and narcotics, by the procedure described above [89,90].
2. *Adsorption of dyes.* The adsorption of dye at oil/water interfaces can also be investigated by the present method. Since dyes are usually soluble in water, experiments are carried out for a system of aqueous dye solution containing various inorganic salts as the builder. The interaction between dye and surfactants can also be investigated by adding inorganic salts to the oil phase [88,92].
3. *Electric emulsification.* One of the applications of the presence of electrocapillary phenomena at oil/water interfaces is the formation of emulsions (see Part III, Chap. 13). When the applied potential becomes sufficiently high, spontaneous emulsification (i.e., a shower of fine droplets) takes place. Emulsions thus formed are in general monodisperse and hence of great practical value. Moreover, this method can be used to clarify the mechanism of emulsification [107-111].

6.4 POINT OF ZERO CHARGE OF SUSPENSION

6.4.1 Introduction

The point of zero charge (PZC) is defined as the state in which the net surface charge of the solid is zero. In other words, at the PZC a solid surface has the same electrical potential as that of the bulk of a contacting liquid phase. The potential difference between a solid surface and a liquid is determined by the activity of potential-determining (PD) ions in the liquid. The PD ions depend on the mechanism of electrification of solid surfaces and are classified into two types. Ions of the first type are the component ions of such solid materials as sparingly soluble salts (e.g., silver halides, silver sulfides, $BaSO_4$, $BaCr_2O_7$, $PbSO_4$, etc.) [112]. Ions of the second type are H^+ or OH^- ions of, for instance, metal oxides with surface hydroxyls, polypeptides with carboxyls and amino groups, and synthetic polymers with sulfate groups which are introduced from potassium persulfate as an initiator of polymerization.

6.4.2 Determination of PZC

Direct Measurement of Surface Charge

Electrokinetic Phenomena. Electrokinetic phenomena, which permit the measurement of the zeta (ζ) potential of the system, are associated with the presence of the diffuse electrical double layer. The PZC is determined by measuring the ζ potential as a function of the concentration of PD ions and by finding the point of zero ζ potential. For the determination of the ζ potential, one can conveniently choose an appropriate type of electrokinetic technique depending on the particle size: for smaller particles, the electrophoretic method is available, and for larger ones the streaming potential, electroosmosis, or sedimentation potential method may be used. Electrokinetic measurements require only a small number of samples compared to the other methods described below. However, the methods mentioned above cannot be used in electrolyte solutions in which the concentration is high and thus the double-layer thickness is depressed. Details of electrokinetic measurements are described in Chap. 7.

Suspension Effect (Sol Concentration Effect or Pallman Effect [113, 114]). This phenomenon is also associated with charge distribution in the diffuse double layer, and can typically be observed when an electrode potential is measured in a sol or a suspension including charged particles; that is, the electrode potential value determined differs depending on whether charged particles are present. This is because the electrode potential in a suspension is affected by the PD ions in the diffuse double layer, whose concentration is different from that in the bulk. The measurement of this suspension effect is carried out by the following techniques.

Method 1. After setting the same types of irreversible electrodes (e.g., normal calomel electrodes) in the supernatant liquid and the bed of a sediment of charged particles, respectively, the potential difference is measured between the two electrodes. Here the PZC is the condition in which this potential difference is zero.

Method 2. The potential determined by method 1 is identical to the Donnan potential, so that if the measurement of the potential difference is carried out by use of the same irreversible electrodes dipped in the suspension and dispersion media, which are separated by a semipermeable membrane, a similar result will be obtained.

Method 3. When each of the same reversible electrodes, which works on the ions in the dispersion medium, is dipped in a suspension and a dispersion medium, respectively, which are connected electrically through a salt bridge, a potential difference due to the suspension effect will be observed. As in the case of electrokinetic methods, it is difficult to obtain a reliable PZC value by this method in systems having higher concentrations of electrolytes.

Adsorption or Ion-Exchange Reaction Measurement of Potential-Determining Ions

Adsorption by Dispersed Particles [115]. When particles having an electrically neutral surface are suspended in a solution that satisfies the condition of PZC, neither adsorption nor desorption of PD ions on the surface occurs. If the activity of PD ions in a solution is different from the PZC of the sample, either adsorption or desorption of the PD ions will occur; the change can be followed potentiometrically if any suitable electrode is available. Accordingly, one can determine the PZC of the sample by introducing a fixed amount of solid material into a series of solutions with various concentrations of PD ions and by finding the condition in which neither adsorption nor desorption of PD ions occurs and in which no potential change appears on the electrode.

Potentiometric Titration. Method 1 [116]. After determining the absolute surface charge density of suspended particles by means of the technique described above, the suspension is titrated potentiometrically with a solution of PD ions which permits calculation of the change in surface charge density of the powder materials through the mass balance of PD ions in the reaction. The condition of the PZC corresponds to the activity of PD ions at which the surface charge density is zero.

Method 2 [117]. The main part of this method is similar to method 1, differing in that the surface charge density of particles is determined experimentally in this method rather than through calculation. That is, the blank titration is carried out on the suspension medium, which has a similar concentration of PD ions and a similar ionic strength as those in the suspension before titration. When the starting point of the titration curve of the suspension is put on the point of the same potential on the blank titration curve, the intersection point of the two curves gives the PZC of the sample.

Method 3 [118]. A suspension that includes an excess amount of PD ions (A) to the PZC and has a low ionic strength is titrated with the PD ions (B) carrying the charge opposite to that of A. By this procedure the change in surface charge density of particles is determined. After enough titration has been attained to reverse the sign of the surface charge, some supporting electrolyte is added to the suspension to increase the ionic strength and accordingly the surface charge density. Next, backward titration is conducted with a solution of the PD ions (A). If the specific adsorption of ions in the supporting electrolyte is zero, a repeat of this experiment will give surface charge density curves which have a common intersection point (i.e., the ZPC). This technique does not require determination of the absolute surface charge density of particles.

Maximum Flocculation [119]

Under the conditions of a PZC the coagulation of particles can easily occur, since the electrostatic repulsion force between particles is absent. With sparingly soluble salts, PZC is also the condition in which the solubility is at a minimum. In this method the turbidity of suspension is measured as a function of the concentration of PD ions. The accuracy of measurement is low in this method, so that only an approximate value for the PZC can be obtained.

6.4.3 Definition of the PZC [120-122]

According to the definition, the PZC is described by the activity of PD ions at which the net surface charge of the solid is zero. In the determination of the PZC, however, one cannot avoid the coexistence of supporting electrolyte and counterions of PD ions in the solution. If either cations or anions other than PD ions are adsorbed specifically in the Stern layer, the PZC value determined by means of electrokinetic or suspension effect methods should shift to a higher concentration of PD ions carrying a charge opposite in sign to that of the specifically adsorbed ions. In contrast, the PZC value obtained by the adsorption method shifts in the direction opposite to that determined by electrokinetic measurement. Therefore, the specific adsorption of the component ions of the supporting electrolyte can be checked through a comparison of PZC values measured by different methods varying in principle. Thus, as the PZC values obtained by different techniques may lead to different interpretations of the double-layer structure, it is necessary to give different notations to each value. The PZC value determined by electrokinetic or suspension effect measurements has been defined as the isoelectric point (IEP) and that determined by the adsorption method as the PZC; to avoid confusion, the present author designates the latter as p.z.c.

6.4.4 Comparison of PZC Values Determined by Different Methods

If the specific adsorption of indifferent ions does not occur, the PZC values determined by different methods should conform with each other. Some examples representing this are introduced briefly here. By using microelectrophoresis and maximum flocculation techniques Healy et al. [119] measured the IEP values of various MnO_2 samples

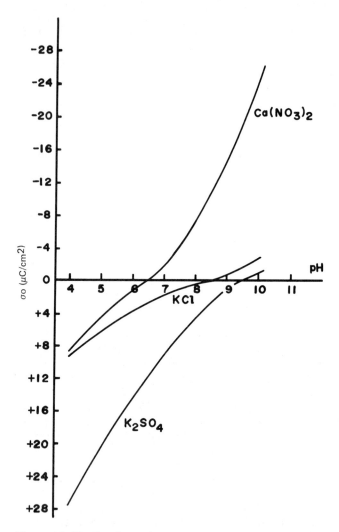

Figure 6.12 Surface charge on hematite as a function of pH in the presence of 10^{-3} M solutions of KCl, $Ca(NO_3)_2$, and K_2SO_4. (From Ref. 125.)

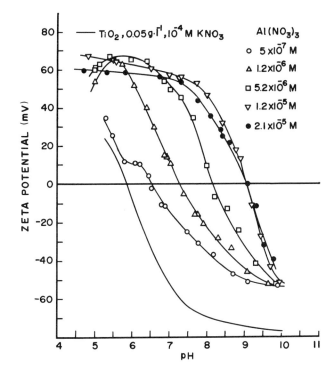

Figure 6.13 Variation of the ζ potential of TiO_2 colloid (0.05 g/liter) in 10^{-4} M KNO_3 with pH at 25°C, as a function of $Al(NO_3)_3$ concentration. The curve determined in the absence of $Al(NO_3)_3$ is included for comparison. (From Ref. 123.)

having different crystal structures and found a good agreement between the experimental results. Blok and de Bruyn [116] obtained similar p.z.c. values of TiO_2 by applying titration method 1 and the adsorption method. Healy observed very close values of IEP and p.z.c. of TiO_2: 5.86 by the electrophoresis method [122] and 5.8 ± 0.1 by the titration method [123]. On samples of AgI, whose colloid chemical properties have been investigated more extensively than those of the other sparingly soluble salts, Overbeek [124] summarized PZC data and obtained a reliable mean value of pI = 5.52, which is very similar to the value of 5.53 determined very recently [112]. Honig and Hengst [112] compared the PZC data on such sparingly soluble salts as silver halides, sulfates and oxalates of Ca, Ba, and Pb, and so on, by various methods with those measured by means of the suspension effect. Agreement among the data was not satisfactory. The main reason for this was ascribed to the difference in the preparation methods of samples, since they might produce samples that differ in surface structures.

In most experiments to determine the PZC, the supporting electrolytes containing such monovalent ions as Na^+, K^+, Cl^-, NO_3^-, and ClO_4^- are used at a concentration lower than 10^{-2} M. Even when the concentration of these ions is increased to 1 M, the shift of PZC values has been found to be very small or zero. In the case of using other kinds of ions, however, it has been found that specific adsorption occurs, which results in a shift of the PZC. Here, one can see some examples of the PZC which are affected by the specific adsorption of ions in the supporting electrolytes. Although examples of determining the p.z.c. as well as the IEP on the same system have not been reported, there are some works treating such effects of specific adsorption on the p.z.c. or IEP values which have been done by different research groups. Figure 6.12 shows the shift of p.z.c. values by the specific adsorption of Ca^{2+} or SO_4^{2-} on hematite, where Ca^{2+} causes the p.z.c. to shift to a lower pH value and SO_4^{2-} to shift to a higher pH value. On the other hand, the specific adsorption of aluminum ions gives higher values for the IEP on TiO_2 (Fig. 6.13). The specific adsorption of monovalent ions is discussed later.

6.4.5 PZC as a Measure of Electrification of Solids

When the electrification of solids is determined by the adsorption of common PD ions, for example H^+ ions for metal oxides, the PZC is a useful parameter with which the electrification mechanism can be elucidated. In 1965, Parks [126] proposed an empirical formula for the PZC of metal oxides and hydroxides:

$$PZC = A - 11.5\left(\frac{Z}{R} - 0.0029a\right) \tag{6.6}$$

where A is a constant that depends on the extent of hydration and the coordination number of surface metal ions in the solid surface: 18.6 for octahedrally coordinated metal ions and 16.2 for tetrahedrally coordinated ones. Z is the valency of metal ions in the solid surface, R the separation between the proton of the surface hydroxyl and the center of the neighboring metal ion, and a the crystal field stabilization energy for hydroxyl ions to coordinate with transition metal ions. Recently, Yoon, [127] modified the term A based on theoretical calculations. In Eq. (6.6) it is assumed that the surface sites which generate the surface charge are homogeneous and that they have the same valence and coordination number. However, actual surfaces are usually heterogeneous, and accordingly an observed value for the PZC is the mean of the PZC values on various sites. Recently, Kittaka and Morimoto [128] found abnormally low IEP values on metal oxides having the inverse spinel structure (Fe_3O_4, Mn_3O_4, $CuFe_2O_4$ etc.); this was attributed to the presence of trivalent cations which have the coordination number 4. This seems to be a strange phenomenon because the ratio of these cations in the bulk solids, including trivalent ions with coordination numbers 4 and 6 and divalent ions with coordination number 6, is only 1/3, and because there is no cleavage plane containing only this kind of metal ion. Thus the data above show that minor parts of surface sites strongly affect the electrification of metal oxides, the other kinds of ions having no effect even if exposed. Few works have been published which are concerned with the effect of semiconductivity and defects in a solid on the mechanism of the electrification of the solid surface. A report by Honig and Hengst [129] describes the relationship between crystal defects of AgBr and its electrification.

6.4.6 Dependence of the PZC on the Temperature of Measurement

Berube and de Bruyn [116] found that the p.z.c. of TiO_2 decreases and finally approaches $(1/2)pKw$ as the temperature is raised to 95°C. A similar result was found by Tewari and Campbell [130] on $Co(OH)_2$ and $Ni(OH)_2$. On Co_3O_4 and NiO, however, the p.z.c. value decreased in parallel with the change in $(1/2)pKw$ as the temperature was increased. These facts indicate that the relative affinity of PD ions H^+ and OH^- to the surface of metal oxides does not change with variation in the temperature, whereas with hydroxides it does. In the case of AgI [116] it was ascertained that the p.z.c. approaches $(1/2)pL$ when the temperature is raised, where L is the solubility product of AgI in water.

6.4.7 Capacitance of Suspensions

In the process of determining the p.z.c., one measures the surface charge density as a function of the surface potential (or the activity of PD ions, e.g., pH); the relationship between them is of great significance because it includes information about the double-layer structure on the solid surface. By differentiating the surface charge density-potential curve, one can obtain the capacitance of the electrical double layer on insulators and semiconductors, which are frequently encountered in the suspensions. This type of data cannot be obtained through direct electrical measurement except for systems involving metals. A number of studies on the electrical double layer in systems containing solid-electrolyte solutions have been reported: On AgI [118,131], ZnO [132], Fe_2O_3 [133,134], Al_2O_3 [135], TiO_2 [136,137], SnO_2 [138], SiO_2 [139,140], and so on. The most widely investigated interfaces for elucidating the double-layer structure in the liquid side are those of Hg-electrolyte solutions and the results obtained are as follows: the specific adsorption of cations on the negatively charged Hg surface increases in the lyotropic order $Cs^+ > K^+ > Na^+ > Li^+$, and anions on the positively charged Hg surface in the order of $I^- > Br^- > Cl^- > NO_3^- \simeq ClO_4^-$. The capacitance measurements for AgI and SiO_2 have shown that the order of adsorbability of ions is similar to that on the Hg surface. This phenomenon has been explained by the fact that the smaller the radius of hydrated ions, the more closely the ions are adsorbed on the surface. In contrast, most of the other metal oxides give the inverse order for the adsorbability of ions; that is, ions having smaller true radii are adsorbed more closely to the surface. This contradiction has not yet been explained satisfactorily [123,137,141,142]. Bérubé and Bruyn [136] and Yates, James, and Healy [123] explained the latter fact by assuming that the surface of, for instance, TiO_2 has a structure-promoting property for water. The ability of the structure-promoting property of alkali ions increases in the order $Li^+ > Na^+ > K^+ > Cs^+$. Therefore, the structure-promoting ions can be adsorbed closely on the surface of hydroxylated metal oxides, similarly to the phenomenon that ions with a structure-promoting property come together.

REFERENCES

1. J. O'M. Bockris and A. K. N. Reddy, *Modern Electrochemistry*, Vol. 2, Plenum Press, New York, 1970, Chap. 7.
2. R. Tamamushi, *Denki Kagaku*, Tokyo Kagaku Dōjin, Tokyo, 1967, Chap. 4.
3. J. Th. G. Overbeek, Electrochemistry of the Double Layer, in *Colloid Science* (H. R. Kruyt, ed.), Vol. 1, Elsevier, Amsterdam, 1952, pp. 115-193.
4. P. Delahay, *Double Layer and Electrode Kinetics*, Wiley-Interscience, New York, 1965, Part 1.
5. M. J. Sparnaay, The Electrical Double Layer, in *The International Encyclopedia of Physical Chemistry and Chemical Physics*, Pergamon Press, Oxford, 1972.
6. D. M. Mohilner, The Electrical Double Layer, Part I: Elements of double-layer theory, in *Electroanalytical Chemistry*, Vol. 1 (A. J. Bard, ed.), Marcel Dekker, New York, 1966, pp. 241-409.
7. R. Parsons, Equilibrium Properties of Electrified Interfaces, in *Modern Aspects of Electrochemistry* (B. E. Conway and J. O'M. Bockris, eds.), No. 5, Plenum Press, New York, 1969, pp. 103-179.
8. R. Payne, The Electrical Double Layer in Nonaqueous Solutions, in *Advances in Electrochemistry and Electrochemical Engineering*, Vol. 7 (P. Delahay and C. W. Tobias, eds.), Wiley-Interscience, New York, 1970, pp. 1-76.
9. R. Payne, The Study of the Ionic Double Layer and Adsorption Phenomena, in *Techniques of Electrochemistry* (E. Yeager and A. J. Salkind, eds.), Vol. 1, Wiley-Interscience, New York, 1972, Chap. 2.
10. D. C. Grahame, *Chem. Rev.*, *41*, 441 (1947).
11. M. A. V. Devanathan and B. V. K. S. R. A. Tilak, *Chem. Rev.*, *65*, 635 (1965).

12. J. O'M. Bockris, K. Müller, H. Wroblowa, and Z. Kovac, *J. Electroanal. Chem.*, *10*, 416 (1965).
13. R. S. Hansen, D. J. Kelsh, and D. H. Grantham, *J. Phys. Chem.*, *67*, 2316 (1963).
14. R. S. Perkins and T. N. Andersen, Potentials of Zero Charge of Electrodes, in *Modern Aspects of Electrochemistry* (B. E. Conway and J. O'M. Bockris, eds.), No. 5, Plenum Press, New York, 1969, pp. 203-290.
15. R. G. Barradas and F. M. Kimmerle, *Can. J. Chem.*, *45*, 109 (1967).
16. Cl. Buess-Herman, L. Gierst, and N. Vanlaethem-Meuree, *J. Electroanal. Chem.*, *123*, 1 (1981).
17. G. Quarin, Cl. Buess-Herman, and L. Gierst, *J. Electroanal. Chem.*, *123*, 35 (1981).
18. J. Lawrence and D. M. Mohilner, *J. Electrochem. Soc.*, *118*, 1596 (1971).
19. H. Nakadomari, D. M. Mohilner, and P. R. Mohilner, *J. Phys. Chem.*, *80*, 1761 (1976).
20. D. M. Mohilner and T. Kakiuchi, *J. Electrochem. Soc.*, *128*, 350 (1981).
21. C. A. Smolders, *Recl. Trav. Chim.*, *80*, 635 (1961).
22. H. Vos and J. M. Los, *J. Colloid Interface Sci.*, *74*, 364 (1980).
23. D. C. Grahame, R. P. Larsen, and M. A. Poth, *J. Am. Chem. Soc.*, *71*, 2978 (1949).
24. D. C. Grahame, E. M. Coffin, J. I. Cummings, and M. A. Poth, *J. Am. Chem. Soc.*, *74*, 1207 (1952).
25. A. N. Frumkin, *Sven. Kem. Tidskr.*, *77*, 300 (1965).
26. L. Campanella, *J. Electroanal. Chem.*, *28*, 228 (1970).
27. R. S. Hansen, R. E. Minturn, and D. A. Hickson, *J. Phys. Chem.*, *60*, 1185 (1956).
28. R. Parsons and R. Peat, *J. Electroanal. Chem.*, *122*, 299 (1981).
29. J. O'M. Bockris, M. A. V. Devanathan, and K. Müller, *Proc. R. Soc., A*, *274*, 55 (1963).
30. R. Guidelli, *J. Electroanal. Chem.*, *123*, 59 (1981).
31. S. Trasatti, *J. Electroanal. Chem.*, *123*, 121 (1981).
32. See, for example, P. Delahay, *Double Layer and Electrode Kinetics*, Wiley-Interscience, New York, 1965, Part 1, pp. 101-112.
33. S. Levine, *J. Electroanal. Chem.*, *123*, 105 (1981).
34. K. Takahashi and R. Tamamushi, *Electrochim. Acta*, *16*, 875 (1971).
35. K. Takahashi and R. Tamamushi, *Bull. Chem. Soc. Jpn.*, *48*, 3540 (1975); *Electrochim. Acta*, *16*, 1157 (1971).
36. R. D. Armstrong and W. P. Race, *J. Electroanal. Chem.*, *33*, 285 (1971).

37. W. Lorenz, *Z. Elektrochem.*, *62*, 192 (1958).
38. W. Lorenz and F. Möckel, *Z. Elektrochem.*, *60*, 507, 939 (1956).
39. R. D. Armstrong, W. P. Race, and H. R. Thirsk, *J. Electroanal Chem.*, *27*, 21 (1970).
40. K. Takahashi, *Denki Kagaku*, *35*, 437 (1967).
41. K. Takahashi, *Electrochim. Acta*, *13*, 1609 (1968).
42. R. D. Armstrong, W. P. Race, and H. R. Thirsk, *J. Electroanal. Chem.*, *16*, 517 (1968).
43. See, for example, J. O'M. Bockris and A. K. N. Reddy, *Modern Electrochemistry*, Vol. 2, Plenum Press, New York, 1970, Chaps. 8-11; P. Delahay, *Double Layer and Electrode Kinetics*, Wiley-Interscience, New York, 1965, Part II; A. J. Bard and L. R. Faulkner, *Electrochemical Methods: Fundamentals and Applications*, Wiley, New York, 1980.
44. D. E. Smith, AC Polarography and Related Techniques, in *Electroanalytical Chemistry*, Vol. 1 (A. J. Bard, ed.), Marcel Dekker, New York, 1966, pp. 1-155.
45. D. E. Smith, *CRC Crit. Rev. Anal. Chem.*, *2*, 248 (1971).
46. M. Sluyters-Rehbach and J. H. Sluyters, Sine Wave Methods in the Study of Electrode Processes, in *Electroanalytical Chemistry*, Vol. 4 (A. J. Bard, ed.), Marcel Dekker, New York, 1970, pp. 1-128.
47. I. Epelboin and M. Keddam, *Electrochim. Acta*, *17*, 177 (1972).
48. M. Keddam, O. R. Mattes, and H. Takenouti, *J. Electrochem. Soc.*, *128*, 257, 266 (1981), and references therein.
49. R. Parsons, *J. Electroanal. Chem.*, *118*, 3 (1981).
50. M. Fleishmann, J. Robinson, and R. Waser, *J. Electroanal. Chem.*, *117*, 257 (1981).
51. G. Vallette, *J. Electroanal. Chem.*, *122*, 285 (1981).
52. T. Takamura and K. Takamura, Application of the Optical Reflection Method to the Study of Electrosorption, Anodic Oxide, and Under-Potential Deposition, in *Surface Electrochemistry* (T. Takamura and A. Kozawa, eds.), Japan Scientific Societies Press, Tokyo, 1978, pp. 179-241.
53. D. D. Macdonald, *Transient Techniques in Electrochemistry*, Plenum Press, New York, 1977, Chap. 8.
54. K. Sugimoto and Y. Sawada, *Boshoku Gijutsu*, *24*, 669 (1975).
55. D. C. Grahame, *J. Am. Chem. Soc.*, *71*, 2975 (1949).
56. R. Payne, The Study of the Ionic Double Layer and Adsorption Phenomena, in *Techniques of Electrochemistry* (E. Yeager and A. J. Salkind, eds.), Vol. 1, Wiley-Interscience, New York, 1972, pp. 71-106.
57. G. J. Hills and R. Payne, *Trans. Faraday Soc.*, *61*, 316 (1965).
58. R. D. Armstrong, W. P. Race, and H. R. Thirsk, *Electrochim. Acta*, *13*, 215 (1968).

59. J. W. Hayes and C. N. Reilley, *Anal. Chem.*, 37, 1322 (1965).
60. T. F. Retajczyk and D. K. Ror, *J. Electroanal. Chem.*, 16, 21 (1968).
61. K. Matsuda, K. Takahashi, and R. Tamamushi, *Sci. Pap. Inst. Phys. Chem. Res. Jpn.*, 64, 62 (1970).
62. P. Delahay, R. de Levie, and A. M. Giuliani, *Electrochim. Acta*, 11, 1141 (1966).
63. F. C. Anson, R. F. Martin, and C. Yarinitzky, *J. Phys. Chem.*, 73, 1835 (1969).
64. C. Yarinitzky and F. C. Anson, *J. Phys. Chem.*, 73, 3123 (1970).
65. J. H. Christie, R. A. Osteryoung, and F. C. Anson, *J. Electroanal. Chem.*, 13, 236 (1967).
66. F. C. Anson, *Anal. Chem.*, 38, 1924 (1966).
67. A. J. Bard and L. R. Faulkner, *Electrochemical Methods: Fundamentals and Applications*, Wiley, New York, 1965, pp. 519-538.
68. J. B. Flanagan, K. Takahashi, and F. C. Anson, *J. Electroanal. Chem.*, 81, 261 (1977); 85, 257 (1977).
69. R. D. Armstrong, M. F. Bell, and A. A. Metcalfe, *J. Electroanal. Chem.*, 77, 287 (1977).
70. J. A. Harrison and C. E. Small, *Electrochim. Acta*, 25, 447 (1980).
71. G. Lauer, R. Abel, and F. C. Anson, *Anal. Chem.*, 39, 765 (1967).
72. J. Guastalla, *J. Chim. Phys.*, 53, 470 (1956).
73. J. Guastalla, *Proc. 2nd Int. Congr. Surf. Activ.*, 3, 112 (1957).
74. M. Blank and S. Feig, *Science*, 141, 1173 (1963).
75. M. Blank, *Proc. IVth Int. Congr. Surf. Activ.*, 2, 233 (1964).
76. M. Dupeyrat and J. Michel, *J. Colloid Interface Sci.*, 29, 605 (1969).
77. M. Dupeyrat, *J. Chim. Phys.*, 61, 316 (1964).
78. M. Dupeyrat and E. Nakache, *J. Colloid Interface Sci.*, 73, 332 (1980).
79. C. Gavach, *Proc. 1st Int. Symp. Biol. Aspects Electrochem.*, May 1971.
80. A. Watanabe, M. Matsumoto, H. Tamai, and R. Gotoh, *Kolloid Z. Z. Polym.*, 220, 152 (1967).
81. A. Watanabe, M. Matsumoto, H. Tamai, and R. Gotoh, *Kolloid Z. Z. Polym.*, 221, 47 (1967).
82. A. Watanabe, M. Matsumoto, H. Tamai, and R. Gotoh, *Kolloid Z. Z. Polym.*, 228, 58 (1968).
83. A. Watanabe and H. Tamai, *Bull. Inst. Chem. Res.*, Kyoto Univ., 47, 283 (1969).

84. A. Watanabe, K. Nishizawa, K. Higashitsuji, and Y. Sakamori, *Colloid Interface Chem. Symp.*, Sendai, Japan, 1969.
85. A. Watanabe and H. Tamai, *Yukagaku*, 20, 101 (1971).
86. A. Watanabe and H. Tamai, *Kolloid Z. Z. Polym.*, 246, 587 (1971).
87. A. Watanabe, *Nippon Kagaku Zasshi*, 92, 575 (1971).
88. A. Watanabe and H. Tamai, *Sen-i Gakkaishi*, 27, 175 (1971).
89. A. Watanabe, A. Fujii, Y. Sakamori, K. Higashitsuji, and H. Tamai, *Kolloid Z. Z. Polym.*, 243, 42 (1971).
90. A. Watanabe, H. Tamai, and K. Higashitsuji, *J. Colloid Interface Sci.*, 43, 548 (1973).
91. A. Watanabe and H. Tamai, *Hyomen*, 13, 67 (1975).
92. H. Tamai, K. Hayashi, and A. Watanabe, *Colloid Polym., Sci.*, 255, 773 (1977).
93. H. Tagaya and A. Watanabe, *Res. J. Living Sci. (Nara)*, 27, 95 (1981).
94. W. D. Harkins and F. E. Brown, *J. Am. Chem. Soc.*, 38, 288 (1916).
95. W. D. Harkins and F. E. Brown, *J. Am. Chem. Soc.*, 41, 499 (1919).
96. J. T. Davies and E. K. Rideal, *Interfacial Phenomena*, 2nd ed., Academic Press, London, 1963.
97. V. V. Subba Rao, R. J. Fix, and A. C. Zettlemoyer, *J. Am. Oil Chem. Soc.*, 45, 449 (1968).
98. J. T. Davies, *Trans. Faraday Soc.*, 43, 1052 (1952).
99. C. Tanford, S. A. Swanson, and W. S. Shore, *J. Am. Chem. Soc.*, 77, 6414 (1955).
100. L. G. Longsworth and C. F. Jacobson, *J. Phys. Colloid Chem.*, 53, 126 (1949).
101. P. Seta and C. Gavach, *C. R. Acad. Sci. Ser. C*, 275(21), 1231 1972).
102. H. Tagaya, A. Watanabe, and M. Kito, *32nd Colloid Interface Chem. Symp.*, Kochi, Japan, 1979.
103. H. Tagaya, A. Watanabe, K. Higashitsuji, and K. Nishizawa, *33rd Colloid Interface Chem. Symp.*, Hokkaido, Japan, 1980 (in press).
104. J. Koryta, P. Vanysek, and M. Brezina, *J. Electroanal. Chem.*, 67, 263 (1976).
105. Z. Samec, V. Marecek, J. Weber, and D. Homolka, *J. Electroanal. Chem.*, 99, 385 (1979).
106. M. Senda and T. Kakutani, *Hyomen*, 18, 535 (1980).
107. K. Higashitsuji, K. Nishizawa, K. Kamada, and A. Watanabe, *Nippon Kagaku Kaishi*, 1974, 995 (1974).
108. A. Watanabe, K. Higashitsuji, and K. Nishizawa, *Colloidal Dispersions and Micellar Behavior* (K. L. Mittal, ed.), ACS Symposium Series No. 9, American Chemical Society, Washington, D.C., 1975, p. 97.

109. A. Watanabe, K. Higashitsuji, and K. Nishizawa, *J. Colloid Interface Sci.*, 64, 278 (1978).
110. K. Higashitsuji and A. Watanabe, *Nippon Kagaku Kaishi*, 1979, 974 (1979).
111. K. Higashitsuji and A. Watanabe, *Nippon Kagaku Kaishi*, 1979, 1287 (1979).
112. E. P. Honig and J. H. Th. Hengst, *J. Colloid Interface Sci.*, 29, 510 (1969).
113. H. Pallmann, *Kolloid Chem. Beih.*, 30, 334 (1930).
114. J. Th. G. Overbeek, in *Colloid Science* (H. R. Kruyt, ed.), Vol. 1, Elsevier, Amsterdam, 1952, p. 184.
115. S. M. Ahmed, *Can. J. Chem.*, 44, 1663 (1966).
116. Y. G. Bérubé and P. L. de Bruyn, *J. Colloid Interface Sci.*, 27, 305 (1968).
117. P. H. Tewari and A. W. McLean, *J. Colloid Interface Sci.*, 40, 267 (1972).
118. J. Lyklema and J. Th. G. Overbeek, *J. Colloid Interface Sci.*, 16, 595 (1961).
119. T. W. Healy, A. P. Herring, and D. W. Fuerstenau, *J. Colloid Interface Sci.*, 21, 435 (1966).
120. *IUPAC Inf. Bull.* (No. 3: Manual of Definitions, Terminology and Symbols in Colloid and Interface Chemistry, 1970).
121. J. Lyklema, *Discuss. Faraday Soc.*, 52, 318 (1971).
122. G. R. Wiese and T. W. Healy, *J. Colloid Interface Sic.*, 51, 434 1975).
123. D. E. Yates, R. O. James, and T. W. Healy, *J. Chem. Soc. Faraday Trans. I*, 76, 9 (1980).
124. J. Th. G. Overbeek, in *Colloid Science* (H. R. Kruyt, ed.), Vol. 1, Elsevier, Amsterdam, 1952, p. 161.
125. A. Breeuwsma and J. Lyklema, *J. Colloid Interface Sci.*, 43, 437 (1973).
126. G. A. Parks, *Chem. Rev.*, 65, 177 (1965).
127. R. H. Yoon, T. Salam, and G. Donnay, *J. Colloid Interface Sci.*, 70, 483 (1979).
128. S. Kittaka and T. Morimoto, *J. Colloid Interface Sci.*, 75, 398 (1980).
129. E. P. Honig and J. H. Hengst, *J. Colloid Interface Sci.*, 31, 545 (1969).
130. P. H. Tewari and A. B. Campbell, *J. Colloid Interface Sci.*, 55, 531 (1976).
131. H. De Bruyn, *Recl. Trav. Chim. Pays-Bas*, 61, 5 (1942).
132. L. Blok and P. L. de Bruyn, *J. Colloid Interface Sci.*, 32, 518 (1970).
133. G. A. Parks and P. L. de Bruyn, *J. Phys. Chem.*, 66, 967 (1962).
134. G. Y. Onoda, Jr., and P. L. de Bruyn, *Surf. Sci.*, 4, 48 (1966).

135. J. B. Peri, *J. Phys. Chem.*, *69*, 211 (1965).
136. Y. G. Bérubé and P. L. de Bruyn, *J. Colloid Interface Sci.*, *28*, 92 (1968).
137. J. A. Davis, R. O. James, and J. O. Leckie, *J. Colloid Interface Sci.*, *63*, 480 (1978).
138. T. Komura, M. Takahashi, and H. Imanaga, *Nippon Kagaku Kaishi*, *1978*, 1596 (1978).
139. G. H. Bolt, *J. Phys. Chem.*, *61*, 1166 (1957).
140. A. Komura, K. Hatsutori, and H. Imanaga, *Nippon Kagaku Kaishi*, *1978*, 779 (1978).
141. M. J. Fuller, *Chromatogr. Rev.*, *14*, 45 (1971).
142. W. Stumm, R. Kummert, and L. Sigg, *Croat. Chem. Acta*, *53*, 291 (1980).

7
Electrokinetic Measurements

Shigeharu Kittaka Okayama University of Science, Okayama, Japan

Kunio Furusawa The University of Tsukuba, Sakura-Mura, Ibaraki, Japan

Masataka Ozaki Yokohama City University, Yokohama, Kanazawa, Japan

Tetsuo Morimoto Okayama University, Okayama, Japan

Ayao Kitahara Science University of Tokyo, Tokyo, Japan

7.1. Electrophoresis 184
    7.1.1. Introduction 184
    7.1.2. Moving boundary method 184
    7.1.3. Microelectrophoresis 188
    7.1.4. Recent developments in electrophoresis 192
7.2. Electroosmosis 199
    7.2.1. Introduction 199
    7.2.2. Measurement 199
7.3. Streaming Potential and Streaming Current 202
    7.3.1. Streaming potential 202
    7.3.2. Streaming current 210
7.4. Surface Conductance 214
    7.4.1. Introduction 214
    7.4.2. Measurement of surface conductivity 214
    7.4.3. Mechanism of surface conductance 215
    7.4.4. Experimental results 216
7.5. Nonaqueous Systems 217
    7.5.1. Microelectrophoresis 218
    7.5.2. Electrodeposition method 219
    7.5.3. Moving boundary method 222
    References 222

7.1 ELECTROPHORESIS

7.1.1 Introduction

According to the well-known Derjaguin-Landau-Verwey-Overbeek (DLVO) theory, the surface or Stern potential of particles plays a major role in the stability of colloidal systems [1]. The surface potential, however, is an absolute potential, and its value is not measurable by any present experimental technique.

The zeta (ζ) potential is the potential at the slipping plane in the electrical double layer and its value is not precisely the same as that of the surface potential. A discussion of its physical meaning and the difference between it and the surface potential appears in Chap. 4. In spite of the uncertainty of its meaning and ambiguous character, the ζ potential has frequently been used for the discussion of colloid stability and its value is still considered useful in relation to the electrical double layer.

The most familiar method for determining the ζ potential is electrophoresis, which consists of setting up a potential gradient in a solution and measuring the rate of movement of particles in the solutions. There are two distinct approaches to measuring this rate, depending on whether or not the moving particles are visible under a microscope. If they are invisible, as in the case of some protein molecules, special techniques to observe the movement are required. These include the familiar Tiselius method. For particles about 0.5 µm in diameter or larger, however, direct observation is possible with a traveling microscope illuminated with a strong light source. An ultramicroscope is useful for suspensions of smaller particles, down to 0.1 µm.

In addition, new techniques have been developed. There are, for example, the laser Doppler method, the rotating grating method, and the rotating prism method. These techniques not only determine the mean mobility value rapidly, but also estimate the mobility distribution about the mean value.

7.1.2 Moving Boundary Method

U-Tube Method

The simplest method used to determine electrophoretic velocity is the U-tube method. The apparatus consists of a U-shaped glass tube called a "Burton tube," fitted with platinum electrodes at its open ends. After making a sharp boundary of the suspension with the supernatant (or dialyzer), a potential difference is applied between the electrodes. The rate at which the boundary moves up one limb of the tube is measured using a cathetometer. After several minutes, the polarity of the applied potential is reversed and the movement of the boundary is measured again. Since this measurement is performed fairly easily, the U-tube method is convenient for determining the sign of the charge of the suspended particles.

Improved U-Tube Method

To avoid disturbances due to reaction products generated at the electrodes in the U tube, Ottewill and Shaw [2] have designed the electrophoretic cell shown in Fig. 7.1, in which there are two pairs of disk-shaped platinum electrodes. One pair is for the application of the potential and the other is for measuring the electric field in the U tube. The stopcock is drilled to give a hole of the same diameter as that of the uniform tube (AB). The colloidal solution is introduced via the limb until the stopcock (E) is completely filled and then the stopcock is closed. After removal of the excess dispersion above the stopcock, dialyzer is introduced to fill the cell. The boundary is formed by slowly opening the stopcock. The movement of the boundary is followed with a cathetometer and the velocity is determined. To check the reliability of the technique, the electrophoretic mobilities determined by this cell for polystyrene latices were compared by Ottewill and Shaw [2] with those obtained by microscopic electrophoresis. Ottewill and Shaw added 0.5% sucrose to the latex dispersion

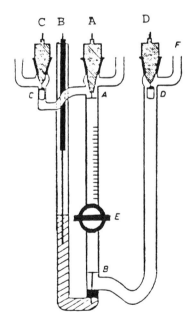

Figure 7.1 Diagram of an improved U tube. A, B, electrodes used for the determination of potential; C, D, electrodes used for the application of potential. (From Ref. 2.)

Table 7.1 Comparison of Electrophoretic Mobilities by the Moving Boundary Method and Microelectrophoresis in the Presence and Absence of Sucrose at pH 9.0

Latex	Method	Sodium chloride concentration: 5×10^{-4} M Mobility (μ sec^{-1} V^{-1} cm) with:		10^{-2} M Mobility (μ sec^{-1} V^{-1} cm) with:	
		0.5% sucrose	Sucrose absent	0.5% sucrose	Sucrose absent
A	Moving boundary	2.82	2.90	2.00	2.02
B	Moving boundary	2.95	2.88	2.25	2.18
C	Moving boundary,	4.07	–	2.57	–
	Microelectrophoresis	–	4.10	–	2.62
E	Moving boundary,	5.12	–	3.62	–
	Microelectrophoresis	–	5.06	–	3.60

to facilitate rapid formation of a sharp boundary. All measurements were carried out as a function of pH at constant ionic strength. It is clear from Table 7.1 that the presence of sucrose had a negligible effect on the mobility determination. The technique using the improved U tube produces reliable electrophoretic data as well as the determination of the sign of the charge on the particle surface.

Tiselius Method

The Tiselius method, when employed as an electrophoretic technique, is especially important in the investigation of proteins and other hydrophilic colloids. The phase boundary of these samples cannot be detected by the usual color differences but are observed by employing the Schlieren method, which is based on the difference between the refractive indices of the colloidal solutions and their dialyzers. The concentration gradient of the colloid is indicated by a maximum in the variation of the refractive index with distance. The position of this maximum is a measure of the position of the boundary between the colloid and the dialyzer. A typical Schlieren optical system is shown in Fig. 7.2a.

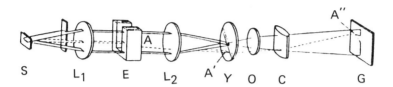

A

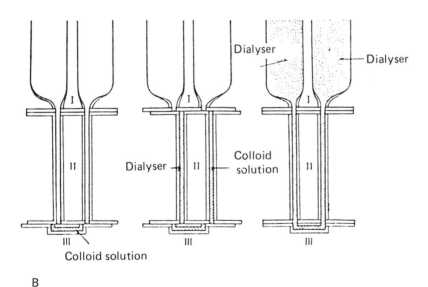

B

Figure 7.2 Electrophoretic measurements using a Tiselius cell. (A) Diagram of the Schlieren method. S, Light source; L_1, L_2, Schlieren lenses; E, cell with concentration gradient; Y, inclined slit; O, camera lens; C, cylindrical lens; G, photographic plate; A, A', A'', hump of refractive index. (B) Method of filling colloidal solution and dialyzer.

The Tiselius electrophoresis cell has a rectangular cross section and is divided into three parts as shown in Fig. 7.2b. The boundary between the colloid and the dialyzer is made by shifting two parts (II, III) of the cell along each other. The lower part (II) contains the colloidal solution and the upper part (II) the dialyzer. The electrodes are usually of the reversible type (Ag/AgCl) and are placed in very large electrode vessels. To avoid the effect of convective flow, the measurements are usually carried out at low temperature (near 4°C), where aqueous solutions have their maximum density.

One of the essential characteristics of this method is the possibility of collecting electrophoretic data on individual components in a mixture. In such cases, the Schlieren picture contains two or more different maxima corresponding to the mobilities of different components.

7.1.3 Microelectrophoresis

Single-Cell Method

For colloidal particles 0.5 μm or larger, direct observation with a standard microscope is possible. Particles of 0.5-0.1 μm are observable with an ultramicroscope. The suspension is placed in a cell of rectangular or circular cross section. One of the standard types of rectangular cells is sketched in Fig. 7.3. The main part of the cell is composed of thin Pyrex glass that is 30-60 mm long, 5-20 mm wide, and 0.1-1.5 mm thick (inner dimensions). Each end of the cell is connected to a glass tube which houses an electrode which is used to apply the potential. The entire apparatus is supported by a metal or resin plate. Another pair of platinum electrodes is inserted just outside the rectangular cell to detect the potential difference between the cell ends. Reversible electrodes such as $Cu/CuSO_4$ or Ag/AgCl are frequently used for application of the potential to avoid the generation of gas bubbles. Also, Neihof [3] employed Pd-H electrodes to avoid gas generation and contamination by the reaction products. For microelectrophoresis, a field strength of 3-4 V/cm is normally applied to the solution.

When the colloid particles are large, but thin and platelike, some difficulties in the measurement of the electrophoretic mobilities are encountered. One problem is that the platelet particles frequently tumble and they disappear out of the view field of the microscope as they tumble. The period during which the particle is not visible is sometimes very long, with the result that the particle under observation becomes lost among other particles. In addition, if the cell is laid horizontally as usual, the rapid sedimentation of the particles makes observation to a definite level impossible. To avoid these difficulties, a very thin rectangular cell is used [4]. Problems arising from sedimentation are avoided by setting the cell vertically.

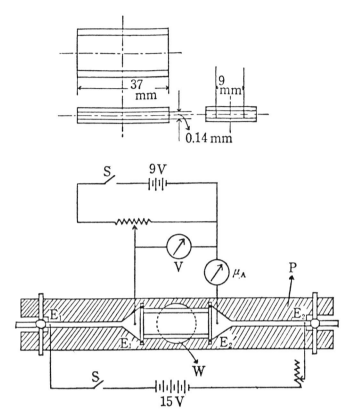

Figure 7.3 Diagram of a typical rectangular microelectrophoretic cell. E_1, E_2, E_1', E_2', platinum electrodes; S, switch; V, voltmeter; μ_A, microammeter; P, acrylic acid resin plate for supporting the cell; W, window for observing.

Stationary Level

When electrophoresis is carried out in a glass cell, there is always an effect due to electroosmosis. This is because the glass wall is negatively charged with respect to the aqueous solution. The electroosmotic flow is superimposed on the electrophoretic movement of the particles. For a cylindrical cell of inner radius r_w (glass capillaries of inner radii 0.1-2.0 mm are usually employed), the following relation holds between the apparent velocity observed microscopically, U_{app}, and the true electrophoretic mobility, U;

$$U_{app} = U + U_{osm}\left(\frac{2r^2}{r_w^2} - 1\right) \tag{7.1}$$

where r is the distance from the axis of the cylindrical cell and U_{osm} the electroosmotic velocity of liquid at the inner surface of the wall ($r = r_w$). It is clear from Eq. (7.1) that the velocity profile of the liquid is a parabolic function of the radius. Also, $U_{app} = U$ for $r = \pm r_w/\sqrt{2}$. This position is called the "stationary level." The actual electrophoretic velocity can be obtained only from observation at this level.

The stationary levels in a rectangular cell can be calculated in the same way. These two levels were given by Komagata [5] as

$$\frac{Y}{b} = \pm \sqrt{\frac{1}{3}\left[1 + \frac{384}{\pi^5 (a/b)}\right]} \qquad (7.2)$$

where a and b are the halves of the width and the height of the cell, respectively. Y is the distance of the stational level from the center of the cell. Y/b is, therefore, the relative depth. Table 7.2 shows some stationary levels calculated from Eq. (7.2) for selected values of the ratio a/b. It is known that as soon as the cell is not flat enough (a/b < 20), the stationary level deviates markedly from the value for infinite width (a/b = ∞). Figure 7.4 shows the mobility profile of the platelet particles of tungstic acid sol observed using vertical placement of the cell in Fig. 7.3. A parabolic curve is obtained as predicted from Eq. (2). On reversing the polarity of the electrodes, the direction of particle movement is reversed: rising particles in the first case begin to sink in the second. There is a small difference in the velocity observed between rising and sinking particles. This is naturally due to the sedimentation of the particles, which can be eliminated by averaging these two velocities. For example, by substituting $a = 4.5$ mm and $b = 0.07$ mm, obtained from Fig. 7.3, into Eq. (7.2), the stationary levels $Y/b = \pm 0.58$ are obtained. From Fig. 7.4, the electrophoretic velocity, the velocity at the stationary levels, is determined to be 5.67×10^{-4} cm^2/sec·V, by taking the average of 4.94×10^{-4} cm^2/sec·V, the rising velocity, and 6.40×10^{-4} cm^2/sec·V, the sinking velocity.

Two-Tube Method

According to Eq. (7.1), the velocity due to electroosmosis is a parabolic function of the radius. The velocity gradient of the liquid at the stationary level is generally large; thus the observed velocity of

Table 7.2 Stationary Level in a Rectangular Cell with Various a/b

a/b	10	15	20	40	100	∞
Y/b	0.612	0.602	0.596	0.586	0.580	0.578

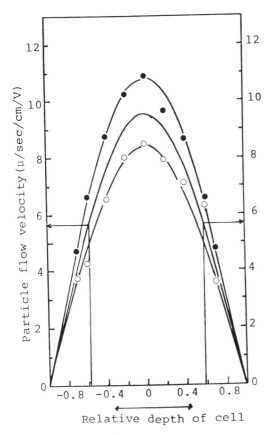

Figure 7.4 Electrophoretic flow curve of a standard tungstic acid sol. ○, Ascending particle flow velocity. ●, Descending particle flow velocity.

the particles changes rapidly with the radius, so that errors in electrophoretic velocities may be substantial.

Smith and Liss [6] have designed an improved form of a single closed cell to overcome the disadvantages of the single tube. It consists of two glass capillary tubes of different diameters parallel to each other; as shown in Fig. 7.5, the ends of the capillary tubes are attached to two larger glass tubes which contain the platinum electrodes. The dimensions of these two glass capillaries should have the relation

$$\frac{L_2}{L_1} = \frac{R_2}{R_1}\left[\left(\frac{R_2}{R_1}\right)^2 - 2\right] \qquad (7.3)$$

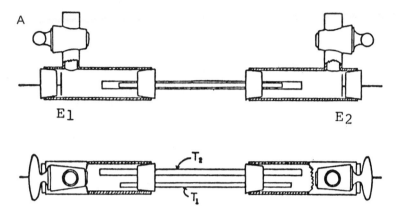

Figure 7.5 Diagram of a two-tube cell for determination of electrophoretic mobility. E_1, E_2, Platinum electrodes; T_1, T_2, glass capillaries with different radii (R_1, R_2) and lengths (L_1, L_2).

where L_1, L_2, R_1, and R_2 are the lengths and radii, respectively, of the narrower and wider capillaries. It is essential that capillaries 1 and 2 should be of the same material, since the development of the theory of liquid flow assumes that the same surface conditions exist in each tube. In this cell, return flow takes place only through the wider tube, T_2, and there is no movement of the liquid along the axis of the narrower tube. Therefore, the velocities observed at one-half depth in the narrower tube are the actual electrophoretic velocities. In this case, electrophoretic velocities are given at the maximum of the parabolic velocity as a function of depth. Hence the following advantages are obtained: (a) slight errors in focusing (i.e., error due to the depth of view field in a microscope) are less important than in the case of a single cell; and (b) the rotational effect of particles is minimized, since the velocity gradient near the level of observation is small.

The two-tube method has also been applied in the case of capillaries of rectangular cross section. According to Hamilton and Stevens [7], such a double-tube flat cell is more convenient in focusing at the stationary level.

7.1.4 Recent Developments in Electrophoresis

Various methods described in Sec. 7.1.3 have been used for the measurement of the electrophoretic mobilities of colloidal particles. In the past 10 years there have been several significant developments in measuring techniques for electrophoresis. Among them are the laser Doppler method [8-15], the rotating prism method [16], and the rotating grating method [17]. These techniques not only permit rapid measurement of the electrophoretic mobility but also permit the mobility distribution to be determined in a multicomponent system.

Laser Doppler Method

When a laser beam is scattered by colloidal particles moving in a definite direction, the frequency of the scattered light changes from that of incident light by the so-called Doppler effect. Since the colloidal particles suspended in a medium are undergoing Brownian movements, a Doppler shift will be also caused by these movements. Thus the frequency distribution of the scattered light will be made up from the superpositions of the Doppler shifts caused by the drift velocity and the random Brownian motion. The random Brownian movements widen the spectral distribution of the scattered light. According to Ware and Flygare [8], the power spectrum density function, $I(\omega)$, of the scattered light from the colloidal particles in a electric field is

$$I(\omega) \propto \frac{DK^2/\pi}{(\omega - \omega_0 \pm Kv)^2 + (DK^2)^2} \qquad (7.4)$$

where D is the translational diffusion coefficient of the particle; ω_0 and ω are the angular frequency of the incident and scattered light; v is the electrophoretic velocity; and K is the scattering vector with the magnitude equal to $(4n\pi/\lambda) \sin(\theta/2)$, in which n is the refractive index of the medium, λ is the wavelength of the incident light in vacuo, and θ is the scattering angle in the medium. In Fig. 7.6, the spectrum of the scattered light is shown schematically. The central frequency of the scattered light is shifted from that of incident light by Kv. Clearly, the measurement of this shift provides a direct measure of the electrophoretic mobility. The positive or negative character of the Doppler shift is dependent on the direction of the drift motion. The magnitude of the Doppler shift due to the Doppler electrophoresis is very small ($\simeq 100$ Hz) in comparison to the visible light ($\simeq 5 \times 10^{12}$ Hz). Such a small difference cannot be detected by using any of the

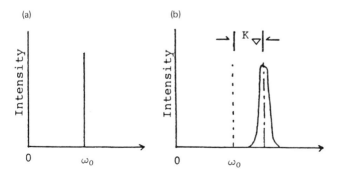

Figure 7.6 Power spectrum of incident light (a) and scattered light (b) from colloidal particles.

usual spectrometers of high resolution such as Fabry-Perot's interferometer, but it can be observed by a method of light beating or heterodyning. The heterodyning method has long been used in electromagnetic radio waves or microwaves. The possibility of using the heterodyning method for visible light was first discussed by Forrester [18] in 1947, and he achieved the detection of beat frequencies in 1955. The first use of the heterodyning method for the measurement of flowing fluid was by Yen and Cummins [19] in 1967. Ware and Flygare [9] applied the method to electrophoresis in 1971.

In the heterodyning method, the scattered light is mixed with a reference beam which is split from the incident beam before introducing the beam to the moving particles. Consequently, output electrical voltages, generated in the photodetector, correspond to the frequencies of the patterns of beating between the scattered light and the incident light. As shown in Fig. 7.6, the beat signals will have a spectral distribution which have a peak at the frequency Kv. By analyzing the frequency components of the beat signals, the electrophoretic mobility can be evaluated instantaneously. The analysis can be achieved by using an autocorrelator, a narrowband audio filter, or an FFT spectrum analyzer, which is based on fast Fourier transform (FFT) theory. The FFT analyzer may be the most useful tool for the analysis, especially for a multicomponent system. Figure 7.7 presents an arrangement of laser Doppler electrophoresis apparatus which was constructed by one of the authors [20]. A 5-mW He-Ne laser is used for the light source and a roof prism is used in conjunction with a focusing lens. The employment of the roof prism ensures the detection of the radiation up to large scattering angles. The scattered light is admitted to a photomultiplier through the pinholes (P_1 and P_2) together with the reference beam, which is reduced by a diffuser placed in the light path. Both beams are detected by the phototube, and beat signals are generated from the phototube. After amplification, the output is fed into the FFT analyzer, which transforms the beat signals and displays the frequency distribution of the signals. The signals may also be stored in the memory of the analyzer. By taking the average of the stored data, it is possible to obtain a smooth spectrum. A hard copy of the spectrum may be obtained with an X-Y recorder.

The sample is placed in a rectangular cell (thickness 3 mm and width 10 mm) having a cross-sectional area similar to the typical Tiselius cell, in which the effect of the electroosmosis is known to be negligible. An electrical potential is applied to the platinum electrodes by electronic switching circuits which send pulses of alternating polarity. This avoids any accumulation of the colloidal particles at the electrodes. The switching frequency is chosen so that the time constant of the pulse is longer than the sampling time of the analyzer. The switching circuits also send pulses to the analyzer for timing. The measurement can be carried out in several seconds to several minutes depending on the sample conditons. The measurement of the electrophoretic mobility

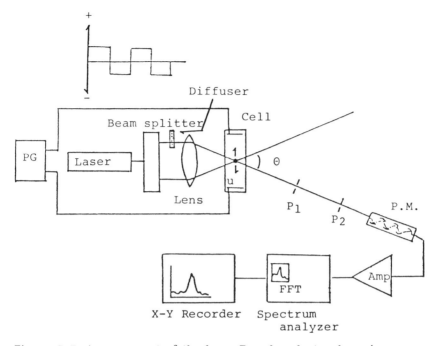

Figure 7.7 Arrangement of the laser Doppler electrophoresis apparatus. Laser, 5-mW He-Ne gas laser; lens, focal length = 30 cm; P_1, P_2, pinholes (200 μm and 100 μm radii); P.M., photomultiplier (Hamamatsu-TV, R-374); FFT, spectrum analyzer; PG, pulse generator.

should be done in a short time, since laser Doppler electrophoresis necessitates a higher electric field than in the usual microelectrophoresis. Figure 7.8 shows an example of the electrophoretic spectrum obtained using the apparatus in Fig. 7.7. The average velocity of the sample can be calculated from the peak frequency of the spectrum. The electrophoretic mobility or ζ potential can be read directly from the spectrum by changing the abscissa to these values. The advantages of the laser Doppler method are the rapid measurement of electrophoretic mobility and the simultaneous determination of the distribution curve of the mobility.

Rotating Prism Method [16]

This method utilizes a rotating cubic prism placed inside a microscope which is used for conventional microelectrophoresis. The prims is rotated electrically. The rotation of the prism causes translational movements of the images of the colloidal particles under observation. The schematic diagram of the rotating prism method is shown in Fig.

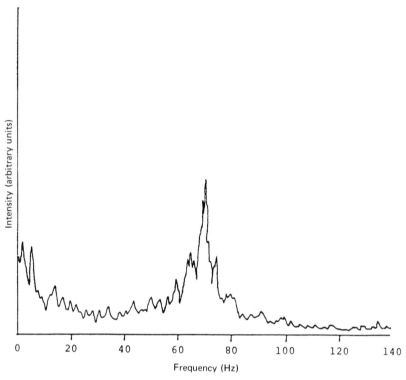

Figure 7.8 Electrophoretic light-scattering spectrum for a latex.

7.9. The rotation speed is controlled by the operator to keep the image of the particle stationary. By adjusting the rotation speed of the prism to match that of an individual particle, the mobility of the individual particle is obtained. The distribution of the electrophoretic mobility can be obtained by measuring the mobilities of many particles in a short time. Each measurement takes only several seconds. The average mobility of the colloid is determined by adjusting the rotating speed so that the images of many particles (i.e., the cloud of particles) appear stationary. Since the operator measures many particles at the same time, the precision is said to be the same as if 10-30 colloidal particles were measured.

Rotating Grating Method [17]

In this technique, a rotating grating is set in the light path of a conventional microelectrophoresis apparatus with a laser being used as the light source. The image of the particle is projected onto a glass disk on which a grating is engraved, and the disk is rotated slowly

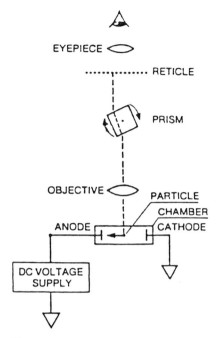

Figure 7.9 Schematic diagram of the rotating prism method. (From Ref. 16.)

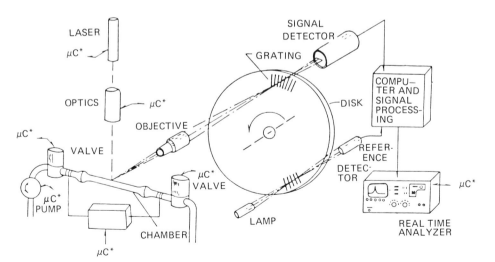

Figure 7.10 Block diagram of the rotating grating method. μC^*, microcomputer input/output. (From Ref. 17.)

at a constant speed. The schematic block diagram of the system is illustrated in Fig. 7.10. The image of the particle alternatively meets with the lines of the grating and the intensity of the transmitted light varies. The width of the engraved lines and the spacing are chosen so that they are both larger than the images of the colloidal particles. Since the photodetector corresponds to the intensity of the transmitted light, the output voltage of the detector will contain signal components which change, depending on the velocity of the particle and the speed of the grating. When the colloidal particles move in the direction of the rotation of the disk, the output signal will have a lower-frequency component than for the case in which the image of the particles is at rest. In the converse case, namely when the images move against the direction of the rotation, it will have higher frequency. Therefore, the electrophoretic velocity of the particle can be evaluated by analyzing the frequency components of the output signals from the detector. The velocity of the particle can be converted to the electrophoretic mobility by taking into account the field strength in the cell chamber for the electrophoresis. The distribution of the mobilities of the particles is represented by the distribution of the frequency component.

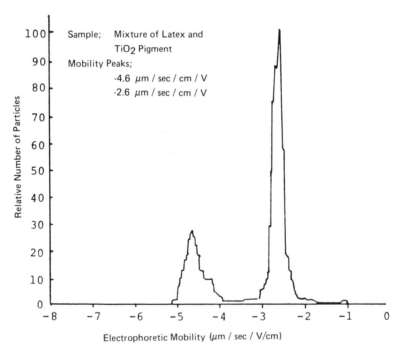

Figure 7.11 Mobility histogram of a mixture of latex and TiO_2 pigment. (From Ref. 17.)

In this method the positive or negative character of the particle motion is obtainable directly from the frequency shift, in comparison with the frequency component when the image is at rest. An example of the mobility distribution for a mixture of latex and TiO_2 dispersions is shown in Fig. 7.11. Two peaks are observed in the distribution due to the two components. Since the magnitude of ζ potentials have a great effect on the stability, detailed and accurate discussion will be possible only by using a histogram of the mobilities.

7.2 ELECTROOSMOSIS

7.2.1 Introduction

Electroosmosis is the flow of liquid along a solid surface resulting from the parallel application of an external electric field. The apparatus for the measurement of this flow is composed of the following parts: a sample cell with attached electrodes which serve to measure the electric conductivity of the liquid, a pair of electrodes for applying the electric field to the sample, a flowmeter or manometer for detecting osmotic pressure, and a DC power supply. By measuring the electroosmotic flow rate V, the specific conductivity of the liquid λ, and the electric current I, the ζ potential can be calculated using the Helmholtz-Smoluchowski equation:

$$\zeta = \frac{4\pi\eta\lambda V}{\varepsilon I} \tag{7.5}$$

where η and ε are the viscosity coefficient and the dielectric constant of the liquid, respectively.

7.2.2 Measurement

Setting of Samples

Figure 7.12 illustrates a typical apparatus for the measurement of electroosmosis. The ideal shape of a sample for the strict treatment of the experimental result is a capillary, which can be made of glass, synthetic polymers, or fibers. In order to apply the Helmholtz-Smoluchowski equation to the experimental data to calculate the ζ potential, the relationship $\kappa a > 1$ must be satisfied between the capillary radius (a) and the thickness of the diffuse double layer, $1/\kappa$. Actual samples for ζ-potential measurements are solids in the form of powders. They are used in the form of a diaphragm composed of large numbers of irregularly shaped capillaries.

DC Power Supply and Electrodes

The DC power supply for the electroosmosis system must supply a potential difference of up to 500 V. Two pairs of electrodes are in-

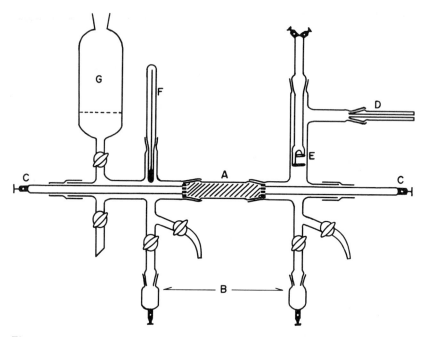

Figure 7.12 Electroosmosis apparatus. A, Diaphragm; B, Ag · Ag_2SO_4/SO_4^{2-} electrodes; C, displaceable electrodes; D, capillary for measuring electroosmotic flow rate; E, electrodes for measuring the conductivity of liquid; F, thermometer; G, reservoir of liquid. (From Ref. 21.)

corporated into the apparatus. One pair is for the application of the external electric field to the electroosmosis system. Because an electric current passes through the system during the course of the measurement, electrolytic reaction occurs on the surface of electrodes. Therefore, reversible electrodes such as Ag·AgCl/Cl⁻ [22], Ag·Ag₂SO₄/ SO₄²⁻ [21], or Hg·Hg₂Cl₂/Cl⁻ [23] should be employed to avoid the evolution of gaseous products. The second pair of electrodes, usually Pt electrodes, are for the measurement of electrical conductivity of the liquid.

Measurement of Electroosmotic Flow

Two methods have been used for the measurement of electroosmotic flow. One method is direct determination of the electroosmotic flow rate by use of the apparatus illustrated in Fig. 7.12. Before applying the potential, the level of the liquid in reservoir G is adjusted to the same height as that of the horizontally placed capillary tube, D. The rate of meniscus movement in the capillary tube is measured after the potential is applied. In this tupe of apparatus, the radius of the capillary D should not be too small to affect the flow rate due to pressure ascribed to the surface tension of the liquid [24]. The pressure is constant when the contact angle of the liquid with the capillary wall is zero. However, the pressure depends on the curvature of the meniscus, and the latter is determined, when the capillary radius is small, by the contact angle of the liquid with the wall, which varies according to whether the liquid advances or recedes. To minimize this effect it is better to choose a capillary with a somewhat larger radius (e.g., 0.25 mm) [21].

The other method of measuring the osmotic flow rate is to follow the velocity of a bubble, introduced into the capillary (Fig. 7.13). Here it should be understood that the bubble moves in the film formed on the glass wall by the liquid. Derjaguin [25] expressed the thickness t of this film by the equation

$$t = 1.32R \left(\frac{U\eta}{\gamma}\right)^{2/3} \tag{7.6}$$

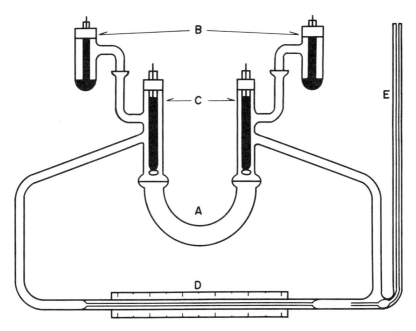

Figure 7.13 Electroosmosis apparatus. A, U tube containing diaphragm; B, calomel electrodes; C, platinum electrodes; D, capillary tube for measuring the velocity of air bubble; E, capillary tube for introducing air bubble. (From Ref. 23.)

where R is the capillary radius, U the tangential velocity of the bubble, and γ the surface tension of the liquid. When the ζ potential of the sample is close to zero and that of the capillary wall is relatively large, the electroosmotic flow resulting from the latter cannot be neglected. This effect may be eliminated by keeping the capillary at a distance from the electrodes thus working.

Instead of direct measurement of the flow velocity, one can observe the pressure required to stop the electroosmotic flow. This pressure is read as the difference in heights of menisci of the liquid in glass tubes which are set vertically on each side of the diaphragm. This method, however, requires knowledge of the structure of the sample plug for the correct calculation of the ζ potential. It thus gives only a relative value for the ζ potential of powder particles.

Some consideration must be given to correction of the Helmholtz-Smoluchowski equation for the effects of surface conductance and double-layer thickness. Since these problems also occur in the measurement of the ζ potential by the streaming potential method, they will be treated in the following section.

7.3 STREAMING POTENTIAL AND STREAMING CURRENT

7.3.1 Streaming Potential

The potential difference developed by liquid flow along a capillary wall is called the streaming potential and can be measured with the electrodes placed at both ends of the capillary. This phenomenon is the reverse of electroosmosis, so the construction of the two measuring systems is similar. Determination of the ζ potential by means of the streaming potential method necessitates measurement of (a) the pressure to force the liquid flow through a capillary or a diaphragm of powders, (b) the streaming potential set up between the electrodes placed at both ends of the sample, and (c) the specific conductivity of the liquid streaming through the capillary system. For an ideal system, in which the thickness of the electric double layer is small and the sample has a simple geometry such as in a capillary or flat-plate slot, the Helmholtz-Smoluchowski equation is valid in calculating the ζ potential:

$$\zeta = \frac{4\pi\eta\lambda E}{\varepsilon P} \qquad (7.7)$$

where λ is the specific conductivity of the liquid, E the streaming potential, P the pressure applied for streaming the liquid. η the viscosity of the liquid, and ε the dielectric constant of the liquid. This phenomenon, however, includes various complex factors which prevent accurate determination of the ζ potential. The main factors are the presence of surface conductance and a relatively large thickness of the electric double layer compared with the capillary radius which appears in a dilute electrolyte solution. These effects are discussed in this section.

Measurement of Streaming Potential

Figure 7.14 illustrates the apparatus for the measurement of the streaming potential. The flow of the liquid is caused by the pressure of N_2 gas. The pressure is measured by either a Hg or a H_2O manometer (A), depending on the pressure range needed. The use of large liquid reservoirs (B) avoids large changes in the liquid level during the measurement. The soda lime trap (C) is used to prevent the contamination by CO_2 from the air, which has a significant influence on the electrification of the surface. The diaphragm of the powder sample (D) must be as dense as possible to keep it firm during the liquid flow. A glass capillary or a polymer tube may also be used as a sample instead of the diaphragm. Electrodes of perforated disks

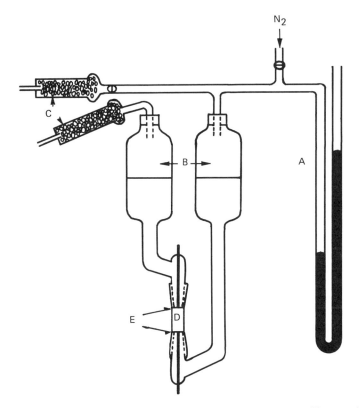

Figure 7.14 Streaming potential apparatus. A, Manometer; B, reservoir of liquid; C, soda lime; D, diaphragm of powder samples; E, electrode.

(E) made of Pt, Au [26,27], and Ag·AgCl-plated Pt [28] are used for determining the streaming potential. Unplated metal electrodes are likely to be markedly polarized to give unreliable results which differ often from those by Ag·AgCl electrodes.

A pH meter is commonly used as an electrometer for measuring the streaming potential. When the resistivity of the measuring system is high (e.g., higher than a few megohms), an electrometer with high input impedance, such as a vibrating reed or chopper type electrometer with impedance higher than 10^{13} Ω, can produce reliable measurements. They have sufficient sensitivity when the full scale ranges from 1 mV through 10 V. The specific conductivity of the liquid of the diaphragm should be measured by use of the usual AC bridge method. Another method [29] that may be employed to obtain the resistivity of the liquid in the sample plug is to use the following relation, together with the streaming potential data:

$$R_1 = R_2 \frac{E_1 - E_2}{E_2} \tag{7.8}$$

where R_1 is the resistivity of the diaphragm, E_1 the streaming potential at a pressure P, and E_2 the streaming potential at the same pressure but with a standard resistor (R_2) connected in parallel with the circuit of the diaphragm. Finally, by introducing a 0.1 M KCl solution into the system and by determining the cell constant, the specific conductivity of the liquid can be calculated.

Comments on Measurements

Figure 7.15 illustrates the relationship between the streaming potential E and pressure P in the following systems: Pyrex glass capillary—I (length 33.3 cm and radius 0.057 cm), capillary—II (length 24.5 cm and radius 0.104 cm), Teflon pieces, and polystyrene beads in water. In every system the E versus P plot is linear at lower pressures and become nonlinear at higher pressures. This suggests that laminar flow disappears at the higher flow rate. The streaming potential must therefore be measured at the lower pressure where the E versus P relation is linear.

Even when a good linear relationship holds between E and P, the plot does not always intersect the origin of the coordinate system. The reason for this is understood to be due to a difference in the polarization of the two electrodes. In such cases, the slope ($\Delta E/\Delta P$) is substituted in Eq. (7.7) instead of the ratio E/P. Some attempts have been made to cancel this polarization. For example, one technique is to apply pressure all at once to force the liquid to stream through the capillary system. The simultaneous response, which is ascribed to the streaming potential, is recorded. When streaming potential data obtained using Pt electrodes were corrected by this method, the re-

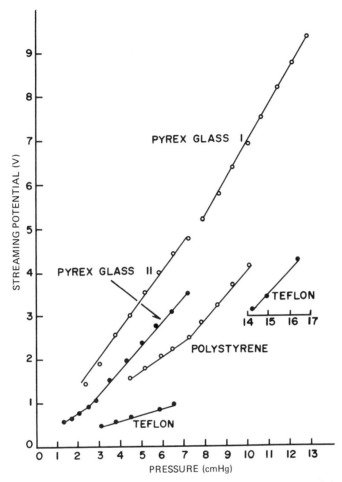

Figure 7.15 Relationship between streaming potential and pressure. (From Ref. 30.)

sults were found to be identical with data obtained by use of reversible Ag·AgCl electrodes [31]. The second method [32] is to remove the polarization potential difference electrically by means of an RC circuit as shown in Fig. 7.16. The closing of switch 1 leads to the charging of the condenser owing to the different polarization potentials of the electrodes, and an electric current is observed in the circuit. After the condenser is charged, switch 1 is opened and switch 2 is closed. Immediately, the liquid is forced to stream in the capillary, which causes an electric current to flow through the circuit. This additional amount of electricity, which corresponds to the stream-

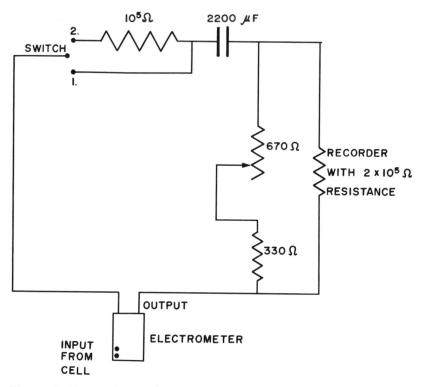

Figure 7.16 RC circuit for canceling the polarization potential of electrodes. (From Ref. 32.)

ing potential, is stored in the condenser. If this circuit is measured within 10 sec, one can determine the streaming potential with an accuracy higher than 99% of the correct value.

Correction for the Helmholtz-Smoluchowski Equation

   A. *Surface Conductance.* In the derivation of Eq. (7.7) the specific conductivity used was that of the bulk liquid. However, when the electrolyte concentration is low or an organic solvent is used, the actual conductivity of the liquid in the diaphragm is higher than that of the bulk liquid due to the contribution of surface conductance. When the sample is a cylindrical capillary with a uniform radius r, this contribution is expressed by $\lambda = \lambda_B + 2\lambda_s/r$, where λ_B and λ_s are specific conductivities of the bulk liquid and the solid-liquid interface, respectively. Thus the ζ potential can be calculated by the following corrected equation:

$$\zeta = \frac{4\pi\eta(\lambda_B + 2\lambda_s/r)E}{\varepsilon P} \tag{7.9}$$

For a diaphragm, however, it is impossible to formulate an analytical equation for the calculation of the ζ potential by taking account of surface conductance, because of the complexity of the geometry of the system. Zhukov and Fridricksberg [33] gave experimental evidence to support the claim that the ζ-potential value is correct when it is calculated by adopting the conductivity in the diaphragm. However, some reports which oppose this result have also been published [27,34,35]. According to the calculation by Overbeek and Wijga [35] in which a simple model system is employed, the value of the ζ potential calculated by this method is always smaller than the correct one if surface conductance is involved in the phenomenon. However, in the system of the powdered glass with the aqueous solution of KNO_3 (2×10^{-4} M), calculation of the ζ potential gives a value approximately 90% that of the ζ potential obtained with the same glass capillary.

B. *Increase in Viscosity at Interface.* In Eq. (7.7) it was assumed that the viscosity is constant throughout the bulk of the liquid and the interior of the electric double layer. The viscosity coefficient for polar molecules, however, increases in an electric field. It is not uncommon for the electric field at the solid-liquid interface to exceed 10^5 V·cm^{-1}. Thus it is not safe to assume that the viscosity coefficient in the double layer is the same as that in the bulk. Theoretical analysis of this problem by Lyklema and Overbeek [36] gave the following relationship between the ζ potential calculated through Eq. (7.7) (ζ_{obs}) and the correct value (ζ) for the system including a Z-Z type electrolyte in solution:

$$\zeta_{obs} = \int_0^\zeta \frac{d\psi}{1 + 32\pi RTcf/\varepsilon \sinh^2 ZF\psi/2RT} \tag{7.10}$$

where ψ is the potential in the electric double layer, F the Faraday constant, c the concentration of electrolyte, and f a constant characteristic of the liquid molecule which is a function of the dipole moment and polarizability of the molecule. The relation between ζ and ζ_{obs} for the 1-1 type of electrolyte at 25.0°C is shown in Fig. 7.17. It is assumed that $f = 10.2 \times 10^{-12}$ V^{-2} cm^2. As can be seen in this figure, ζ_{obs} deviates to values lower than the correct value ζ with an increase in the electrolyte concentration and in the ζ potential itself.

C. *Electroviscous Effect.* It should be noted that the development of a streaming potential induces electroosmosis, which retards the streaming of the liquid. This is called the electroviscous effect, which becomes significant in the case of large values for the ζ poten-

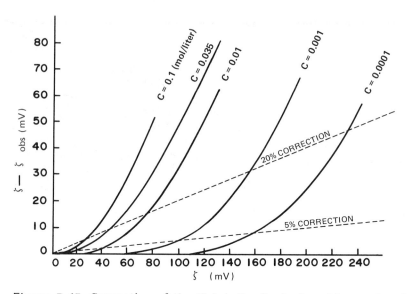

Figure 7.17 Correction of the Helmholtz-Smoluchowski equation for the increase of viscosity in the electric double layer at various concentrations of 1-1 electrolytes. (From Ref. 36.)

tial, large concentrations of counter ions in the diffuse double layer, or greater thickness of the double layer. According to the theoretical calculation on the parallel-plate slot [37], the condition of $\zeta = \zeta_d = 118$ mV and $\kappa h = 2$ increases the viscosity by about 12%, where h is half of the width of the slit. When $\kappa h > 10$, the increase in viscosity is less than 2%. Furthermore, when $\zeta = 30$ mV, the maximum increase amounts to only 1.5%. For measurement using a capillary, this effect will become marked. However, when the overlapping of the facing double layers is small, the effect is minimal and may be disregarded.

D. *Curvature and Overlapping of Electric Double Layers*. When the double layer thickness $1/\kappa$ is not much smaller than the capillary radius, the Helmholtz-Smoluchowski equation cannot be used, because the curvature of the double layer and the extent of overlapping must be taken into account for calculating the correct ζ potential. Oldham, Young, and Osterle [38] investigated these effects theoretically and the result obtained is shown in Fig. 7.18, where the reduced value of the ζ potential, $Z = F\zeta/RT$, is employed. On the basis of this curve and the ζ potential ($Z_w = F\zeta_w/RT$) calculated through Eq. (7.9), one can obtain the correct ζ potential ($Z = F\zeta/RT$). The correction made on the calculated ζ potential becomes marked as t (κr; r is the radius of the capillary) or the ζ potential decreases.

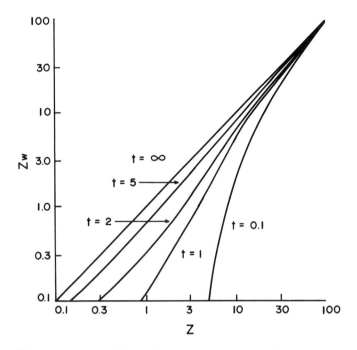

Figure 7.18 Relationship between $Z_w(F\zeta_w/RT)$ and $Z(F\zeta/RT)$ for various $t(\kappa r)$ values. (From Ref. 38.)

In conclusion, it is noted that the most serious factor which obstructs the correct calculation of ζ potential is the thickness of the electric double layer. The original reports should be referred to for details of the corrections made.

Experimental Results Obtained by the Streaming Potential Method

Figure 7.19 displays the ζ potentials of glass capillaries of various radii filled with water, where the effects of corrections A and D on the ζ potential were taken into account [38]. The contribution of correction B is disregarded, because of the low ionic strength of the system studied. ζ_c is the value calculated by using Eq. (7.9) and the bulk conductivity of the liquid, ζ_w is the value corrected for the effect of surface conductance through Eq. (7.9), and ζ is the correct value obtained by using methods A and D. As can be seen in Fig. 7.19, a constant value will be reached after corrections are made.

With exercising the effect of surface conductance, Rutgers and de Smet [39] measured the ζ potential of Jena 16 III glass capillaries filled with electrolyte solutions by means of streaming potential and electroosmosis techniques. They found no difference between the

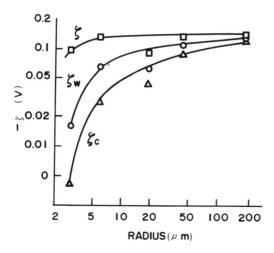

Figure 7.19 Corrections of the ζ potential of small glass capillaries contacting with water. (From Ref. 38.)

values obtained with the two different methods. The ζ potential decreased at concentrations higher than 10^{-3} M of 1-1 electrolytes and decreased more rapidly in the presence of 2-1 electrolytes, as shown in Fig. 7.20. The variation of ζ-potential values is understood to be due to the decrease in the thickness of the diffuse double layer, which is caused by the increase in ionic strength. In the presence of 3-1 electrolytes, the ζ potential decreased rapidly even at low concentrations, changed sign from negative to positive, increased to a maximum value, and then decreased slowly. These complicated phenomena can be explained in terms of the specific adsorption of positive ions, that is, the neutralization of surface charge, reversal of the charge, saturated adsorption of the ions, and depression of the double layer.

The streaming potential method is useful to follow the change of ζ potential with time. Figure 7.21 represents the change in the ζ potential of 1200°C-treated Fe_2O_3 in aqueous NaCl solution with time [40]. The result can be explained as the hydration of the solid surface. This indicates that the electrification of the metal oxide surface is a function of the extent of hydration of the surface.

7.3.2 Streaming Current

The streaming current method can be conveniently used when a sample material has a high electric conductivity. Neither the electroosmosis method nor the streaming potential method are suitable for such material because the back current in the streaming potential measurement

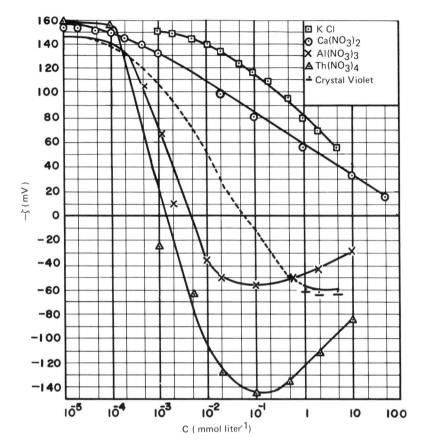

Figure 7.20 Zeta potential of glass capillary contacting with electrolyte solutions. (From Ref. 39.)

flows not only through the liquid phase but also through the solid one, and an electric field cannot effectively be applied to the solid-liquid interface in the measurement of electroosmosis.

Measurement of the Streaming Current

Figure 7.22 shows an electric circuit for measuring the streaming current of a capillary system. The streaming current flows through the outer circuit, whose resistivity R is much less than that of the interior of the sample cell. The streaming current I is determined by measuring the potential difference E between both ends of the cell by use of a sensitive electrometer; $I = E/R$. The ζ potential of the system is calculated from Eq. (7.11):

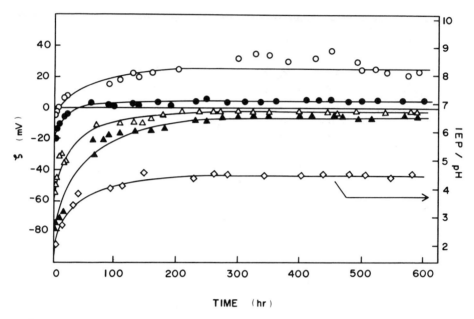

Figure 7.21 Zeta potential and IEP of 1200°C-treated Fe_2O_3 as a function of hydration time. ○, pH 3; ●, pH 4; △, pH 5; ▲, pH 6; ◇, IEP. (From Ref. 40.)

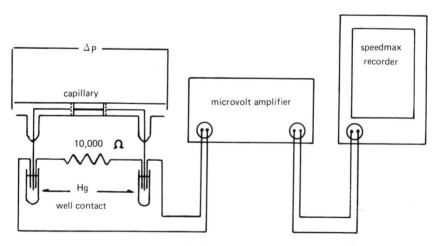

Figure 7.22 Electronic circuit for streaming current measurement. (From Ref. 41.)

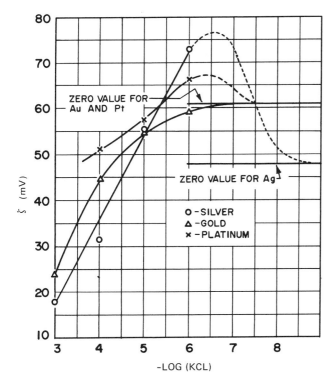

Figure 7.23 Zeta potential of metal capillaries contacting with dilute KCl solution and pure water (zero value). (From Ref. 42.)

$$\zeta = \frac{4\pi\eta CI}{\varepsilon P} \tag{7.11}$$

where η is the viscosity of the liquid, C the cell constant of the system, and ε the dielectric constant of the liquid. This method is useful for cylindrical metal samples, but is of no use for metal powders, since the cell constant can neither be calculated nor measured. Figure 7.23 shows the ζ potential of the system of Ag, Au, or Pt cylinder-KCl solution determined by means of the streaming current method [42]. This technique is, of course, applied to semiconductor or insulator materials. Buchanan and Heyman [43] measured the ζ potential of $BaSO_4$ by both streaming potential and streaming current methods, and obtained very similar results.

7.4 SURFACE CONDUCTANCE

7.4.1 Introduction

Surface conductance at the solid-liquid interface plays an important role in electrokinetic phenomena and in the conductance of suspensions. Bikerman [44-46], Rutgers [47,48], Street [50-52], and Dukhin and Derjaguin [53] have examined this subject theoretically and experimentally. In this section some experimental problems of surface conductance at the solid-liquid interface will be described.

7.4.2 Measurement of Surface Conductivity

Just as in the cases of streaming potential measurements or electroosmosis measurements, solid samples having a simple geometry, such as a capillary, a rod, or a flat-plate slot, give reliable results for the surface conductivity through the use of simple equations. In practice, the surface conductance causes significant problems in the electrokinetic investigation of systems containing powder samples, and it is difficult to obtain reliable data except for the special systems described below.

Surface Conductivity of Capillary or Flat-Plate Systems

A capillary or flat-plate slot has the best geometry for rigorous investigation of surface conductance of glass or synthetic polymer samples. To obtain reliable data, a capillary with a small radius or a parallel-plate slot with narrow spacing is required. A capillary radius of less than 10^{-2} cm is usually used. The equation for calculating the surface conductivity λ_s in the capillary system is as follows:

$$\lambda_s = \frac{(\lambda_c - \lambda_B)r}{2} \qquad (7.12)$$

where λ_c is the specific conductivity of liquid in the capillary with radius r and λ_B is the bulk conductivity of liquid. The equation for the parallel-plate system is

$$\lambda_s = \frac{(\lambda_c - \lambda_B)h}{2} \qquad (7.13)$$

where h is the spacing between two flat plates. The whole construction of the measuring cell may be the same as that for the measurement of electroosmosis or streaming potential. The electrodes, however, are made of platinized platinum like those employed in the conductivity measurement of ordinary solutions. The conductivity is measured by a DC method because of the low conductivity of the system.

Surface Conductivity of a Powder Sample System

In this type of system it is impossible to obtain the correct surface conductivity except when a dilute suspension of spheres, spheroids, or rods is used, because the electric field distribution and the polarization on the particle surface are too complicated to be estimated for other geometries.

Usually, the surface conductivity of powders is measured when the sample is in the form of a suspension or a diaphragm. The basic equation for the calculation of surface conductivity is expressed as follows [51]:

$$\lambda_s = \frac{\phi}{S}(f\lambda_I - \lambda_B) \quad (7.14)$$

where ϕ is the volume fraction of liquid, S the surface area per unit volume of suspension, λ_I the specific conductivity of the suspension, and f the formation factor, which is a function of particle shape and density. The formation factor (f) has been obtained theoretically [53]:

$$f = 1 + \frac{k(1-\phi)}{2} \quad (7.15)$$

where k is a constant determined by the shape of particles. For spherical particles, Eq. (7.15) is reduced to the form

$$f = \frac{3-\phi}{2\phi} \quad (7.16)$$

For particles other than spheres, k may be obtained from the original reference [53].

The f value can be determined experimentally [54] by applying Eq. (7.14) to the data from the specific conductivity of the suspension. When the concentration of an electrolyte is large, the effect of surface conductance can be neglected, and $f\lambda_I - \lambda_B = 0$, that is, $f = \lambda_B/\lambda_I$.

For a concentrated suspension of spherical particles, the equation derived by Fricke and Curtis [49] can be used:

$$\lambda_s = \frac{(\lambda_I/\lambda_B - 1) + (\rho/2)(\lambda_I/\lambda_B + 2)}{\rho(\lambda_I/\lambda_B + 2) - (\lambda_I/\lambda_B - 1)} \lambda_B a \quad (7.17)$$

where a is the radius of a particle and $\rho = 1 - \phi$.

7.4.3 Mechanism of Surface Conductance

Bikerman [44], using the assumption that the ions distributed in the diffuse double layer act as carriers, derived an equation for the surface conductivity of the system including a 1-1 electrolyte:

$$\lambda_s = \sqrt{\frac{RTc}{2\pi F^2}} \left\{ \left[\exp\left(\frac{-F\zeta}{2RT}\right) - 1\right] \left(\lambda_+^\circ + \frac{\varepsilon RT}{2\pi\eta}\right) + \left[\exp\left(\frac{F\zeta}{2RT}\right) - 1\right] \left(\lambda_-^\circ + \frac{\varepsilon RT}{2\pi\eta}\right) \right\} \quad (7.18)$$

where c is the electrolyte concentration, F the Faraday constant, and $\lambda°$ the equivalent conductivity of the ions. If this type of surface conductance plays a major role, there must be a negative surface conductivity when $\lambda_+^° \neq \lambda_-^°$ and the ζ potential is low (e.g. $\lambda_+^° > \lambda_-^°$, $\zeta \geq 0$, and vice versa). Such results, however, have never been reported. The observed values of surface conductivity are positive and much larger than the calculated ones. Accordingly, it must be assumed that ions other than those in the diffuse double layer are involved. One possibility is that there is a contribution due to ions adsorbed in the Stern layer [55,56]. Another possibility [57,58] is that in interfaces such as glass-water, the solid surface swells, and the component ions of the solid and/or the ions adsorbed in the swollen layer participate in the electric conduction. These possibilities should be considered in the explanation of experimental results. At present, the concentration and the mobility of these ions have not been established.

7.4.4 Experimental Results

Table 7.3 [59-61] gives surface conductivity values for glass-KCl solution interfaces determined by the use of capillaries, flat-plate slots, and powder samples. The observed values are larger than the values calculated ($1.4 \times 10^{-9} \ \Omega^{-1}$) with Eq. (7.18) by assuming that $\zeta = 120$ mV (5×10^{-4} M KCl [57]).

Table 7.3 Surface Conductance of Pyrex Glass in the Solution of KCl

Form of Pyrex glass	Concentration of KCl (mol/liter)	λ_s at 25.0°C ($10^{-9} \ \Omega^{-1}$)	References
Flat plates	10^{-3}	100	59,60
	10^{-2}	608	
	10^{-1}	2640	
Spheres	5×10^{-4}	4.30	55
	10^{-3}	7.39	
	10^{-2}	37.2	
Capillary	10^{-5}	1.5	61
	10^{-4}	2.93	
	10^{-3}	13.6	
	10^{-2}	16.3	

Table 7.4 Variation of λ_S with the Volume Fraction of Particles in the Solution of Barbiturate-Acetate Adjusted to pH 7 and Ionic Strength I = 0.01

$\lambda_B \times 10^7$ (Ω^{-1} cm^{-1})	$\lambda_I \times 10^7$ (Ω^{-1} cm^{-1})	$\rho = 1 - \phi$	$\lambda_S \times 10^{7a}$ (Ω^{-1}) 1	2
10,440	10,180	0.0109	1.30	−3.45
10,440	10,060	0.0258	1.21	0.41
10,440	9,847	0.0477	1.38	1.31
10,440	9,599	0.0688	1.33	1.31

[a]Column 1 determined through the experimental formation factor; column 2 determined through the theoretical formation factor shown as Eq. (7.16).
Source: Ref. 54.

James and Carter [54] calculated the surface conductivity of model systems of spherical polystyrene latices as a function of the volume fraction of the particles (Table 7.4) and obtained a constant value of surface conductivity by using the experimentally determined formation factor.

Kittaka and Morimoto [62] determined the activation energy for surface conductance of a silica-electrolyte solution interface in order to obtain information on the ionic mobility (Fig. 7.24). For the system, incorporating a 1-1 or 2-1 electrolyte, the activation energy for surface conductance is less than that of the bulk, whereas in other systems it exceeds that of the bulk. The latter result suggests a large contribution due to strongly adsorbed ions in the Stern layer.

7.5 NONAQUEOUS SYSTEMS

Electrokinetic measurements in nonaqueous systems are not, in principle, different from those in aqueous systems. Electroosmosis and streaming potential methods are used in nonaqueous systems as well as the electrophoresis method. Because columns packed with particles are used as cells in the former two measurements, we meet the problem of the overlapping of double layers in nonpolar media as described in Secs. 5.2.1 and 5.3. Thus the results of the considerable difference between the values of the ζ potential obtained from electrophoresis and

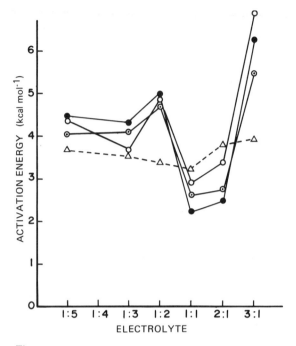

Figure 7.24 Activation energy of surface conductance of three kinds of SiO_2 samples with surface area 20.2 m^2/g (○), 12.2 m^2/g (●), and 3.78 m^2/g (⊙) in the electrolyte solutions: 1-5 $Na_5P_3O_{10}$; 1-3 $Na_3C_6H_5O_7$; 1-2 Na_2SO_4; 1-1 NaCl; 2-1 $MgCl_2$; 3-1 $AlCl_3$. △, activation energy of bulk conductivity of solution. (From Ref. 62.)

the streaming potential or electroosmosis as discussed in Sec. 5.3 are shown there. A reliable electrophoresis method for nonaqueous systems will be described.

Microelectrophoresis, the moving boundary method, and the electrodeposition method are used to measure electrophoretic velocity in nonaqueous dispersions. Several comments will be made about obtaining reliable data in nonaqueous systems.

7.5.1 Microelectrophoresis

A high strength of electric field is necessary to obtain measurable electrophoretic velocity, because the dielectric constant is lower in nonpolar solvents. A strength of several tens of volts per centimeter is usually applied. Hence a short electrode distance of several centimeters is suitable in a cell. In addition an electric source of constant voltage capable of several hundred volts is necessary. But electric power is not necessary because of low conductivity (lower than 1 mA).

To prevent current leakage along the wall of a cell due to the higher resistivity of the media, quartz or polycarbonate resin should be used as the cell material. However, the latter cannot be used for aromatic solvents. Platinum wire or perforated plate can be utilized as electrodes, because polarization is negligible due to low conductivity. Field strengths higher than 200 V/cm should not be used, because regular movement of particles would be prevented by the dielectric polarization and uneven potential distribution between electrodes which may occur [63].

The condition that $\kappa a < 0.1$ holds very often in nonaqueous dispersion, because κ is very low, as discussed in Sec. 5.2.1. Hence the following Hückel equation can usually be applied to calculate the ζ potential in nonaqueous spherical particle systems:

$$\zeta = \frac{6\pi\eta U}{\varepsilon} \tag{7.19}$$

where U is the electrophoretic mobility. The coefficient 6 of the equation must be substituted by 4 or 8 in the case of a cylindrical particle depending on its orientation (4 if parallel to the field and 8 if perpendicular to the field).

Retardation and relaxation effects must be corrected to obtain a precise ζ-potential value from electrophoretic mobility if $0.1 < \kappa a < 100$. Wiesema, Leob, and Overbeek [64] have proposed a method which takes both effects into consideration.

Commercial apparatus for microelectrophoresis is available for measurement in nonaqueous dispersions as well as in aqueous ones. For example, model 3000 manufactured by the Pen Kem Co. in the United States is very suitable for nonaqueous systems, and model 500 is used with a substituted high-field electric source.

7.5.2 Electrodeposition Method

An electrodeposition method which measures the deposition current was developed as a modified electrophoresis by Kondo and Yamada [65], although a deposition method measuring the weight of the depositing substance had been published previously [66]. The former measures the time dependence of the discharging current I carried by moving particles under an applied field. The apparatus is shown schematically in Fig. 7.25. Electrophoretic velocity was experimentally proportional to the applied field strength up to 40 V/cm for a nonaqueous carbon black dispersion (a developing liquid in electrophotoreproduction). The relation between I and time t was theoretically formulated as follows:

$$I = I_1 + I_0 \exp\left(\frac{-t}{T_r}\right) \tag{7.20}$$

220 / Kittaka et al.

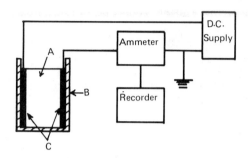

Figure 7.25 Schematic diagram of deposition current measurement arrangement. A, Sample of dispersion; B, polyacrylate vessel; C, copper electrodes (area, 20 cm^2; distance, 1 cm). (From Ref. 65.)

where I_1 and I_0 are leakage current and initial current, respectively, and T_r is the mean time for particles to traverse between both electrodes.

Electrophoretic mobility U is related to T_r by the equation

$$U = \frac{L}{T_r} / \frac{V}{L} = \frac{L^2}{VT_r} \tag{7.21}$$

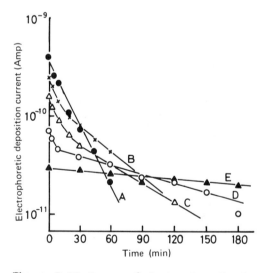

Figure 7.26 Decay of electrophoretic deposition current. Applied field (V/cm); A, 50; B, 30; C, 20; D, 10; E, 5. (From Ref. 65.)

where L is the distance between both electrodes and V the applied voltage.

If the relation of log $(I - I_1)$ and t were linear, T_r, and in turn U, can be obtained. The experimental result on deposition current in the carbon black dispersion showed the linearity of log $(I - I_1) \sim t$, except for an initial stage in which the leak current appears as seen in Fig. 7.26. The ζ potential was calculated from the Hückel equation. Furthermore, the total surface charge Q_0, the charge per particle q, and particle radius a can be obtained from the following equations:

$$Q_0 = 2I_0 T_r \tag{7.22}$$

$$q = \left(\frac{Q_0}{M_0}\right) m \tag{7.23}$$

$$a = 3\left(\frac{M_0}{Q_0} \frac{\eta U}{2\rho}\right)^{1/2} \tag{7.24}$$

where M_0, m, and ρ are the total mass of particles in the cell, the mass of a particle, and the density of a particle, respectively.

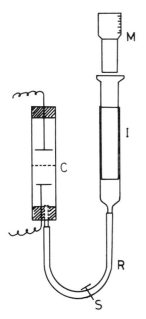

Figure 7.27 Arrangement of moving boundary method in nonaqueous electrophoresis. C, Cell; I, injector; M, micrometer; R, rubber tube; S, screw cock. (From Ref. 69.)

It is to be noted that this method involves the following assumptions: Electrophoretic velocity is much slower than diffusion velocity of particles, and all of the particles reaching the electrode discharge perfectly. The author compared the electrodeposition method with microelectrophoresis using carbon black or titanium oxide dispersed in Aerosol OT-cyclohexane solutions. Both methods showed a similar trend in the change in ζ potential with Aerosol OT concentration, but the former gave a lower ζ potential [67]. A different approach to the deposition current in nonaqueous media was proposed by Mohn [68].

7.5.3 Moving Boundary Method

For dispersions of smaller particles which cannot be detected with an optical microscope, we can use an ultramicroscope or a laser optical source. Particles of about 0.01 μm are detected with the latter.

A moving boundary method is suitable for electrophoresis measurement of much smaller particles. A cylindrical cell made of quartz is recommended in nonaqueous media. Electrodes of spiral wires are used to protect laminar flow and they are located several centimeters apart. An example of a cell and accessory is shown in Fig. 7.27. A sharp boundary between the medium and the dispersion is necessary to obtain more accurate data with the apparatus. It is recommended that the dispersion in the cell C be slowly added from a syringe I using a micrometer M to form a sharp boundary, as shown in Fig. 7.27.

REFERENCES

1. E. J. Verwey and J. Th. G. Overbeek, *The Theory of the Stability of Lyophobic Colloids*, Elsevier, Amsterdam, 1948.
2. R. H. Ottewill and J. N. Shaw, *Kolloid Z. Z. Polym.*, *218*, 34 (1967).
3. R. Neihof, *J. Colloid Interface Sci.*, *30*, 128 (1969).
4. K. Furusawa and S. Hachisu, *Sci. Light*, *15*, 115 (1966).
5. S. Komagata, *Kaimen denki kagaku gaiyo*, Shokodo, Tokyo, 1953, p. 63.
6. M. E. Smith and M. W. Lisse, *J. Phys. Chem.*, *40*, 399 (1936).
7. J. D. Hamilton and T. J. Stevens, *J. Colloid Interface Sci.*, *25*, 519 (1967).
8. B. R. Ware and W. H. Flygare, *Chem. Phys. Lett.*, *12*, 81 (1971).
9. B. R. Ware and W. H. Flygare, *J. Colloid Interface Sci.*, *39*, 670 (1972).
10. B. R. Ware and W. H. Flygare, *Adv. Colloid Interface Sci.*, *4*, 1 (1974).
11. E. E. Uzgiris, *Opt. Commun.*, *6*, 55 (1972).

12. E. E. Uzgiris, *Rev. Sci. Instrum.*, *45*, 47 (1974).
13. E. E. Uzgiris and D. H. Cluxton, *Rev. Sci. Instrum.*, *51*, 44 (1980).
14. T. Yoshimura and A. Kikkawa, *Japan J. Appl. Phys.*, *11*, 1797 (1972).
15. V. Novotny and M. L. Hair, in *Polymer Colloids*, Vol. II (R. M. Fitch, ed.), Plenum Press, New York, 1979, p. 37.
16. P. J. Goetz and J. G. Peniman, Jr., *Am. Lab.*, Oct. 1976.
17. P. J. Goetz, in *Cell Electrophoresis: Clinical Application and Methodology*, INSERM Symposium No. 11, (A. W. Preece and D. Sabolovic, eds.), Elsevier/North-Holland, Amsterdam, 1979.
18. A. T. Forrester, W. E. Parkins, and E. Gerjuoy, *Phys. Rev.*, *72*, 728 (1947).
19. Y. Yen and H. Z. Cummins, *Appl. Phys. Lett.*, *4*, 176 (1964).
20. M. Ozaki and K. Kojima, *33rd Symp. Colloid Interface Chem.*, The Chemical Society of Japan, Sapporo, Japan, 1980.
21. M. von Stackelberg, W. Kling, W. Benzel, and F. Wilke, *Kolloid-Z.*, *135*, 67 (1954).
22. D. Stigter, *J. Colloid Sci.*, *19*, 252 (1964).
23. A. J. Ham and W. Hodgson, *Trans. Faraday Soc.*, *38*, 217 (1942).
24. W. Rose and R. W. Heins, *J. Colloid Sci.*, *17*, 39 (1962).
25. B. V. Derjaguin, *Dokl. Akad. Nauk SSSR*, *39*, 11 (1943).
26. L. H. Reyerson, I. M. Kolthoff, and K. Coad, *J. Phys. Chem.*, *51*, 321 (1947).
27. H. C. Li and P. L. de Bruyn, *Surf. Sci.*, *5*, 203 (1966).
28. A. S. Brown, *J. Am. Chem. Soc.*, *56*, 646 (1934).
29. M. D. Robinson, J. A. Pash, and D. W. Fuerstenau, *J. Am. Ceram. Soc.*, *47*, 516 (1964).
30. D. R. Stewart and N. Street, *J. Colloid Sci.*, *16*, 192 (1961).
31. G. K. Korpi and P. L. de Bruyn, *J. Colloid Interface Sci.*, *40*, 263 (1972).
32. J. M. Horn, Jr., and G. Y. Onoda, Jr., *J. Colloid Interface Sci.*, *61*, 272 (1977).
33. P. N Zhukov and D. A. Fridriksberg, *Kolloidn. Zh.*, *12*, 25 (1950).
34. H. W. Douglas and J. Burden, *Trans. Faraday Soc.*, *55*, 350 (1959).
35. J. Th. G. Overbeek and P. W. Wijga, *Recl. Trav. Chim. Pays-Bas*, *65*, 556 (1946).
36. J. Lyklema and J. Th. G. Overbeek, *J. Colloid Sci.*, *16*, 501 (1961).
37. S. S. Dukhin and B. V. Derjaguin, in *Surface and Colloid Science*, Vol. 7 (E. Matijević, ed.), Wiley, New York, 1974, p. 121.
38. J. B. Oldham, F. J. Young, and J. F. Osterle, *J. Colloid Sci.*, *18*, 328 (1963).
39. A. J. Rutgers and M. De Smet, *Trans Faraday Soc.*, *41*, 758 (1945).

40. T. Morimoto and S. Kittaka, *Bull. Chem. Soc. Jpn.*, 46, 3040 (1973).
41. M. Hurd and N. Hackerman, *J. Electrochem. Soc.*, 102, 594 (1955).
42. M. Hurd and N. Hackerman, *J. Electrochem. Soc.*, 103, 316 (1956).
43. A. S. Buchanan and E. Heyman, *J. Colloid Sci.*, 4, 157 (1949).
44. J. J. Bikerman, *Z. Phys. Chem.*, 163, 378 (1933).
45. J. J. Bikerman, *Kolloid-Z.*, 72, 100 (1935).
46. J. J. Bikerman, *Trans. Faraday Soc.*, 36, 154 (1940).
47. A. J. Rutgers, *Trans. Faraday Soc.*, 36, 69 (1940).
48. A. J. Rutgers and M. De Smet, *Trans. Faraday Soc.*, 43, 102 (1947).
49. H. Fricke and H. J. Curtis, *J. Phys. Chem.*, 40, 715 (1936).
50. D. J. O'Conner, N. Street, and A. S. Buchanan, *Aust. J. Chem.*, 7, 245 (1954).
51. N. Street, *Aust. J. Chem.*, 9, 333 (1956).
52. N. Street, *J. Phys. Chem.*, 62, 889 (1958).
53. S. S. Dukhin and B. V. Derjaguin, in *Surface and Colloid Science*, Vol. 7 (E. Matijević, ed.), Wiley, New York, 1974, p. 49.
54. A. M. James and M. N. A. Carter, *J. Colloid Interface Sci.*, 29, 696 (1969).
55. F. Urban, H. L. White, and E. A. Strassner, *J. Phys. Chem.*, 39, 311 (1935).
56. A. Watillon and R. De Backer, *J. Electroanal. Chem.*, 25, 181 (1970).
57. J. Th. G. Overbeek, in *Colloid Science* (H. R. Kruyt, ed.), Vol. 1, Elsevier, Amsterdam, 1952, p. 235.
58. S. D. James, *J. Phys. Chem.*, 70, 3447 (1966).
59. J. W. McBain, C. R. Peaker, and A. M. King, *J. Am. Chem. Soc.*, 51, 3294 (1929).
60. J. W. McBain and C. R. Peaker, *J. Phys. Chem.*, 34, 1033 (1930).
61. J. A. Schufle, C. T. Huang, and W. Drost Hansen, *J. Colloid Interface Sci.*, 54, 184 (1976).
62. S. Kittaka and T. Morimoto, *J. Colloid Interface Sci.*, 55, 431 (1976).
63. S. Kuo and F. Osterle, *J. Colloid Interface Sci.*, 25, 421 (1967).
64. P. H. Wiesema, A. L. Leob, and J. Th. G. Overbeek, *J. Colloid Interface Sci.*, 22, 78 (1966).
65. A. Kondo and J. Yamada, *Nippon Kagaku Kaishi*, 1972, 716 (1972).
66. M. J. B. Franklin, *J. Oil Colour Chem. Assoc.*, 51, 499 (1968).
67. A. Kitahara, unpublished data.
68. E. Mohn, *Photogr. Sci. Eng.*, 15, 451 (1971).
69. A. Kittahara and S. Kobayashi, unpublished data.

8
Stability Measurement of Disperse Systems

Kunio Furusawa   The University of Tsukuba, Sakura-Mura, Ibaraki, Japan

Matsuo Matsumoto   Kyoto University, Uji, Kyoto, Japan

8.1. Coalescence of Droplets   225
    8.1.1. Introduction   225
    8.1.2. Colloid stability   226
    8.1.3. Coalescence of liquid droplets   229
    8.1.4. Exceptional stabilization   232
8.2. Stability of Colloids   233
    8.2.1. Introduction   233
    8.2.2. Kinetic methods   233
    8.2.3. Static methods   241
8.3. Latex Characterization and Stability   248
    8.3.1. Introduction   248
    8.3.2. Preparation of monodisperse emulsifier-free latices and their characterization   249
    8.3.3. Charge distribution in electrical double layer on polystyrene latices   253
    8.3.4. Coagulation into secondary minimum   255
    8.3.5. Summary   262
    References   262

8.1 COALESCENCE OF DROPLETS

8.1.1 Introduction

When liquid droplets or air bubbles approach each other in a continuous liquid medium, many forces act to thin or thicken the intervening liquid film in the process of drainage. If the film on relatively large droplets is thick, of the order of a few micrometers, dimple formation between confronting surfaces takes place due to the surface deformation. This results in capillary pressure which thins the film. When the film is thinned, the confronting liquid surfaces have a high radius of curvature, resulting in a minor capillary pressure in the film.

The presence of ions in a continuous liquid medium brings about the formation of diffuse double layers around the drop surfaces. The overlapping of the layers followed by the thinning of the medium gives rise to excess pressure (i.e., electrostatic repulsion based on osmotic pressure thickens the liquid film). At the same time, an attractive force always acts between the droplets through the liquid film. This force is mainly the van der Waals dispersion force. Hence, in the absence of other forces, such as capillary pressure and mechanical force, the thinning of the intervening liquid film is affected by the balance between electrostatic repulsion and attraction. A predominance of attractive force results in an extremely thin film and then gives rise to coalescence between liquid droplets, provided that stable and rigid adsorption layers of the surfactant or hydrated layers are absent on the drop surfaces; those layers closely overlapping each other exert osmotic pressure to prevent further thinning of the liquid film.

The forces mentioned above are the same as those observed in the interaction between colloid particles. Hence the work on colloid stability will be briefly summarized in regard to the theoretical and experimental studies of the electrostatic, attractive, steric, and other forces. Following that, general research in colloid stability will be illustrated by the experimental work on liquid droplet coalescence.

8.1.2 Colloid Stability

The interaction force of an electrical double layer based on the Derjaguin-Landau-Verwey-Overbeek (DLVO) theory is described by the constant potential model of the Stern layer assuming thermodynamic equilibrium in the electrical double layer [1-5]. This model, however, is no longer applicable when the relaxation time of the Stern layer for maintaining the constant potential is larger than the time of Brownian collision. This model should be replaced by the constant charge model, which considers no thermodynamic equilibrium in the double layer. The theory of the constant charge model proposed by Overbeek [1] was considered by Frens [6] and related approximations were derived by Wiese and Healy [7], Gregory [8], and Usui [9].

The attractive forces acting between colloid particles in addition to the electrostatic repulsion were described by Hamaker [10], who derived an equation by the simple summation of intermolecular attraction [11] assuming a uniform density of molecules in particles and a continuous liquid medium. Lifshiz [12] and Dzyaloshinskii, Lifshiz, and Pitaevskii [13] indicated in their theories that the attractive force between particles (i.e., condensed bodies in a free space and a liquid medium) is not a simple summation as described by Hamaker, but that the force should be expressed by the permittivity which reflects a bulk property of the particle and of the liquid, which includes the intervening effect, the permittivity being the function of the imaginary frequency. Their theories, which were difficult to apply in practice, have been developed by Ninham and Parsegian [14,15], who derived an approximated equation called the macroscopic theory. The equation consists of two terms divided into temperature-dependent and temperature-independent terms. When the particle has a single and strong characteristic absorption in the ultraviolet region, the equation takes the simple form and depends primarily on the temperature-independent term except in the case of colloid particles and intervening liquid having values similar to the permittivity. A similar macroscopic equation has also been derived by Israelachvili using the mirror image method [16,17].

The repulsive and attractive forces of DLVO theory depend on the separation between colloid particles. A direct proof of the theory was given initially by Derjaguin and Titijevskaya [18], Scheludko and Exerowa [19-21], and Lyklema and Mysels [22]. They measured interferometrically the equilibrium thickness of free liquid film made of an aqueous solution containing the surface-active material as a function of the ionic strength, and obtained qualitative and quantitative agreement with the theory. Roberts and Tabor [23,24] extended the interferometrical method to the measurement of the double-layer force between two solids, glass and rubber. The pioneering work of measurement of the double-layer force was done by Israelachvili and Adams [25]. They developed a new apparatus for the direct measurement of the force between molecularly smooth mica surfaces in simple electrolyte solutions. The separation between the surfaces mounted on a piezoelectric ceramics tube and a spring could be adjusted to an accuracy of 0.7 nm/V using a tube with two other step adjustments by cantilever operation. The separation was measured to an accuracy of 0.1 nm by multibeam interferometry, the force being obtained directly by measuring the deflection of the spring of known constant. The experimental results were in excellent agreement with DLVO theory in the relationship of the forces to the distance between mica surfaces.

Prior to the Israelachvili's work, the attractive force between mica surfaces which is always present in colloid interaction was investigated directly by Tabor and co-workers [26,27]. The experimental method used was the same in principle to that in Ref. 25. When the separation between the surfaces was small, the separation was determined by multibeam interferometry by measuring the jump position of different springs. At large separations the attractive force became weak, which led to lower accuracy in the measurement of separation due to the influence of vibration. In this case a new dynamic method was applied. The force measured in the free space remained nonretarded up to 2-12 nm, and at a separation of more than 50 nm the force became retarded with a clear transition range of 12-50 nm.

According to the macroscopic theory the sign of the attractive force becomes positive (repulsive) when permittivities ε_1 and ε_2 of two colloid particles 1 and 2 surrounded by the continuous medium 3 of permittivity ε_3 are in the order $\varepsilon_1 > \varepsilon_2 > \varepsilon_3$ or $\varepsilon_1 < \varepsilon_2 < \varepsilon_3$. The system of gas bubble-hydrocarbon-high dispersion energy solid satisfies this condition. Derjaguin et al. [28] found the presence of negative attraction (repulsion) in systems of gas bubble-liquid hydrocarbon-sapphire, stainless steel, and mica. The equilibrium thickness of hydrocarbon film measured elipsometrically was found to depend on the negative nonretarded or retarded attraction, which was balanced by the applied hydrostatic pressure. The negatively nonretarded attraction appeared in the range of separation of 6-20 nm between gas bubble and mica in tetradecane, but a negatively retarded attraction was found at separations greater than about 30 nm when stainless steel was used instead of mica.

In addition to the DLVO force, colloidal or liquid dispersions are also stabilized sterically by the adsorbed film of surface-active materials. Ottewill and Walker [29] explained the stabilization of latex particles in the presence of adsorbed film in terms of osmotic pressure resulting from the overlapping of adsorbed layers. In fact, the steric force between the optically smooth and macroscopic surfaces covered by the adsorbed polymer, which is balanced by the applied pressure, increased markedly with decreasing separation between the surfaces [30]. Hesselink and co-workers [31] proposed that the steric force consists not only of the osmotic effect but also of the volume restriction effect owing to the decrease in configurational entropy of adsorbed segments. This theory was confirmed by the study of drainage and equilibrium thickness of the free film using poly(vinyl alcohol) [32]. The other finding of the steric force was also due to Israelachvili, Tandon, and White [33]. They measured directly the steric force between poly(ethylene oxide) (MW 1.48×10^5) films adsorbed on mica surfaces in an aqueous solution and found a monotonical increase of steric repulsion on approach at a separation less than about 250 nm. Using a similar method, Klein [34] studied directly the steric force as the function of the separation between mica surfaces having adsorbed polystyrene (MW 6×10^5) films in cyclohexane. No force in this case was observed up to a separation of 60 nm. But at less than this separation the attraction was initially observed between adsorbed films, and then after passing through a minimum force at about 20 nm a repulsive (steric) force was evident.

The dispersion of liquid droplets is also affected by the interfacial instability such as vibrational propagation and the Marangoni effect which is a gradient in the interfacial tension due to the mass transfer across the interface. Scheludko [21,35,36] and Vrij [37] separately indicated how the thin liquid film between liquid droplets or air bubbles is ruptured by interfacial turbulence; the rupture of the thin film takes place at a certain thickness of the film when the work necessary to deform the interface is equal to the attraction, neglecting the double-layer force.

The drainage of thin liquid film containing a surfactant is remarkably decreased by the increase in effective viscosity of the film with thinning [38]. The increase in the viscosity of the thin liquid film between rubber and a glass place calculating by measuring the frictional force commences from a film thickness of about 50 nm. At less than 30 nm the viscosity becomes much higher than that of the bulk liquid. The increased viscosity acts to delay the approach of droplets, leading to a decay in the rate of coalescence.

8.1.3 Coalescence of Liquid Droplets

According to the DLVO theory, the colloid stability in an aqueous solution is a function of the Stern potential and ionic strength. Hence mercury, which is used as an ideally polarized electrode, is one of the best liquids to use to study the effect of the Stern potential on the coalescence. Using twin dropping mercury electrodes, Watanabe and Gotoh [39] have studied the effect of the polarizing potential on coalescence between mercury droplets formed at the confronted end tips of glass capillaries. The mercury droplets immersed in aqueous solutions containing an inorganic electrolyte do not coalesce over a critical potential (rational potential). This potential is equivalent to the critical Stern potential in the absence of specific adsorption. The critical potential of coalescence is not only dependent on the concentration of ions but also on the valence of the ions. The latter dependence on ionic valencies is what is known as the Schulze-Hardy rule. The condition of rapid coagulation in the DLVO theory predicts a straight line of slope 1/4 in a plot of the logarithm of critical potentials against the logarithm of ion concentrations. The coalescence of mercury droplets conforms exactly with this prediction. The same experiment was extended to the system of water-methanol mixed solvents [40]. When the mixed solvent contained the simple electrolyte KCl, the critical potential increased with increasing concentration of KCl and methanol in the mixed solvent. This result is apparent from the theory because the thickness of the diffuse layer is compressed with increasing concentration of electrolyte and decreasing dielectric constant of the solvent.

Usui, Yamasaki, and Shimoizaka [41] applied Watanabe's method to the study of asymmetrical double-layer interaction in the heterocoagulation of colloid particles. When potentials of the same sign but different magnitude were applied to each droplet formed at the tips of the twin dropping mercury electrodes in an aqueous solution, the product of critical potentials of two mercury droplets was constant at fixed ionic strength. Values of the critical potential thus obtained were entirely the same as those of the theoretical potential of heterocoagulation given by Devereux and de Bruyn [4]. Usui and Yamasaki [42] further extended their work to the study of attachment between glass and mercury in an aqueous solution.

The study of the coalescence of mercury droplets could be extended to investigate the protective action of surfactants as well. Watanabe and co-workers [43] measured the critical potential of coalescence of mercury droplets in aqueous solutions containing nonionic surfactants together with a supporting electrolyte (Fig. 8.1). When the concentration of surfactant of the pluronic type was low, the critical potential was the same as that obtained for surfactant-free solution. With increasing surfactant concentration, the potential was independent of the surfactant concentration, and no coalescence was observed even at a potential for charge zero. The critical concentration which prevents coalescence is associated with HLB values of the surfactant and corresponds to the concentration of the formation of a saturated adsorption layer; this was confirmed by the measurement of differential capacitance at the mercury/solution interface. The stabilization of mercury was also found in liquid hydrocarbon containing a surfactant such as stearic acid. Sonntag [44] measured the equilibrium separation

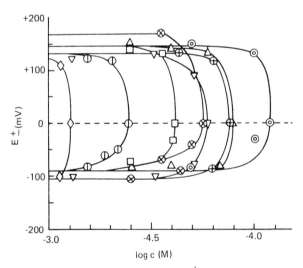

Figure 8.1 Critical potential E_-^+ of coalescence of Hg droplets versus the logarithm concentration c of nonionic surfactant in 0.1 M KCl. The molecular formula of the surfactant is $HO(CH_2CH_2O)_a CH_3(CHCH_2 O)_b (CH_2CH_2O)_c H$. HLB values: △, 5.57; ▽, 11.92; □, 16.82; ◊, 86.91; ◎, 3.95; ⊗, 5.20; ⊕, 11.97; Φ, 53.95.

between mercury surfaces in the solution by the capacitance method. The separation was about 7-8 nm, corresponding to the bilayer thickness of the acid, exhibiting steric stabilization.

It was pointed out that the coalescence of aqueous droplets in oil phases [45] is an example of double-layer interaction between liquid droplets other than mercury. Electrocapillary phenomena at the oil/water interface which were found by Guastalla [46] and developed by Watanabe et al. [47-49] took place when either the oil or aqueous phase contained surface-active ions and the other phase or both included a supporting electrolyte. In this case the electric current flows across the oil/water interface owing to ion transfer by the application of a potential difference between the aqueous phase and the oil phase. Hence the effective polarization at the oil/water interface is steady and ill defined, depending on the applied potential. This indicates that the interfacial tension obtained at the applied potential is not in equilibrium but in a stationary tension.

Droplets of aqueous KCl solution formed at the tips of twin glass capillaries, where they had been immersed in the oil phase containing cationic surfactant and supporting ions, were prevented from coalescing over a critical value of applied potentials (Fig. 8.2). This is similar to the phenomenon found in the case of coalescence of mercury droplets. The critical value of the coalescence potential of aqueous drop-

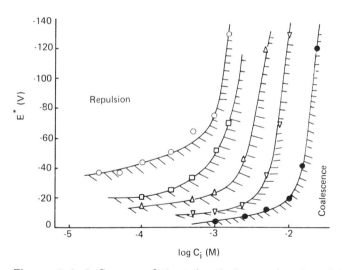

Figure 8.2 Influence of the trimethylammonium bromide concentration C_i in the oil phase on the critical potential of coalescence of 0.1 M KCl aqueous droplets. The concentration of cetylpyridinium chloride in the oil phase: ○, 2.5×10^{-5} M; □, 5.0×10^{-5} M; △, 1.0×10^{-4} M; ▽, 2.5×10^{-4} M; ●, 5.0×10^{-4} M.

lets increased with increasing concentration of supporting ions in the oil phase, which suggests a compression of the diffuse double layer and that the relative Stern potential has an effect on the double-layer interaction.

8.1.4 Exceptional Stabilization

The coalescence of mercury droplets could be a good model experiment to study the DLVO theory. However, there is one coalescence which cannot be explained by the DLVO force. Lessard and Zieminski [50] studied the percentage of coalescence of air bubbles in an aqueous solution as the function of the concentration of inorganic electrolytes. A smaller percentage of the coalescence was obtained by increasing the ionic strength of the solution; the higher the valence of electrolyte, the lower the concentration to prevent the coalescence was. The stabilizing ability of the electrolytes did not depend on the type of electrolytes but on either the valence of anion or cation. This suggests that the coalescence of air bubbles was not affected by the double-layer interaction. Similar phenomena were found in the attachment between gas bubble and mercury in aqueous KF solution [51], or perfluorohexane and mercury in aqueous KCl or KBr solution [52]. In both cases, critical potentials of mercury attachment given by the Stern potential decreased monotonically with increasing concentration of electrolytes. Similarly, the decrease in critical Stern potentials was observed in the case of coalescence between mercury droplets in the KBr solution when the concentration of KBr was higher than 3×10^{-2} M [52]. However, the potentials increased with increasing KBr concentrations at less than the concentration above showing the maximum potential. In this case, critical values of the Stern potential did agree fairly well with those predicted by the rapid coagulation condition up to a concentration of 2×10^{-2} M, where the Stern potentials were calculated from the differential capacitance using Devanathan's model [53], which considers the specific adsorption of Br ions.

Pashley [54] carried out direct measurements of the force and separation between mica surfaces in aqueous solutions of alkaline halides using Israelachvili's method [25]. He showed the presence of an extra force in addition to the DLVO force at small separations of less than about 8 nm. But this extra force was not found in the aqueous solution of HCl, which forms less hydrated hydronium ions, indicating that there is no dependence on the anion. The extra force was thus shown to be due to the solvation effect of alkaline ions. Relating to this work, the theoretical works of Lane and Spurling [55] and van Megen and Snook [56] are worth citing. They considered the effect of the solvent structure and the solvation force on the van der Waals force between two bodies. Mitchell, Ninham, and Pailthorpe [57] also pointed out in their theory that the excess adsorption of solvent molecules at the interface leads to an additional term which acts as a repulsive force in addition to the conventional van der Waals attraction assuming the zero excess adsorption. In fact, direct measurement of the force and the separation between mica surfaces in octamethylcyclosiloxane [58] supports the foregoing theory by the fact that the force exhibits a periodical oscillation corresponding to the size of the solvent molecule up to about 7-10 molecules thickness.

The interaction between air bubbles or liquid droplets in aqueous solutions of high ionic strength is treated at a very small separation, usually less than 10 nm. Hence the exceptional stabilization found in the case of gas bubble-mercury, fluorocarbon-mercury, and mercury-mercury in aqueous solutions of high ionic strength would be explained by considering the solvent structure and the solvation force at the interface, together with the possible presence of a negative attraction in the gas bubble-mercury system.

8.2 STABILITY OF COLLOIDS

8.2.1 Introduction

Hydrophobic sols, such as silver iodide sol, silica sol, metal oxide sol, and others, are very sensitive to inorganic salts; addition of relatively small amounts of a salt can cause them to flocculate. The concentration of a salt required to flocculate the sol is called the critical flocculation concentration (CFC) or flocculation value (C_{CFC}). C_{CFC} is an important measure of the stability of a hydrophobic sol.

Flocculation values depend on a number of factors such as the method of measurement, the nature of the sol, the surface charge and double-layer thickness of the sol particles, and the valence and nature of the counterions. The procedures for measuring flocculation values can be divided into two types, the kinetic method and the static method. In the kinetic method, the rate of flocculation is measured terbidimetrically as a function of the concentration of the coagulation agent.

In the static method, a series of glass tubes including the sol is prepared, increasing amounts of salt are added, and the flocculating state is determined by some suitable methods of alternating shaking and waiting.

8.2.2 Kinetic Methods

In the flocculating rate method, developed by Reerink and Overbeek [59] based on previous work of Smoluchowski [60] and Fuchs [61], the rate of coagulation is measured as a function of the concentration of a coagulation agent. As compared with the static method, this procedure has the advantage of offering a less ambiguous criterion, but no equilibrium is attained between the sol and the solution.

Turbidity and Stability

Turbidity τ is defined by the Lambert-Beer law,

$$I_{tr} = I_{in} \exp(-\tau \ell) \tag{8.1}$$

where I_{in} and I_{tr} are the intensities of the incident and transmitted light, respectively, and ℓ is the length of the optical path of the cell. According to Rayleigh's theory of light scattering, the turbidity is related to the number of particles per cubic centimeter, n, and their individual volume V_s by the relation

$$\tau = A_r N V_s^2 \tag{8.2}$$

with

$$A_r = \frac{24\pi^3 n_0^4}{\lambda^4} \frac{n^2 - n_0^2}{n^2 + 2n_0^2} \tag{8.3}$$

where n_0 and n are the refractive indices of the solvent and the particle, respectively, and λ is the wavelength of the light. These equations hold only for spherical particles with radii smaller than 0.05λ and in the absence of consumptive light adsorption.

An equation of the turbidity change as a function of time for the coagulation process has been derived by Troelstra and Kruyt [62] and Oster [63]. Since the total turbidity of a dilute sol is the sum of the contribution of particles of various sizes, Eq. (8.2) becomes

$$\tau = A_r \sum_{i=1}^{\infty} N_i (V_{si})^2 \tag{8.4}$$

where the subscript i refers to the i-fold particles. Substituting

$$N_i = \frac{N_b (t/T_{tr})^{i-1}}{1 + (t/T_{tr})^{i+1}}$$

for the number of i-fold particles at time t, and considering $V_{si} = iV_s$, where V_s is the volume of a primary particle, we obtain

$$\tau = A_r V_s^2 N_b \sum_{i=1}^{\infty} \frac{i^2 (t/T_{tr})^{i-1}}{[1 + (t/Tr)]^{i+1}}$$

$$= A_r V_s^2 N_b \left(1 + \frac{2t}{T_{tr}}\right) \tag{8.5}$$

where $T_{tr} = W/4\pi DRN_b$ and has the dimensions of time. This gives the time when the total number of particles halves. W is the stability factor, D is the diffusion coefficient, and $N_b = \Sigma_{i=1}^{\infty} N_i$ for t = 0. Here i − 1 and i + 1 are the power indices. Differentiating Eq. (8.5) with respect to t, we obtain

$$\frac{d\tau}{dt} = \frac{8\pi DR(V_s N_b)^2 A_r}{W} \quad (8.6)$$

Equations (8.5) and (8.6) apply only to the initial stages of coagulation. They show that the turbidity changes linearly with time and that the initial slope is proportional to the square of the sol concentration, $V_s N_b$, and inversely proportional to the stability factor, W. If the sol concentration is expressed by C_s in g cm^{-3}, Eq. (8.5) reads

$$\frac{d\tau}{dt} = \frac{k'C_s^2}{W}$$

or

$$\log W = \log\left(\frac{dt}{dA}\right)_{t\to 0} - \log \frac{2.303}{k'C_s^2} \quad (8.7)$$

where

$$k' = \frac{A_r k_0}{d^2} \quad (g^{-2} \text{ cm}^5 \text{ sec}^{-1}) \quad (8.8)$$

$$k_0 = 8\pi DR = \frac{8kT}{3\eta} \quad (\text{cm}^3 \text{ sec}^{-1}) \quad (8.9)$$

Here we have introduced the density of the sol particles, d, and the absorbance, A, which is related to the turbidity by the relation

$$\tau = 2.303A \quad (8.10)$$

Stopped-Flow Measurements

A schematic picture of the stopped-flow spectrophotometer is shown in Fig. 8.3. In this apparatus, the solutions to be mixed are contained in two syringes (D_1 and D_2). By using a pressure actuator, equal volumes (0.2 ml) of each solution are reproducibly mixed within a very short time (10 msec for dilute aqueous solutions). The turbidity of the mixture can be recorded on a storage oscilloscope or on an attached recorder. Usually, measurements are carried out at room temperature, 20 ± 1°C, and at a wavelength of 546 nm (mercury green). At this wavelength, no consumptive light absorption occurs.

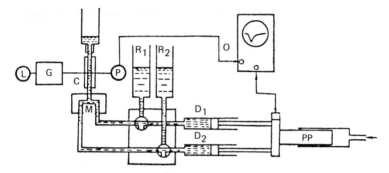

Figure 8.3 Schematic representation of stopped-flow spectrophotometer. L, Light source; G, spectrophotometer; C, sample cell; M, mixing vessel; P, photomultiplier; O, oscilloscope; R_1, R_2, sample reservoir; D_1, D_2, cylinder for charging the sample; PP, pressure actuator.

Curves of Absorbance versus Time

Coagulation of Negative Silver Iodide Sol by Inorganic Ions. Figure 8.4 shows some typical curves of absorbance versus time obtained by adding KNO_3(aq) to negative AgI sols [64]. In this figure, the

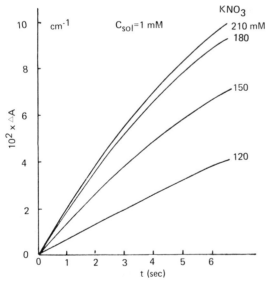

Figure 8.4 Increase of the absorbance of AgI sols with time due to coagulation by KNO_3. ΔA is the increase in absorbance per centimeter path length. The abscissa scale is in seconds. Sol concentration 1 mM. The final KNO_3 concentration is indicated. (From Ref. 64.)

salt concentrations indicated are the final concentrations after mixing by the stopped-flow method. It is clear from these curves that the initial portions are sufficiently linear to be in agreement with theoretical expection [Eq. (8.4)]. The deviation from linearity is due to the nonvalidity of the assumption that the scattering from aggregates is the same as that from spherical particles with the same total volume, $Vs_i = iV_s$.

With increasing salt concentration, the initial slope increases until finally a concentration region is attained where the coagulation rate becomes independent of the salt concentration. This is the region of fast coagulation beyond the C_{CFC}. The C_{CFC} is easily determined by plotting the logarithm of the reciprocal slope (log dt/dA) against log C_s, where C_s is the salt concentration.

In Fig. 8.5, plots of log $(dt/dA)_{t \to 0}$ versus log C_s are given for three sol concentrations and three salts with counterion valences $z = 1$, 2, and 3. As can be seen from Eq. (8.7), the quantity $(dt/dA)_{t \to 0}$ is proportional to the stability ratio W. The curves obtained are of the form predicted by theory. They are linear in the slow-coagulation range and become parallel to the abscissa in the rapid-coagulation range. The value of C_{CFC} for each curve is obtained from the intersection of these two sections of the curves [i.e., $C_{CFC} = 174$, 4.0, and 0.11 mM for KNO_3, $Ca(NO_3)_2$, and $La(NO_3)_3$, respectively]. These values are higher than those reported for classical static experiments [65]. This should be expected since in the kinetic exper-

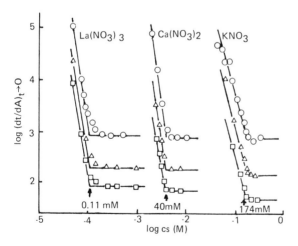

Figure 8.5 Dependence of $(dt/dA)_{t \to 0}$ on the electrolyte concentration for three salts and three sol concentrations. The breaks (indicated by arrows) give the critical coagulation concentrations. ○, C_{sol} 0.2 mmol/liter; △, C_{sol} 0.5 mmol/liter; □, C_{sol} 1.0 mmol/liter. (From Ref. 64.)

iments, the coagulation times are on the order of seconds and orthokinetic coagulation is absent.

A detailed discussion of the slopes of the fast-coagulation branches in these plots was given in Chap. 3. At a given sol concentration, the fast coagulation rate is roughly independent of the counterion valency, although a slightly lower coagulation rate at higher valency can be detected. It is due to a different distribution of charge in the Stern layer, which was initially suggested by Kruyt and Troelstra [62].

If rapid coagulation were a bimolecular process, its rate should be proportional to the squares of the sol concentration. It can be seen from the horizontal branches in Fig. 8.5 that the measured coagulation rates at high C_{sol} are slightly low in comparison to those at low C_{sol}. According to Fleer and Lyklema [64], this behavior is due to an experimental artifact of the stopped-flow spectrophotometer. Because of too wide an angle of acceptance of the photomultiplier, the unmodified instrument records a part of the secondarily scattered light, thus giving rise to too low an absorbance, especially at higher sol concentrations.

Allowing for this instrumental artifact, the results may be interpreted in agreement with the bimolecular nature of the coagulation process, as indicated in Fig. 8.5.

Coagulation of Positive Silver Iodide Sols by Anionic Surface-Active Agents. The change of absorbance with time that occurred after the addition of various concentrations of aqueous solutions of long-chain sulfates to the positive AgI sols was measured [66]. In Fig. 8.6, plots of $\log (dt/dA)_{t \to 0}$ versus log C are given. As can be seen, the curves of stability for the surface-active agents have different shapes from those of Fig. 8.5. Instead of being parallel to the abscissa, the stabilities reach a minimum and increase again with higher concentrations. Figure 8.7 shows curves of the zeta (ζ) potential versus log molar concentration for the same surface-active agents. Generally, as the concentration of the surface-active agent is increased the ζ potential remains practically constant until a certain concentration is reached. Afterward it gradually decreases to zero with increasing slope, becomes negative, and finally approaches a limiting negative value. These electrophoretic curves suggest that the phenomena in Fig. 8.6 are due to the reversal of the particle charge which has occurred by the combination of the surface-active agents with the positive silver iodide particle surfaces.

The curves of ζ versus log C are shown to be of the general shape predicted by the theory developed by Watanabe [67]. That is, the values of the maximum number of available adsorption sites, N_1, and the adsorption constant, k_2, can be determined by using the assumption that the monolayer adsorption of the surface-active agent is taking place up to the zero point of ζ. By substituting $\psi_d = \zeta$ and $z = -1$, the following relations are obtained [66]:

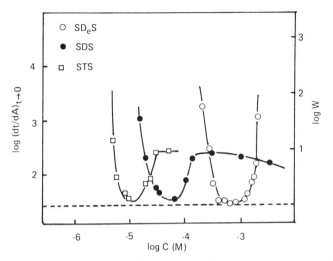

Figure 8.6 Log stability versus log molar concentration for sodium alkyl sulfates. (From Ref. 66.)

$$\left(\frac{d\zeta}{d \log C}\right)_{\zeta=0} = 2.303 \left[\frac{\zeta\varepsilon(1+\tau)}{4\pi aeN_1} - 1\right] \tag{8.11}$$

$$\frac{1}{C°} = k_2 \left[\frac{4\pi aeN_1}{\zeta\varepsilon(1+\tau)} - 1\right] \tag{8.12}$$

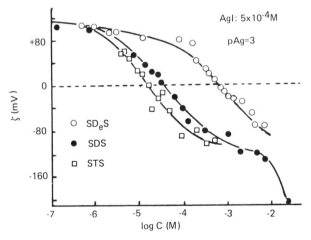

Figure 8.7 Zeta potential versus log molar concentration for sodium alkyl sulfates. (From Ref. 66.)

Table 8.1 Parameters for Adsorption of Anionic Surface-Active Agents on Positive Silver Iodide Sol

Surface-active agent	$\left(\dfrac{d\zeta}{d \log C}\right)_{\zeta=0}$ (V)	$C°$ (mol)	N_1 (cm^{-2})	k_2 (liter/mol)	CMC in water at 25°C (mol)
SDeS (Na decyl sulfate)	−0.096	3.98×10^{-4}	1.94×10^{13}	6.80×10^{3}	3.6×10^{-2}
SDS (Na dodecyl sulfate)	−0.096	4.21×10^{-5}	1.94×10^{13}	6.41×10^{4}	7.6×10^{-3}
STS (Na tetradecyl sulfate)	−0.096	1.68×10^{-5}	1.94×10^{13}	1.61×10^{5}	2.3×10^{-3}

Source: Ref. 66.

where τ and a are the turbidity and the radius of sol particles, respectively, and $C°$ is the concentration of surfactants at the zero of ζ.

Values of various parameters, calculated from the experimental data of the ζ potential versus logC curves, using Eqs. (8.11) and (8.12), are given in Table 8.1. For comparative purposes, the critical micelle concentration (CMC) values of the surface-active agents in water are also given. The values of k_2 appear to increase regularly as the CMC values decrease.

8.2.3 Static Methods

This method has the drawback that the stability criterion is somewhat arbitrary, and it has the advantage that if properly done, a time scale is used to attain adsorption equilibrium between the particles and the liquid.

The practical methods used for estimating C_{CFC} depend on monitoring some suitable property, either of the total system (absorbance, rheology, or electrical conductivity) or of the aggregates themselves (sedimentation volume, sedimentation rate, or filtration rate).

Some examples of the static method for estimating stability are described below.

Coagulation of Polystyrene Latices by an Inorganic Salt [68]

In the static method, 1.0 ml latex of approximately 0.5% w/w is mixed with 4.0 ml of KCl solutions of various concentrations in glass-stoppered tubes and the mixtures are kept at 25°C in a water bath. After 2 hr the contents of the tubes are mixed and the tubes replaced in the bath. After an equilibration time of 18 hr, the transmission coefficient of the supernatant liquid is measured against water at 540 nm with a spectrophotometer. The transmission coefficients are plotted against the electrolyte concentration. An example of these plots is shown in Fig. 8.8. The intersection of the two linear parts of the curve yields an electrolyte concentration that is taken as the C_{CFC}. The results for the C_{CFC} of latices with different surface charges are presented in Table 8.2. It is evident that the C_{CFC} increases with increasing surface charge as expected from DLVO theory. But there is only a small difference between the C_{CFC} values for the latex with largest surface charge and that with the smallest surface charge. It is evident that colloid stability depends on the potential at the Stern layer over which the diffuse double-layer charge is distributed, not directly on the surface charge of the sample.

Loose Aggregation of Tungstic Acid Sol Particles [69]

Platelet particles of tungstic acid sol form loose and reversible aggregates which are caused by the second minimum in the potential energy curve of the interaction between particles. Critical ionic concen-

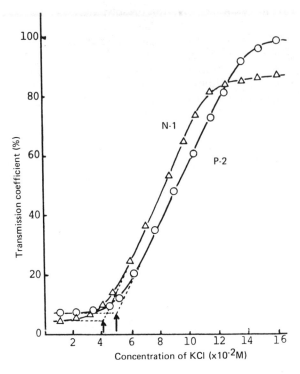

Figure 8.8 Transmission coefficient of the supernatant of polystyrene latices (N-1, P-2) as a function of concentration of KCl. The arrows indicate the coagulation concentration. (From Ref. 68.)

Table 8.2 Coagulation Concentration of KCl for Latices with Different Surface Charges

Sample	Surface charge ($\mu C/cm^2$)	Coagulation concentration (mol $\times 10^2$)
N-1	1.1	4.02
K-1	3.2	4.35
C-2	6.9	4.71
P-2	8.7	4.97

Source: Ref. 68.

trations for this aggregation are determined by observing the aggregating state using an inverse-type microscope and measuring the sedimentation state. The microscopic observation is performed as follows. A cell filled with 30 ml of a standard suspension is placed on the microscope and maintained at a temperature of 25°C by circulating water from a thermostat through a jacket. The salt concentration of the suspension is varied by adding a concentrated salt solution dropwise and stirring thoroughly. The aggregating state of the suspension is examined after each addition. The salt concentration is determined by means of a conductivity measurement using electrodes present in the cell.

Upon addition of various kinds of electrolytes, the suspension forms loose and reversible aggregates at a concentration characteristic of each electrolyte. These values of the critical concentration are shown in Table 8.3. At the critical concentration, the particles, which are separate at first, aggregate in pairs, triplets, and in larger groups. It is remarkable that every particle in these aggregates is still undergoing Brownian motion, although rotation around the spindle axis is suppressed (Fig. 8.9). This shows that particles in the aggregates exist with enough space to allow them to move freely. Another re-

Table 8.3 Flocculation Values for the Loose Agglomeration of Tungstic Acid Sol

Electrolyte	Flocculation Value (M)	
	By microscopy	By settling time
LiCl	2.27×10^{-3}	1.9×10^{-3}
NaCl	2.00×10^{-3}	1.7×10^{-3}
KCl	1.71×10^{-3}	1.6×10^{-3}
$(1/2)Na_2SO_4$	1.60×10^{-3}	—
$CaCl_2$	4.61×10^{-4}	4.5×10^{-4}
$BaCl_2$	4.52×10^{-4}	4.4×10^{-4}
$LaCl_3$	6.42×10^{-5}	—

Source: Ref. 69.

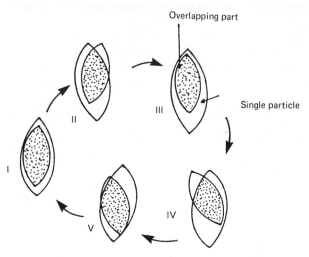

Figure 8.9 Diagrammatic picture showing the feature of relative Brownian motion of closely neighboring particles. Relative situation changes with time, say, from state I → state II → ··· → state V → state I.

markable property of this aggregation is its perfect reversibility. Also, the thinner the platelet particles, the easier the formation of aggregates.

The interference color of two overlapped particles is always yellowish green, regardless of the type of added electrolytes. This color is characteristic of the critical point for pair formations and indicates that, at the critical electrolyte concentration, the distance between the aggregating particles is almost the same and equal to about 120 nm [70]. The color is very sensitive to changes in the electrolyte concentration and, with an increase in the concentration, it shifts toward shorter wavelengths. The Brownian motion of the particles diminishes gradually and finally subsides at high electrolyte concentrations, where all the particles aggregate into a few large islands.

The sedimentation experiments are performed as follows. Relatively thick suspensions (about 5% by weight of particle in solutions) of various electrolyte concentrations are placed in sedimentation tubes and the time required to achieve complete sedimentation is measured. The temperature is controlled at 25°C using a water bath. In Fig. 8.10, a typical curve obtained from sedimentation experiments is presented. This demonstrates the influence of the electrolyte concentration on the settling time. The settling time is the time required for the particles to descend from a height of 30 cm to the bottom of the tube. There is a fairly sharp threshold on the curve indicating the occurrence of some aggregation. Threshold concentrations are listed in Table 8.3 together

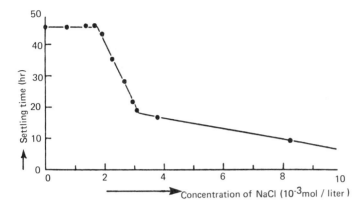

Figure 8.10 Sedimentation curve of tungstic acid sol when NaCl was used as a flocculation salt. (From Ref. 69.)

with the values of critical pair formation obtained by microscopic examination. Fairly good agreement between them indicates that pair formation will occur at the sedimentation threshold.

Flocculation of Tungstic Acid Sol by Polyacrylamide [71]

In flocculation experiments with polymeric substances, careful attention must be paid to the method of mixing the polymer solution and the colloid suspension, because inhomogeneous mixing may lead to nonreproducible results. Two methods of mixing, the one-portion mixing method and the two-portion mixing method, are compared to determine the effect of the order of mixing on the results of flocculation experiments.

In the one-portion mixing method, 5 ml of $BaCl_2$ solution is carefully added to 2 ml of tungstic acid sol (t-sol) in such a way that a sharp boundary between the sol and the electrolyte solution is formed. Then 2 ml of a polyacrylamide (PAAm) solution is pipetted into the tube. After mixing by end-over-end rotation at 10 rpm for 45 min, the system is allowed to stand for 1 hr to allow the flocculated particles to settle out. Then the extinction coefficient of the supernatant is measured using a spectrophotometer at a wavelength of 550 nm.

In the two-portion mixing method, a portion of t-sol (v milliliters), 5 ml of the electrolyte solution, and 2 ml of the PAAm solution are carefully placed in a cylinder in the same way as in the one-portion method. After mixing these components for 15 min by end-over-end rotation, the second portion of the sol (2 − v ml) is added. The cylinder is then rotated for 30 additional minutes to mix both portions. According to Fleer and Lyklema [72], maximum flocculation was produced when the volumes of the first and second portions of the sol were equal.

The time is chosen in such a way that the total mixing time is the same for both the one- and the two-portion methods. After allowing the flocculated particles to settle out for 1 hr, the extinction coefficient of the supernatant is measured.

The critical electrolyte concentration for flocculation, C_{CFC}, is defined as the concentration at which the absorbance of the supernatant is reduced to 50% of its original value. In Figs. 8.11 and 8.12 the dependence of C_{CFC} on the polymer concentration, obtained by the two types of mixing processes, is given for various molecular weight PAAm fractions. For all the fractions, a decrease in the value of C_{CFC} is initially observed and, after attaining a minimum, the C_{CFC} increases beyond the critical polymer concentration. The critical polymer concentration is dependent on the molecular weight of the PAAm. In the flocculation curves for PAAm values in the range $M_w = 1.2 \times 10^5$ to 5×10^6, the minimum range becomes broader and the region of flocculation extends to higher polymer concentrations with increasing molecular weight. On the other hand, for the fractions having $M_w = 5\text{-}10 \times 10^6$, the flocculating power becomes less and the minimum in the flocculation curve becomes narrower with increasing molecular weight. It is striking that the molecular weight dependence of the critical polymer concentration for the onset of stabilization depends strongly on the method used to mix the sol and the polymer.

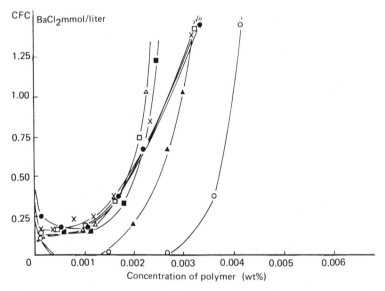

Figure 8.11 Dependence of C_{CFC} on the polymer concentration (one-portion mixing). –●–, MW 120,000; –x–, MW 590,000; –□–, MW 1,120,000; –▲–, MW 3,000,000; –○–, MW 5,000,000; –■–, MW 7,000,000; –△–, MW 9,290,000. (From Ref. 71.)

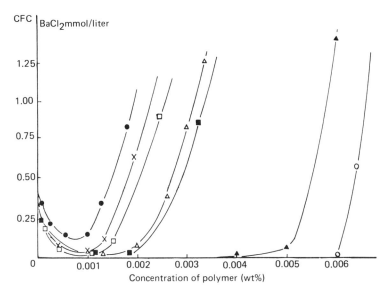

Figure 8.12 Dependence of C_{CFC} on the polymer concentration (two-portion mixing). —●—, MW 120,000; —x—, MW 590,000; —□—, MW 1,120,000; —▲—, MW 3,000,000; —○—, MW 5,000,000; —■—, MW 7,000,000; —△—, MW 9,280,000. (From Ref. 71.)

The results obtained for the two-portion mixing method show a strong molecular weight dependence (Fig. 8.12), whereas the results for the one-portion method do not depend significantly on the molecular weight (Fig. 8.11). In particular, the flocculation power of the 3 and 5×10^6 PAAm fractions is very strong in the two-portion mixing method.

Several factors are responsible for these results. In Fig. 8.13 a schematic representation is given to explain the effects of molecular weight on flocculation behavior. The factors determining the stability of a colloid system, if the particles carry an adsorbed polymer layer, are the bridging attraction energy (V_{bridge}) and the steric repulsion energy (V_{st}). These are in addition to the "classical" van der Waals attraction (V_a) and electric double-layer repulsion ($V_{e\ell}$) terms.

It is usually considered that over the whole molecular weight range of adsorbed polymers, V_a is small in comparison to the other energy contributions. In contradistinction, $V_{e\ell}$ still plays a significant role in controlling the stability. For interparticle distances larger than about twice the adsorbed polymer layer thickness, the total interaction energy (V_t) is determined only by $V_{e\ell}$. Thus the effect of the addition of electrolyte is to reduce the distance of closest approach to about twice the adsorbed polymer layer thickness, and only then does polymer bridging between the two particles become possible. In this

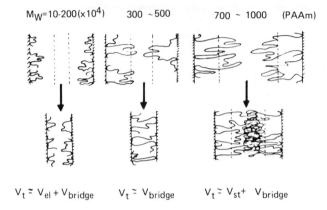

Figure 8.13 Schematic representation to show the molecular weight effect on flocculation power.

connection, it may be expected that the amount of electrolyte needed for flocculation will decrease with increasing thickness of the polymer layer. This is the tendency observed in the flocculation results using PAAm of $M_w = 1.2 \times 10^5$ to 2×10^6.

It is probable that the decrease in flocculation power for the molecular weight range higher than 5×10^6 may be due to the steric repulsion between polymer layers. In this molecular weight region, the layer thickness of the adsorbed polymer may be so large as to be beyond the distance of closest approach for the particles. V_{st} increases rapidly as the amount of adsorption increases and the effective range of the steric repulsion extends to larger distances as the layer thickness increases [73]. One concludes that the optimum conditions for flocculation result when the adsorbed polymer layer formed on the platelet particles is thicker than that of the electrical double layer, but the polymer layer is not so thick that steric repulsion, resulting from the overlapping of the polymer layers on adjacent particles, becomes effective. For the explanation of the efficiency of the method of mixing, a bridging mechanism for the flocculation and irreversibility of the adsorbed polymer layer is essential.

8.3 LATEX CHARACTERIZATION AND STABILITY

8.3.1 Introduction

In recent years, monodisperse latices have been recommended as the model dispersion for the study of a number of fundamental phenomena in the field of colloid science [74]. The main reasons for this are the high monodispersity of the particles and the well-defined distribution of surface charge [75]. The behavior of latices depends on whether they are prepared in the presence or absence of an emulsifier. If no emulsifier is used in the preparation, the surface charge consists of covalently bound polar groups originating from the initiator used in the polymerization reaction [76]. If an emulsifier is used in the preparation of the latices, adsorbed ionic groups of the emulsifier, in addition to the covalently bound polar groups, can contribute to the surface charge. So the surface charge of the latices polymerized in the presence of an emulsifier depends on the adsorption-desorption equilibrium of the emulsifier. According to Stone-Masui and Watillon [77], the adsorbed emulsifier cannot be eliminated completely from the latex surface.

In this section the preparation and characterization of monodisperse polystyrene latices with a wide range of surface charge prepared in the absence of emulsifiers, and the structure of electrical double layers on their surfaces, are discussed. Finally, their stability on coagulation into secondary minima is described.

8.3.2 Preparation of Monodisperse Emulsifier-Free Latices and Their Characterization

The surface charge of emulsifier-free polystyrene latices with potassium persulfate as the initiator is due to sulfonate groups originating from the initiator. To increase the surface charge, the number of sulfonate groups on the final particles should be maximized [78]. This can be achieved by utilizing a high concentration of potassium persulfate and adjusting the pH of the polymerization to a higher value with the addition of $KHCO_3$ or K_2HPO_4 buffers. The purpose of this condition is to suppress the Kolthoff reaction [79],

$$OSO_3^- + H_2O \rightarrow HSO_4^- + OH \cdot$$

which would otherwise enhance the number of OH radicals acting as initiators at the expense of sulfate radicals.

Data from the preparation and characterization of emulsifier-free latices are shown in Table 8.4. Columns 5 and 6 of this table give the average particle size determined by electron microscopy. The presence of KCl in the polymerizing system leads to larger particles (compare N-2 with K-1 or K-2). It is seen that in the absence of added electrolyte, an increase in the initiator concentration leads to larger particles. Addition of $KHCO_3$ or K_2HPO_4, however, does not alter the radius significantly (compare N-2 with C-1 or N-3 with P-2). The size distribution is very narrow, with $D_w/D_n < 1.01$ for all samples.

The surface charge in column 8 has been obtained by potentiometric or conductometric titration of the H form of the latex with NaOH in an argon atmosphere. The latex is converted to the H form by ion exchange for 3 hr using Dow mixed-bed resins according to Van den Hul

Table 8.4 Preparation and Characterization of Polystyrene Latices with a Wide Range of Surface Charge

Sample	Polymerization conditions $K_2S_2O_8$ conc. (mol/liter $\times 10^4$)	Electrolyte (mol/liter $\times 10^2$)	pH	Particle size Radius (nm)	Spec. area (m^2/g)	σ_0 ($\mu C/cm^2$) Before i.e.[a]	After i.e.
1	2	3	4	5	6	7	8
N-1	3.71	–	3.2	212	13.5	3.1	1.1
N-2	19.07	–	2.6	260	11.0	6.9	3.5
N-3	34.21	–	2.3	290	9.9	8.7	4.1
		KCl					
K-1	19.2	0.27	2.6$_5$	350	8.5	–	3.2
K-2	19.2	1.0	2.6$_6$	528	6.4	–	3.5
K-3	37.1	0.5	2.6$_2$	381	7.9	–	4.3
		$KHCO_3$					
C-1	18.50	1.0	7.6	255	11.2	8.4	4.5
C-2	27.70	1.0	7.5	290	9.9	17.2	6.9
C-3	37.40	0.5	6.2	280	10.2	12.3	7.4
		K_2HPO_4					
P-1	29.3	0.5	7.4	325	8.7	15.7	8.1
P-2	36.9	0.5	7.8	300	9.6	15.0	8.7

[a]i.e., ion exchange.
Source: Ref. 81.

and Vanderhoff [80]. The equivalent points determined from each technique are always very close. The titration curves for latex C-3 are shown in Fig. 8.14. The shapes of these curves are due to the titration of a strong acid. The acid groups are certainly $-OSO_3^-$ groups originating from $K_2S_2O_8$ [75]. Such titration curves, however, are observed only for the fresh samples. For the latices left for a long period after preparation, there are two breakpoints on the titration curves, which suggests that the surface sulfonate groups would be hydrolyzed gradually and change to the hydroxyl groups or

Table 8.4 (cont'd.)

Weight fraction of sulfur $\times 10^2$				Surface/bulk sulfur ratio	
Total		Surface			
Before i.e.	After i.e.	Before i.e.	After i.e.	Before i.e.	After i.e.
9	10	11	12	13	14
3.0	2.1	1.4	0.5	0.47	0.24
4.5	3.3	2.5	1.3	0.56	0.39
5.0	3.4	2.9	1.4	0.58	0.41
–	–	–	–	–	–
–	–	–	–	–	–
–	–	–	–	–	–
9.0	7.5	3.1	1.7	0.34	0.23
11.0	7.6	5.6	2.3	0.51	0.30
9.6	8.0	4.2	2.5	0.44	0.31
8.0	5.8	4.6	2.4	0.58	0.41
10.0	8.0	4.8	2.8	0.48	0.35

carboxyl groups during storage. The rate of hydrolysis will depend on the conditions of storage [81].

The surface charge σ_0 increases with $K_2S_2O_8$ concentration in each of the series. This is predictable from the mechanism of emulsion polymerization. Buffering the system during polymerization leads to higher values of σ_0 (compare the N series with the C or P series). As mentioned before, this is attributed to suppression of the Kolthoff reaction. The highest surface charge attained after ion exchange is 8.7 $\mu C/cm^2$. In the absence of specific adsorption it would lead to

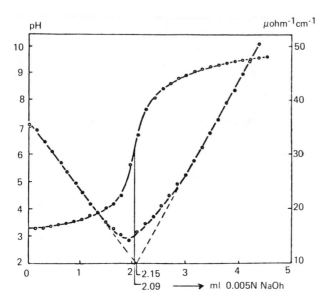

Figure 8.14 Conductometric and potentiometric titration of latex C-3 after ion exchange.

diffuse double-layer potentials of 190 mV and 80 mV in 10^{-3} M or 10^{-1} M (1-1) electrolytes, respectively. However, due to specific adsorption in the Stern layer, lower values of the ζ potential are found from electrokinetic study.

The last columns of Table 8.4 contain data on the sulfur content S of the latices. The absolute sulfur content is expressed as weight fraction of the polymer and has been determined by x-ray fluorescence. Comparison between S before and after ion exchange (columns 9 and 10) indicates the loss of up to 30% of the sulfur due to ion exchange. This is perhaps due to desorption of oligomers. Columns 11 and 12 give the amount of sulfur calculated from σ_0. The amount of sulfur on the surface before ion exchange has been found by using the assumption that the weight loss due to ion exchange is entirely due to a loss of surface sulfur;

column 11 = (12) + (9) − (10)

In turn, the surface charge before ion exchange (column 7) follows directly from the amount of adsorbed sulfur [11]. Obviously, the value of σ_0 before ion exchange is considerably higher than that after ion exchange.

The last two columns give the percent of sulfur that can be titrated (i.e., the surface sulfur). Considering the low ratio of surface volume to bulk volume, the figures are very high after ion exchange.

They indicate that during polymerization, the hydrophilic sulfate groups remain predominantly at the outside of the growing latex particle.

8.3.3 Charge Distribution in the Electrical Double Layer on Polystyrene Latices

It has been established [82] that in the neighborhood of a charged particle, there is an ionic atmosphere which is composed of two kinds of layers, the Stern layer and the Gouy diffuse layer, which are distinguishable by the distribution of ions in each. Few experimental studies of the composition and properties of these layers have been undertaken despite the fact that they are essential for understanding colloid stability, electrokinetic phenomena, and so on.

The surface charge of emulsifier-free polystyrene latices comes from the dissociation of surface sulfonate groups originating from the potassium persulfate used as the initiator. Since the conductivity of a latex suspension is determined by the number and distribution of counter ions in the double layer, the conductometric titration curves obtained using certain bases provide direct information on the structure of the double layer. Figure 8.15 shows examples of the titration curves. The titration was conducted as follows. About 30 ml of the latex, after ion exchange, was titrated under an argon atmosphere with two basic solutions, NaOH and Ba(OH)$_2$, at 25°C. In these titrations it is necessary to adjust the concentration of solid latex to 3-4% by weight, because the experimental error at less than 3% is large, while the titration patterns are distorted at higher than 4% due to the surface conductivity at the latex-solution interface.

In the case of titration with NaOH, both the descending and ascending branches of the conductometric titration curve are linear, and the shape of the curve is typical of the titration of a strong acid with a strong base. This is the result of an equal preference of the strong-acid surface groups for Na$^+$ and H$^+$ ions. However, the negative slope of the descending branch is much smaller than that for a strong acid in solution and it depends on the surface charge of the latex. This is due to the limited mobility of the counterions in the double layer. Usually, the conductivity of a system depends on the product of the number of movable ions and their mobilities. Therefore, the descending slope in the present system is influenced by the following three factors: (a) the concentration of latex solid in the sample, (b) the amount of surface charge, and (c) the degree of binding of the counterions to the surface, which in turn depends on the structure of the electrical double layer. If factors (a) and (b) would be defined from the experimental conditions and be controlled for the apparent titration slope, the residual slope x' gives the part coming from factor (c), which provides useful information on the structure of the electrical double layer. Figure 8.16 shows the plot of slopes obtained by the foregoing process versus surface charge density σ_0, which is de-

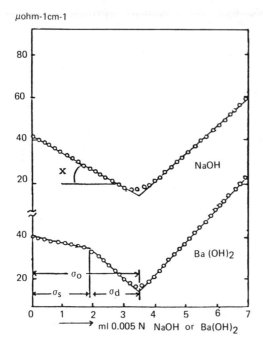

Figure 8.15 Conductometric titration curves using NaOH and Ba(OH)$_2$. σ_0, Surface charge; σ_s, Stern layer charge; σ_d, diffuse-layer charge; X, slope of the descending branch of the titration curve.

termined from the equivalent point of the same titration curve. As shown in the figure, the slope x' decreases rapidly with an increase in σ_0. This indicates that the higher the surface charge, the more tightly the counter ions are bound to the surface.

The lower conductometric titration curve in Fig. 8.15, in which an inflection point is found on the descending branch, shows the results obtained using Ba(OH)$_2$. According to Van den Hul and Vanderhoff [80], the occurrence of the inflection point is associated with the fixation of the counterions in the Stern layer. First, the counterions, which are attached to the surface sulfonate ions, are exchanged preferentially for Ba^{2+}, and then the dissociated ions in the diffuse double layer are exchanged. The counterions in the diffuse double layer contribute substantially to the conductivity through their high mobilities. According to this concept, the number of counterions located in the Stern layer and in the diffuse layer can be determined from the position of the inflection point in the descending branch. Figure 8.17 shows the results obtained from analysis of the inflection data [81]. It gives the charge distribution in the Stern layer σ_s and in the diffuse

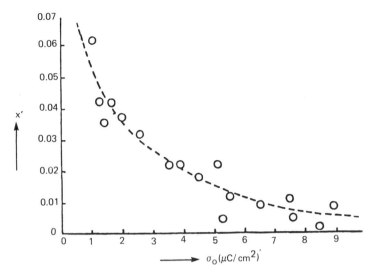

Figure 8.16 Relation between the corrected slope of the titration curve x' and the surface charge σ_0 of latices. (From Ref. 71.)

layer σ_d as a function of the surface charge σ_0. It is obvious that, at high σ_0 values, σ_s increases at about the same rate as σ_0, leading to a virtually constant value of σ_d. This is identical to the result obtained using NaOH.

A similar result has been reported by Norde [83] for polystyrene latices prepared using the same emulsifier-free system. His result is also plotted in Fig. 8.17. He calculated σ_d from the ζ potential for latices of various surface charges using the Gouy-Chapman equation. The ζ potential was determined by the method of Wiersema, Loeb, and Overbeek [84] from electrophoretic mobilities of the latices in 0.01 M KNO_3 solution. As can be seen in the figure, his data are located on the same line as that obtained from analysis of the inflection points. This suggests that the region divided by the inflection point of the conductometric titration curve coincides with the slipping plane of electrophoresis in 0.01 M KNO_3.

8.3.4 Coagulation into Secondary Minimum

One of the most remarkable properties of the colloid theory proposed by Derjaguin-Landau-Verwey-Overbeek [85] (DLVO theory) is that the interaction potential between two colloid particles must have a shallow "secondary minimum" which is located outside the maximum acting as a potential barrier for usual coagulation. Many experiments have been performed to prove its existence: for example, studies on

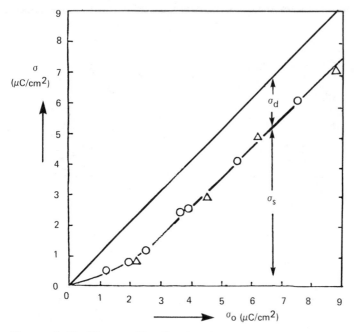

Figure 8.17 Charge distribution in the Stern layer σ_s and the diffuse layer σ_d as a function of the surface charge σ_0. △: data by Norde. (From Ref. 71.)

tactoid formation, thixotropic phenomena [86], adherence of glass particles to the surface of glass plate [87], and reversible agglomeration of plate-shaped tungstic acid particles [88]. The experimental evidence, however, does not come from spherical colloid particles but from anisometric particles such as rod-shaped and plate-shaped particles. This would probably be due to experimental difficulties in the study of large spherical colloids, where the particles sediment quickly because of their weight.

Schenkel and Kitchener [89] exhibited experimentally the existence of a "secondary minimum" using spherical particles of cross-linked polystyrene with diameters of 10 μm. This article describes how variation in the particle size of emulsifier-free polystyrene latices influence the secondary minimum. Polystyrene latices have a low density (1.05 g/cm^3) compared to the usual inorganic colloids and the gravitational effect resulting from their weight is relatively small.

The emulsifier-free polystyrene latices used were prepared by the Kotera-Furusawa method [76] and their properties were characterized in detail [90]. The variation of critical flocculation concentration (CFC) with latex particle size was examined to demonstrate the existence of a secondary minimum. For the case of coagulation into a secondary minimum, a lower CFC would be expected for larger particles under otherwise equal conditions. It would be quite inconsistent with the prediction followed in the coagulation into primary deep minimum.

The CFC of each sample was determined by the static method. The transmission coefficient for light of wavelength 500 nm was measured for latex suspension of various salt concentrations after 20 hr and the results were plotted against the salt concentration. The CFC was determined as an inflection point on the graph. An example of the experimental results is shown in Fig. 8.18.

It is interesting to notice that the slope of the steep portion of each curve depends on the particle size of the sample. The CFC values determined from the curve are listed in Table 8.5 together with the ζ potentials measured at the CFC of each salt. As can be seen from the table, the value of CFC increases with particle size and reaches a maximum for SL-107 ($2a = 758$ nm) and then decreases gradually. This tendency cannot be explained by a simple theoretical analysis of the stability of the suspension, which shows that the larger the particle, the higher the CFC value. This decrease is probably due to the existence of a secondary minimum in the interaction potential energy curve. It is believed that the origin of coagulation would shift from the usual primary minimum to a shallow secondary minimum at a particle diameter of about 700-800 nm. The latter effect causes a decrease in CFC values with particle size.

The experimental data obtained were analyzed using DLVO theory, in which the interaction potential energy V between two interacting spheres is expressed as the superposition of two parts, V_a and V_r. V_a is the van der Waals attraction and V_r is the double-layer repulsion potential energy. They are expressed as follows:

$$V = V_a + V_r$$
$$= -\frac{Aa}{12H} + 3.469 \times 10^{19} \varepsilon (kT)^2 \frac{a\gamma^2}{z^2} \exp(-\kappa aH) \quad (8.13)$$

where

$$\kappa \left(\frac{8\pi nz^2 e^2}{\varepsilon kT} \right)^{1/2} \quad (8.14)$$

$$\gamma = \frac{\exp(ze\psi_0/2kT) - 1}{\exp(ze\psi_0/2kT) + 1} \quad (8.15)$$

In the equations above,

A = Hamaker constant
a = radius of the spherical particle
H = shortest distance between two particles

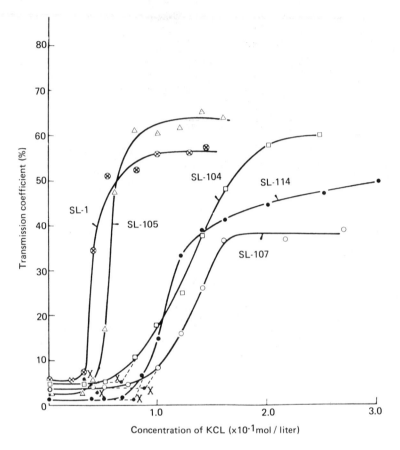

Figure 8.18 Transmission coefficient versus concentration of KCl plots for the various latices. X marked on each curve indicates the critical flocculation concentration of the salt. (See Table 8.5 for the particle size.) (From Ref. 90.)

ε = dielectric constant of the medium (78.5 for water at 20°C)
k = Boltzmann constant
T = absolute temperature of the suspension
z = valence of the ion
n = ion density
e = ionic charge
ψ_0 = surface potential

Equation (8.13) still has an unknown constant, A. The value of CFC (mol/liter) is substituted for n and the ζ potential is measured at the CFC for ψ_0. To determine the constant A, the following graphical

Table 8.5 Critical Flocculation Concentration of Various Salts for the Different Polystyrene Latices and the ζ Potentials at These Conditions

Sample	Particle diameter (mμ)	KCl		BaCl$_2$		LaCl$_3$	
		Flocc. conc. (M)	ζ poten. (mV)	Flocc. conc. (M)	ζ poten. (mV)	Flocc. conc. (M)	ζ poten. (mV)
SL-1	350	3.4×10^{-2}	−23	3.6×10^{-3}	−23	9.2×10^{-4}	−23
SL-105	416	4.8×10^{-2}	−25	4.6×10^{-3}	−25	13.6×10^{-4}	−25
SL-107	758	8.8×10^{-2}	−27	8.2×10^{-3}	−27	9.0×10^{-4}	−24
SL-114	1078	8.0×10^{-2}	−28	7.4×10^{-3}	−25	5.2×10^{-4}	−23
SL-104	1374	6.8×10^{-2}	−29	6.5×10^{-3}	−27	3.0×10^{-4}	−21

Source: Ref. 90.

method was employed. Assuming numerical values for A, the potential curves were drawn at the CFC of the small latex SL-2 (2a = 378 nm) and they were examined to see which curve was adequate for the critical condition of the flocculating colloidal system. This was actually done by judging whether the maximum of each potential energy curve was large or small compared to kT. The results are shown in Fig. 8.19. The curves in the figure were drawn for KCl electrolyte taking 3.5×10^{-2} mol/liter for the CFC, -25 mV for the surface potential ψ_0, and various A values. Curves 1-4 (1: $A = 0.7 \times 10^{-13}$ erg; 2: $A = 1.0 \times 10^{-13}$ erg; 3: $A = 1.3 \times 10^{-13}$ erg; 4: $A = 1.8 \times 10^{-13}$ erg) represent the potential maxima, having heights of 39.0, 27.5, 16.5, and 2.0 kT, respectively. Assuming that the potential maximum for the critical condition is probably about 15-20 kT [75], curve 3 with the value 1.3×10^{-13} erg for A is chosen as the acceptable one. This A value is close to the ones proposed by Ottewill and Shaw [91], 1.1-0.1×10^{-13} erg, and Watillon and Joseph-Petit [92], 2.2-0.4×10^{-13} erg.

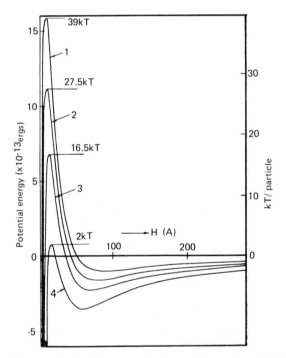

Figure 8.19 Curves of interaction potential between two particles for monovalent ion. c = 0.035 mol/liter, $\psi_0 = -25$ mV, 2a = 378 nm. 1: $A = 0.7 \times 10^{-13}$ erg; 2: $A = 1.0 \times 10^{-13}$ erg; 3: $A = 1.3 \times 10^{-13}$ erg; 4: $A = 1.8 \times 10^{-13}$ erg.

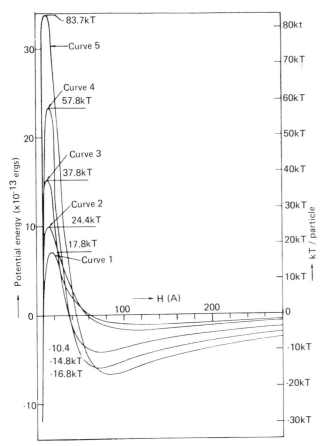

Figure 8.20 Curves of interaction potential between two particles of the various size latices for divalent ion. Curve 1: $2a = 350$ nm, $c = 3.6 \times 10^{-3}$ mol/liter, $\psi_0 = -23$ mV, $A = 1.3 \times 10^{-13}$ erg; Curve 2: $2a = 416$ nm, $c = 4.6 \times 10^{-3}$ mol/liter, $\psi_0 = -25$ mV, $A = 1.3 \times 10^{-13}$ erg; Curve 3: $2a = 758$ nm, $c = 8.2 \times 10^{-3}$ mol/liter, $\psi_0 = -27$ mV, $A = 1.3 \times 10^{-3}$ erg; Curve 4: $2a = 1078$ nm, $c = 7.4 \times 10^{-3}$ mol/liter, $\psi_0 = -25$ mV, $A = 1.3 \times 10^{-3}$ erg; Curve 5: $2a = 1374$ nm, $c = 6.5 \times 10^{-3}$ mol/liter, $\psi_0 = -27$ mV, $A = 1.3 \times 10^{-3}$ erg.

It is interesting to draw the potential energy curves for the various CFC values of each sample using this A value. An example of the results is illustrated in Fig. 8.20 for divalent ions. Curves 1 and 2 of Fig. 8.20 are the net potential energy curves corresponding to SL-1 ($2a = 350$ nm, $c = 3.6 \times 10^{-3}$ mol/liter, and $\psi_0 = -23$ mV) and SL-105 ($2a = 416$ nm, $c = 4.6 \times 10^{-3}$ mol/liter, and $\psi_0 = -25$ mV) at the CFC value of $BaCl_2$. It is seen that these two curves exhibit the usual primary minimum of the interaction potential. Curves 3, 4, and 5 correspond to SL-107 (2a = 758 nm, c = 8.2×10^{-3} mol/liter, and ψ_0 = −27 mV), SL-114 (2a = 1078 nm, c = 7.4×10^{-3} mol/liter, and ψ_0 = −25 mV), and SL-104 (2a = 1374 nm, c = 6.5×10^{-3} mol/liter, and ψ_0 = −27 mV). It can be seen that these curves still have a large potential energy barrier (40-80 kT) at the CFC and do not explain coagulation. For the suspension of large particles, the existence of a potential minimum is evident (e.g. curve 5 of Fig. 8.20, V_{min} = −17 kT at 90 A). The depth of this secondary minimum on the potential energy curve would be sufficient to cause the coagulation of particles. From this point of view, it is concluded that the CFC for large colloids is expected to have a lower limit of salt concentration due to trapping of particles in the secondary minimum.

It has been shown that the potential energy curves for monovalent and trivalent ions [90] are similar to the curves for the divalent ion discussed above. These findings support the interpretation based on a secondary minimum.

8.3.5 Summary

Polystyrene latices were prepared in the absence of emulsifiers with a wide range of surface charge by controlling the concentration of potassium persulfate as the initiator and by performing the polymerization in $KHCO_3$ or K_2HPO_4 buffer. Conductometric titration of these latices with NaOH and $Ba(OH)_2$ showed that the charge distribution in the Stern layer increases linearly with an increase in the surface charge of the latices, σ_0. The charge distribution in the diffuse layer becomes relatively constant as the surface charge of the latices increases.

The critical flocculation concentration (CFC) and ζ potential at the CFC were determined using a series of "emulsifier-free" polystyrene latices from 350 to 1400 nm in diameter. The CFC values increased with particle size, reaching a maximum value with particles of 750 nm diameter and then decreased gradually in line with the diameter. This result could not be explained by simple theoretical consideration of the stability of suspensions. A satisfactory explanation was given only after using the secondary minimum term in the interaction potential energy curve. This means that the origin of coagulation shifted from the usual primary minimum to the shallow secondary minimum at a particle diameter of about 750 nm.

REFERENCES

1. E. J. W. Verwey and J. Th. G. Overbeek, *Theory of the Stability of Lyophobic Colloids*, Elsevier, New York, 1948.

2. B. V. Derjaguin and L. D. Landau, *Acta Phys. Chim.*, *14*, 633 (1941).
3. B. V. Derjaguin, *Discuss. Faraday Soc.*, *18*, 85 (1954).
4. O. F. Devereux and P. L. de Bruyn, *Interaction of Plane-Parallel Double Layers*, MIT Press, Cambridge, Mass., 1963.
5. R. Hogg, T. W. Healy, and D. W. Fuerstenau, *Trans. Faraday Soc.*, *62*, 1638 (1966).
6. G. Frens, Thesis, University of Utrecht, 1968.
7. G. R. Wiese and T. W. Healy, *Trans. Faraday Soc.*, *66*, 490 (1970).
8. J. Gregory, *J. Chem. Soc. Faraday Trans. II*, *69*, 1723 (1973).
9. S. Usui, *J. Colloid Interface Sci.*, *44*, 107 (1973).
10. H. C. Hamaker, *Physica's Grav.*, *4*, 1058 (1937).
11. F. London, *Z. Physik.*, *63*, 245 (1930).
12. E. M. Lifshiz, *Sov. Phys.*, *2*, 73 (1956).
13. I. E. Dzyaloshinskii, E. M. Lifshiz, and L. P. Pitaevskii, *Adv. Phys.*, *10*, 165 (1961).
14. B. W. Ninham and V. A. Parsegian, *Biophys. J.*, *10*, 646 (1970).
15. V. A. Parsegian and B. W. Ninham, *Biophys. J.*, *10*, 664 (1970).
16. J. N. Israelachvili, *Proc. R. Soc. Lond. A*, *331*, 39 (1972).
17. J. N. Israelachvili, *Q. Rev. Biophys.*, *6*, 341 (1973).
18. B. V. Derjaguin and A. S. Titijevskaya, *Proc. 2nd Int. Congr. Surf. Activ.*, *I*, 211 (1957).
19. A. Scheludko and D. Exerowa, *Kolloid-Z.*, *165*, 148 (1959).
20. A. Scheludko and D. Exerowa, *Kolloid-Z.*, *168*, 24 (1960).
21. A. Scheludko, *Adv. Colloid Interface Sci.*, *1*, 391 (1967).
22. J. Lyklema and K. J. Mysels, *J. Am. Chem. Soc.*, *87*, 2539 (1965).
23. A. D. Roberts and D. Tabor, *Special Discuss. Faraday Soc.*, *1*, 243 (1970).
24. A. D. Roberts, *J. Colloid Interface Sci.*, *41*, 23 (1972).
25. J. N. Israelachvili and G. E. Adams, *J. Chem. Soc. Faraday Trans. I*, *4*, 975 (1978).
26. D. Tabor and R. H. S. Winterton, *Proc. R. Soc. Lond. A*, *312*, 435 (1969).
27. J. N. Israelachvili and D. Tabor, *Proc. R. Soc. Lond. A*, *331*, 19 (1972).
28. B. V. Derjaguin, Z. M. Zorin, N. V. Churaev, and V. A. Shishin, *Wetting, Spreading and Adhesion* (J. F. Paddy, ed.), Academic Press, London, 1978, p. 201.
29. R. H. Ottewill and T. Walker, *Kolloid-Z.*, *227*, 108 (1968).
30. F. W. Cain, R. H. Ottewill, and J. B. Smitham, *Faraday Discuss. Chem. Soc.*, *65*, 33 (1978).
31. F. Th. Hesselink, A. Vrij, and J. Th. G. Overbeek, *J. Phys. Chem.*, *75*, 2094 (1971).

32. J. Lyklema and T. van Vliet, *Faraday Discuss, Chem. Soc.*, 65, 25 (1978).
33. J. N. Israelachvili, R. K. Tandon, and L. R. White, *J. Colloid Interface Sci.*, 78, 430 (1980).
34. J. Klein, *Nature*, 288, 248 (1980).
35. A. Scheludko, *Proc. K. Ned. Akad. Wet.*, Ser. B, 65, 87 (1962).
36. A. Scheludko and E. Manev, *Trans. Faraday Soc.*, 64, 1123 (1968).
37. A. Vrij, *Discuss. Faraday Soc.*, 42, 23 (1966).
38. A. D. Roberts, *J. Phys. D: Appl. Phys.*, 4, 423 (1971).
39. A. Watanabe and R. Gotoh, *Kolloid-Z.*, 191, 36 (1963).
40. M. Matsumoto, *J. Chem. Soc. Jpn.*, 91, 708 (1970); *Bull. Inst. Chem. Res., Kyoto Univ.*, 48, 171 (1970).
41. S. Usui, T. Yamasaki, and J. Shimoiizaka, *J. Phys. Chem.*, 71, 3195 (1967).
42. S. Usui and T. Yamasaki, *J. Colloid Interface Sci.*, 29, 629 (1969).
43. A. Watanabe, M. Matsumoto, and R. Gotoh, *Kolloid-Z.*, 201, 147 (1965).
44. H. Sonntag, *Z. Phys. Chem.*, 221, 365, 373 (1962).
45. M. Matsumoto, *Bull. Inst. Chem. Res., Kyoto Univ.*, 47, 361 (1969).
46. J. Guastalla, *Proc. 2nd Int. Conf. Surf. Activ.*, 3, 112 (1957).
47. A. Watanabe, M. Matsumoto, H. Tamai, and R. Gotoh, *Kolloid-Z.*, 220, 152 (1967).
48. A. Watanabe, M. Matsumoto, H. Tamai, and R. Gotoh, *Kolloid-Z.*, 221, 47 (1967).
49. A. Watanabe, M. Matsumoto, H. Tamai, and R. Gotoh, *Kolloid-Z.*, 228, 59 (1968).
50. R. R. Lessard and S. A. Zieminski, *Ind. Eng. Chem. Fundam.*, 10, 260 (1971).
51. S. Usui, S. Sasaki, and F. Hasegawa, *Jpn. Discuss. Colloid Interface Chem.*, 33, 162 (1980).
52. M. Matsumoto, A. G. Gaonkar, and T. Takenaka, *Jpn. Discuss. Colloid Interface Chem.*, 34, 314 (1981).
53. M. A. V. Devanathan and B. V. K. S. R. A. Tilak, *Chem. Rev.*, 65, 635 (1965).
54. R. M. Pashley, *J. Colloid Interface Sci.*, 80, 153 (1980); 83, 531 (1980).
55. J. E. Lane and T. H. Spurling, *Chem. Phys. Lett.*, 67, 107 (1979).
56. W. van Megen and I. K. Snook, *J. Chem. Soc. Faraday Trans. II*, 77, 181 (1981).
57. D. J. Mitchell, B. W. Ninham, and B. A. Pailthorpe, *J. Chem. Soc. Faraday Trans. II*, 74, 1098, 1116 (1978).

58. R. G. Horn and J. N. Israelachvili, *Chem. Phys. Lett.*, 71, 192 (1980).
59. H. Reerink and J. Th. G. Overbeek, *Discuss. Faraday Soc.*, 18, 74 (1954).
60. M. Smoluchowski, *Phys. Z.*, 17, 557, 586 (1916).
61. N. Fuchs, *Z. Phys.*, 89, 736 (1934).
62. S. A. Troelstra and H. R. Kruyt, *Kolloid-Z.*, 101, 182 (1946); H. R. Kruyt and S. A. Troelstra, *Kolloidchem. Beih.*, 54, 225 (1943).
63. G. Oster, *J. Colloid Sci.*, 2, 290 (1947).
64. G. J. Fleer and J. Lyklema, *J. Colloid Interface Sci.*, 55, 228 (1976).
65. J. Th. G. Overbeek, in *Colloid Science* (H. R. Kruyt, ed.), Vol. 1, Elsevier, Amsterdam, 1952, p. 307.
66. A. Watanabe, *Bull. Inst. Chem. Res. (Kyoto Univ.)*, 38, 179 (1960).
67. A. Watanabe, *Bull. Inst. Chem. Res. (Kyoto Univ.)*, 38, 158 (1960).
68. K. Furusawa, *Bull. Chem. Soc. Jpn.*, 55, 48 (1982).
69. K. Furusawa and S. Hachisu, *Sci. Light*, 16, 91 (1967).
70. K. Furusawa and S. Hachisu, *Sci. Light*, 12, 157 (1963).
71. K. Furusawa, Y. Tezuka, and N. Watanabe, *J. Colloid Interface Sci.*, 73, 21 (1980).
72. G. J. Fleer and J. Lyklema, *J. Colloid Interface Sci.*, 46, 1 (1974).
73. F. Th. Hesselink, A. Vrij, and J. Th. G. Overbeek, *J. Phys. Chem.*, 75, 2094 (1971).
74. W. Norde and J. Lyklema, *J. Colloid Interface Sci.*, 66, 257, 285 (1978).
75. H. J. Van den Hul and J. W. Vanderhoff, *J. Electroanal. Chem.*, 37, 161 (1972).
76. A. Kotera, K. Furusawa, and Y. Takeda, *Kolloid Z. Z. Polym.*, 239, 677 (1970).
77. J. Stone-Masui and A. Watillon, *J. Colloid Interface Sci.*, 52, 479 (1975).
78. K. Furusawa, W. Norde, and J. Lyklema, *Kolloid Z. Z. Polym.*, 250, 908 (1972).
79. I. M. Kolthoff and I. K. Miller, *J. Am. Chem. Soc.*, 73, 3055 (1956).
80. H. J. Van den Hul and J. W. Vanderhoff, *J. Colloid Interface Sci.*, 28, 336 (1968); J. W. Vanderhoff, H. J. Van den Hul, R. J. M. Tausk, and J. Th. G. Overbeek, in *Clean Surfaces: Their Preparation and Characterization for Interfacial Studies* (G. Goldfinger, ed.), Marcel Dekker, New York, 1970, p. 155.
81. K. Furusawa, *Bull. Chem. Soc. Jpn.*, 55, 48 (1982).
82. O. Stern, *Z. Elektrochem.*, 30, 508 (1924).

83. W. Norde, Thesis, Agricultural University, Wageningen, The Netherlands, 1976.
84. P. H. Wiersema, A. L. Loeb, and J. Th. G. Overbeek, *J. Colloid Interface Sci.*, 22, 78 (1966).
85. E. J. W. Verwey and J. Th. G. Overbeek, *Theory of the Stability of Lyophobic Colloids*, Elsevier, Amsterday, 1948.
86. H. R. Kruyt, ed., *Colloid Science*, Vol. 1, Elsevier, Amsterdam, 1958, p. 324.
87. A. von Buzagh, *Kolloid-Z.*, 47, 370 (1929); 51, 105, 230 (1930).
88. K. Furusawa and S. Hachisu, *Nippon Kagaku Zasshi*, 87, 695 (1966); 90. 947 (1969); *Sci. Light*, 16, 91 (1967).
89. J. H. Schenkel and J. A. Kitchener, *Trans. Faraday Soc.*, 56, 161 (1960).
90. A. Kotera, K. Furusawa, and K. Kudo, *Kolloid Z. Z. Polym.*, 240, 837 (1970).
91. R. H. Ottewill and J. N. Shaw, *Discuss. Faraday Soc.*, 42, 154 (1966).
92. A. Watillon and A. M. Joseph-Petit, *Discuss. Faraday Soc.*, 42, 143 (1966).

PART III: APPLICATIONS

9
Detergency

Fumikatsu Tokiwa   Kao Corporation, Tokyo, Japan
Tetsuya Imamura   Kao Corporation, Tochigi, Japan

9.1. Dirt and Detergency   269
9.2. Interfacial Electrical Phenomena in Removal of Soil   270
    9.2.1. Removal of soil   270
    9.2.2. Potential energy curves   272
    9.2.3. Factors governing removal of soil   276
9.3. Soil Removal and Redeposition   277
    9.3.1. Shape models of fiber and soil particle   277
    9.3.2. Interfacial electrical effect   278
    9.3.3. Stability constant for heterocoagulation   281

9.1 DIRT AND DETERGENCY

The term "detergency" often means the ability to remove dirt adhering to fiber in a solution containing a detergent. Dirt is generally composed of oily dirt, such as squalene, fatty acids, and glycerides, and solid dirt, such as soot, mud, and metal oxides [1]. The latter type of dirt is generally called "soil." The mechanism for the removal of dirt from fiber is very different for solid dirt and for oily dirt. Oily dirt, which generally spreads over an entire fiber surface, is usually first deformed to make droplets when the penetration of a surfactant occurs before the dirt is removed from the fiber. Thus oily dirt can be dispersed as small droplets (i.e., an emulsion) and can also be solubilized in surfactant micelles by "solubilization." On the other hand, solid soil mainly adheres to fiber primarily by means of the van der Waals forces; in this case the charge effect due to the adsorption of surfactant and builder ions onto the soil and the fiber plays an important role in the removal of the soil.

In a detersive system, fiber and soil have their own charges with the adsorption of surfactant and/or builder ions and are surrounded by electrical double layers. The soil that is removed from the fiber by certain forces may be dispersed in the detergent solution as charged colloidal particles. Furthermore, since the short axis of the fiber has a size of nearly colloidal dimensions, the fiber can be regarded as being composed of colloidal particles with an infinite length. Therefore, the detersive system can be regarded as a colloid-dispersion system. As has been indicated by Tachibana [2], the first detersive step is to peptize the coagulation by which the soil adheres to the fiber. Then it is necessary to disperse the soil well in a detergent solution and also to prevent the redeposition of the soil onto the fiber.

The previous approaches concerned with the relation between soil removal and interfacial electrical phenomena were based on the Dergaguin-Landau-Verwey-Overbeek (DLVO) theory [3] of the stability of hydrophobic colloids. However, as has been indicated by Kitahara [4] and Tokiwa [5], it would be better to discuss this problem from the viewpoint of the heterocoagulation theory. In this chapter the characteristics of the heterocoagulation theory and the results of the relation between detergency and interfacial electrical phenomena and, on the other hand, detergency and the heterocoagulation theory are discussed in comparison with the results of these approaches.

9.2 INTERFACIAL ELECTRICAL PHENOMENA IN REMOVAL OF SOIL

9.2.1 Removal of Soil

The electron microscopic investigation of the adhesion of solid soil on fiber shows that almost all of the soil is deposited on the surface of the fiber, while no considerable amount of soil is occluded in the interior of the fiber [6]. The adhesion of soil is caused mainly by the dispersion forces or the van der Waals forces. In addition to these forces, which extend over relatively large distances, there are specific bonds [7] between the atoms or ions located on two adhering surfaces, such as calcium bridging or hydrogen bonding. These are effective over a few angstrom units and contribute significantly to the adhesion only if the soil and the fiber are in very intimate contact and if the soil particles are small enough. The forces of these specific bonds are at least one order of magnitude larger than the van der Waals forces [7]. More detailed information about the strength of adhesion has recently been obtained by direct measurements. (As an example of the van der Waals forces working alone, an adhesion force of 2.3×10^{-5} dyn has been reported using the hydrodynamic method [8] for carbon black particles with a radius of 0.105 μm on a cellophane substrate.)

The adhesion forces of soil on substrates are greatly diminished after immersion in water, decreasing to about one-fourth of their original strength [9]. This change can be explained by the new Hamaker constant, A_{123}, which is given by an equation derived by Hamaker [10]:

$$A_{123} = A_{12} + A_{33} - A_{13} - A_{23} \tag{9.1}$$

where A_{12} is the Hamaker constant for the soil (1) and the fiber (2); A_{33} for water molecules (3); A_{13} for the soil and water, and A_{23} for the fiber and water. Although soil particles generally adhere to the substrate by van der Waals forces, the adhesion forces are also greatly influenced by the outer diffuse parts of the electrical double layers formed at the soil-liquid and fiber-liquid interfaces.

The potential energies of the van der Waals attraction, V_A, the electrical double-layer repulsion, V_E, the Born repulsion force, V_B, and the total energy, V_T, as a function of the distance between fiber and soil lead to curves such as those shown in Fig. 9.1. V_B is evolved only by an overlapping of the electron clouds of soil and fiber [3] when they approach very closely. V_A tends to make a soil particle adhere to fiber, while V_E and V_B tend to keep them away. The removal of soil from fiber is largely conditioned by the energy depth, ΔV_T (= $V_{T,max} - V_{T,min}$), and the energy barrier, $V_{T,max}$, as will be described in detail below. The figure shows schematically that those values are affected by the values of V_E.

From consideration of the V_T versus x curve reported by Lange [11], where x is the distance from the soil particle, the removal of soil adhering to fiber is explained by the step-by-step process shown in Fig. 9.2. In state I, the soil adhering to the fiber is in a detergent solution, but the solution does not yet penetrate between the two sub-

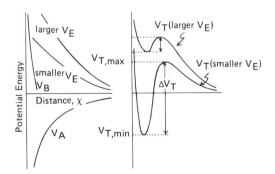

Figure 9.1 Potential energy curves between soil and fiber.

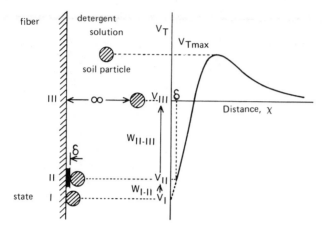

Figure 9.2 Removal of a soil particle from fiber.

stances in the zone of contact. In state II, the detergent solution penetrates the zone by means of the adsorption, penetration, and wetting phenomena of surfactants, separates the two substances to a distance of only δ, and forms an electrical double layer simultaneously by the solvation and adsorption of ions. In state III, the soil overcomes the barrier of $V_{T,max}$ and is separated far from the fiber; thereafter, there is no interaction between them. The soil removal proceeds from state I to state II and then to state III. The work involved in moving from state I to state II, W_{I-II}, is equal to the sum of the van der Waals attractive forces, W, and the adhesion force of the detergent solution to the fiber and the soil, q (i.e., $W_{I-II} = W - q$.). On the other hand, the work involved in moving from state II to state III, W_{II-III}, is determined by the relation between the van der Waals attraction and the repulsive force of the electrical double layers. That is, W_{II-III} is equal to the total energy at the distance δ, $V_{T,\delta}$ (i.e., $W_{II-III} = -V_{T,\delta}$). The total work done for the soil removal process is, therefore, given by the sum of W_{I-II} and W_{II-III}. The main ingredients of the detergent, both surfactants and builders, decrease the work by making the value of $V_{T,\delta}$ smaller. Lange has also studied the contents of q and has indicated that the surface-active properties of anionic and nonionic surfactants are important in reducing the work needed for the soil removal process.

9.2.2 Potential Energy Curves

If it is assumed that the shapes of the soil and fiber are a sphere and a plate, respectively (the model of the shape will be described later in detail), the potential energy of the van der Waals attraction between them, V_A, can be given by the following equation in kT units (all of the following energies are expressed in kT units):

$$V_A = -\frac{A_{113}}{6kT}\left[\frac{2a(x+a)}{x(x+2a)} - \ln\frac{x+2}{x}\right] \quad (9.2)$$

where a is the radius of the sphere, A_{113} is the Hamaker constant between fiber and soil in medium 3 (in the case, fiber and soil are considered to be an identical substance, 1), x is the distance between them, k is the Boltzmann constant, and T is the absolute temperature. The interaction energy between symmetrical electrical double layers of the sphere and the plate, V'_E, is given by Eq. (9.3) (the prime used in V'_E means that it is of a symmetrical type):

$$V'_E = a\frac{D}{kT}\psi_1^2 \ln(1 + \exp(-\kappa x)) \quad (9.3)$$

$$\kappa = \left(\frac{8\pi e^2 N\mu}{1000 DkT}\right)^{1/2} \quad (9.4)$$

where ψ_1 is the surface potential (in this case, the fiber and the soil are considered to be an identical substance 1, i.e., these surface potentials are identical, D is the dielectric constant, κ is the Debye-Hückel parameter, e is the charge per electron, N is Avogadro's number, and μ is the ionic strength. According to Eq. (9.3), the value of V'_E is always positive, and it monotonously decreases with increase in distance. In this respect, V'_E is very different from what is observed with V_E by the use of the heterocoagulation theory. The total energy working between the fiber and the soil, V_T, is given by the sum of V_A and V'_E:

$$V_T = V_A + V'_E \quad (9.5)$$

For calculation of the following V_T values, it is assumed that a soil particle adheres to fiber at the closest distance (i.e., 5A) as a result of the Born repulsive force.

By the DLVO theory, in the approach discussed above the relationship between fiber and soil is treated as a problem of dispersion and coagulation among the same type of colloids. However, in practice, the electrical double layers of fiber and soil are different because of the specific adsorptivities of surfactant or builder ions due to their differences in surface properties. The dispersion and coagulation between different colloid particles have been discussed by Hogg, Healy, and Fuerstenau [12] and Derjaguin [13]; it is called heterocoagulation. The characteristics of the heterocoagulation theory according to Hogg et al. will be discussed next, together with the results of our own experiments.

According to the heterocoagulation theory, the interaction energy between asymmetrical electrical double layers of a sphere and a plate, V_E, is given by the following equation, unlike the V'_E of Eq. (9.3):

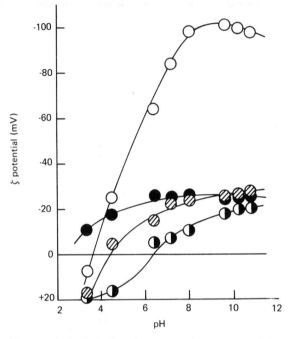

Figure 9.3 Relation between the ζ potentials of cotton (●) polyester (○), nylon (◐) fiber, and Fe_2O_3 particles (◑), and pH: ionic strength, 1×10^{-3}; 25°C. (Radius of Fe_2O_3 about 1 μm.)

$$V_E = \frac{a}{4}\frac{D}{kT}\left[(\psi_1^2 + \psi_2^2)\ln\frac{\exp(2\kappa x) - 1}{\exp(2\kappa x)} + 2\psi_1\psi_2 \ln\frac{\exp(\kappa x) + 1}{\exp(\kappa x) - 1}\right] \quad (9.6)$$

where ψ_1 and ψ_2 are the surface potentials of the fiber and the soil (2), respectively. In Fig. 9.3, the ζ potentials of cotton, polyester, nylon fiber, and ferric oxide particles as a function of pH are shown [14]. The ζ potentials of these substances differ greatly and are dependent on the pH of the solution. Since fibers and soil particles form different electrical double layers, their mutual interactions should be dealt with by means of heterocoagulation theory.

The combination of polyester fiber and Fe_2O_3 particle is chosen from the ζ versus pH curves shown in Fig. 9.3; Fig. 9.4 exemplifies the relation between the interaction energy of their electrical double layers, V_E, and the distance, x, at different pH values according to Eq. (9.6). As the true surface potentials of the fiber and the particle cannot be measured by ordinary methods, the ζ potential is

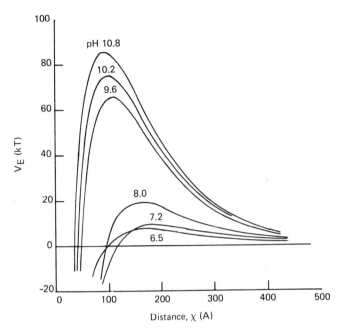

Figure 9.4 V_E versus x curves for asymmetrical electrical double layers between polyester fiber and Fe_2O_3 particles at various pH values.

used instead. The potential curves in the figure show cases when the ζ potentials of the fiber and the particle have the same sign at pH > 6.5. The first term on the right-hand side of Eq. (9.6) is always negative, but the second term is positive. Therefore, the value of V_E obtained from Eq. (9.6) reaches a maximum at a certain distance; it is sometimes negative at a distance smaller than the distance at which the maximum V_E value is observed. In this case, V_E works as an attractive energy [12]. This is characteristic of V_E as viewed in terms of the heterocoagulation theory, and it is clearly different from the interaction energy [Eq. (9.3)] between symmetrical double layers as obtained by DLVO theory.

Although the interaction energy due to the van der Waals forces, V_A, can also be given by Eq. (9.2), in this case the Hamaker constant between fiber 1 and soil 2 in a medium 3, A_{123}, must be used. The Hamaker constant here is given by the geometrical mean of A_{113} and A_{223} [15]:

$$A_{123} = \sqrt{A_{113} A_{223}} \qquad (9.7)$$

9.2.3 Factors Governing Removal of Soil

As shown in Fig. 9.1, the removal of soil is affected by the values of $V_{T,max}$ and ΔV_T (= $V_{T,max} - V_{T,min}$). These values are determined from the total potential energy curve (V_T versus x curve), which is the sum of the V_A versus x curve and the V_E versus x curve. In the curves of V_A versus x obtained from Eq. (9.2) and of V_E versus x from Eq. (9.6), only the V_E versus x curve can be controlled. If the repulsive energy, V_E, can be increased, the value of ΔV_T will become smaller and that of $V_{T,max}$ will become larger, which is preferable to washing; the following three factors are considered to determine the value of V_E.

1. Effect of Soil-Particle Size. As expressed in Eq. (9.6), when the radius of a particle, a, decreases, the value of V_E also decreases. At the same time, this makes the values of $V_{T,max}$ and $V_{T,max}/\Delta V_T$ decrease, which is not preferable for washing. Generally, it is known that smaller particles are more difficult to remove from fiber, especially particles with a radius of less than 0.1 μm [16, 17]. For removing such small particles, the surface potential, ψ, has to be greatly increased; however, in practice it is difficult to achieve such an increment in ψ [16].
2. Effect of Ionic Strength. This effect is included in the Debye-Hückel parameter of Eq. (9.4). In general, if ψ is not affected by ionic strength, the increment in ionic strength makes the value of V_E decrease and the value of V_T decrease. This is not preferable to washing. However, in the case where any ionic surfactants exist in the detergent solution, the adsorption of the surfactants onto soil and/or fiber is greatly affected by the addition of electrolytes, and this changes the value of ψ. In the case of a low surfactant concentration, the addition of an electrolyte promotes the adsorption of the surfactant on the soil and the fiber, and the soil removal is increased due to the increment in ψ. After the adsorption of the surfactant reaches a saturation level, the soil removal is decreased with the electrolyte concentration [18]. This result can be explained in terms of the decrease in $V_{T,max}$ due to the decrease in ψ. On the other hand, in the case of a high surfactant concentration, the addition of an electrolyte decreases $V_{T,max}$ because the adsorption is already at the saturation level, so the soil removal is decreased with the electrolyte concentration.
3. Effect of Surface Potential. In general, it is preferable in detergency to increase V_E by the increment in ψ. Some results concerned with this effect are shown in the next section.

9.3 SOIL REMOVAL AND REDEPOSITION

9.3.1 Shape Models of Fiber and Soil Particle [19]

Before discussing detergency phenomena in terms of the heterocoagulation theory, it is necessary to examine the shape models of the fiber and soil particle. The cylinder model of the fiber and the sphere model of the soil particle may soon be understood. However, the interaction energy between asymmetrical electrical double layers of the sphere-cylinder model cannot be solved analytically. Therefore, it is assumed that the cylinder shares the effects of the sphere and the plate equally, and the following equation is obtained [19]:

$$V_E = \frac{a_1 a_2}{2(a_1 + 2a_2)} \left[(\psi_1^2 + \psi_2^2) \ln \frac{\exp(2\kappa x) - 1}{\exp(2\kappa x)} + 2\psi_1 \psi_2 \ln \frac{\exp(\kappa x) + 1}{\exp(\kappa x) - 1} \right]$$

(9.8)

where a_1 and a_2 are the radii of the sphere and the cylinder, respectively. When Eq. (9.6) of the V_E of the sphere-plate model is compared with E. (9.8), the terms concerned with the potential and distance are identical, and only the term concerned with the shape is different. On the other hand, in the case of the sphere-sphere model with radii of a_1 and a_2, the equation of V_E has the shape term of $a_1 a_2/4(a_1 + a_2)$; this term is the only difference between Eqs. (9.6) and (9.8); the other parts are identical.

Table 9.1 Effect of Models on the Value of the Shape Terms[a]

Radius of sphere or cylinder[b] (μm)		Shape term		
		Sphere 1 -plate, $\frac{a_1}{4}$	Sphere 1 -cylinder, $\frac{a_1 a_2}{2(a_1 + 2a_2)}$	Sphere 1 -sphere 2, $\frac{a_1 a_2}{4(a_1 + a_2)}$
a_1	a_2			
0.1	10	1	0.995	0.989
1	10	10	9.52	9.10
10	10	100	66.8	50.0

[a]The value for a plate–a sphere with 0.1 μm radius is taken as 1.
[b]a_1 is the radius of sphere 1 and a_2 is that of a cylinder or sphere 2.
Source: Ref. 19

The interaction energies of V_E calculated by the sphere-plate, sphere-cylinder, and sphere-sphere models are different regarding the term concerned with the shape effect, shown in Table 9.1 for comparison. Since the diameters of soil particles in actual washing systems are 0.1-1.0 µm, and since the fibers have a diameter of about 10 µm [17,20], the difference among the shape terms of the three models is not so considerable. Therefore, the sphere-plate model is used in the following discussion.

9.3.2 Interfacial Electrical Effect

The relation between the redeposition of removed soil particles onto fiber and the interfacial electrical effect will be considered first. Using Eq. (9.6), based on the heterocoagulation theory, and Eq. (9.5) (in this case, V_E is used instead of V_E'), the maximum values $V_{T,max}$ of the total energy between the fiber and Fe_2O_3 particle, which are calculated from the ζ potentials shown in Fig. 9.3, are plotted against the pH, as shown in Fig. 9.5. The amount of redeposition of Fe_2O_3, which was measured experimentally, is also shown in the figure. The amount of redeposition corresponds well to $V_{T,max}$; that is, it is more difficult for the soil to redeposit onto fiber because of the larger values of $V_{T,max}$ in the alkaline region. In a washing system [21,22], a correlation similar to Fig. 9.5 is observed between $V_{T,max}$ and the

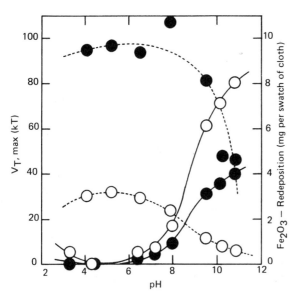

Figure 9.5 Effects of pH on $V_{T,max}$ between fiber and Fe_2O_3 (solid line) and on the amount of Fe_2O_3 redeposition on fiber (dashed line): ●, on cotton; ○, on polyester.

Table 9.2 Values of $V_{T,max}$ and ΔV_T Between Soiled Cloth and Carbon Particles in Detergent Solutions[a]

Detergent solution	Builder	Carbon black[b]/soiled cloth[c]	
		$V_{T,max}$ (kT)	ΔV_T (kT)
1	NaCl	28	58
		(0)	(111)
2	$NaBO_2$	56	43
3	Na_2SO_4	32	53
4	Na_2CO_3	61	47
5	Na_3PO_4	134	40
6	NTA·3Na	174	28
7	Na-citrate	52	49
		(0)	(89)
8	$Na_4P_2O_7$	210	26
9	$Na_5P_3O_{10}$	236	11
		(10)	(59)

[a]Detergent solutions are composed of 0.04% sodium dodecyl sulfate and 0.08% builder. Water of 4°DH was used, and the ionic strength was adjusted to 3.62×10^{-2} with sodium sulfate.
[b]The carbon particles removed from the soiled cloth were suspended in the detergent solution.
[c]Values in parentheses are for carbon particles obtained from "de-oiled" soiled cloth.

amount of soil redeposition and there is no redeposition if $V_{T,max}$ is in the range 80-100 kT.

When soil particles are removed from fiber in a washing system, the effect of the energy depth, ΔV_T, is important as well as the $V_{T,max}$, as shown in Fig. 9.1. Using an artificially soiled cloth made with oily dirt and carbon black, the ζ potentials of the cloth and the soil particles in the detergent solution were measured under conditions similar to those in the actual washing [23]. Table 9.2 summarizes the values of $V_{T,max}$ and ΔV_T between the cloth and the soil particle which were calculated by the heterocoagulation theory with the ζ potentials of both substances. This table shows the effect of builders in the detergent. Generally speaking, with polyanion electrolytes such as sodium tripolyphosphate ($Na_5P_3O_{10}$), ΔV_T is relatively small and $V_{T,max}$ is large. On the other hand, with sodium chloride and sodium sulfate, ΔV_T is larger than $V_{T,max}$. These results are obtained when the artificially soiled cloth made with both oily dirt and carbon black particles is washed. In the case of the soiled cloth made only with carbon black, there is a tendency for ΔV_T to increase and for $V_{T,max}$ to decrease compared to the oily soiled cloth. In the presence of oily substances, the ζ potentials of the cloth and the soil particle are increased, because their surfaces become lipophilic [23]. Consequently, the interfacial electrical effect on V_T becomes larger, as shown in Fig. 9.1.

Thus $V_{T,max}$ and ΔV_T can be controlled to promote detergency, soil removal and soil anti-redeposition by choosing appropriate detergent ingredients, especially builders. For detergency, larger values of $V_{T,max}$ and $V_{T,max}/\Delta V_T$ are preferable. In Fig. 9.6, the detergency for the soiled cloth is plotted against $V_{T,max}/\Delta V_T$. The detergency first increases in proportion to $V_{T,max}/\Delta V_T$, but with a further increase in this ratio, it levels off. In any case, there is a clear correlation between the detergency and the ratio of $V_{T,max}/\Delta V_T$ in the region of smaller $V_{T,max}/\Delta V_T$. No correlation between detergency and $V_{T,max}/\Delta V_T$ is found if the DLVO theory is applied [22].

The detergency in Fig. 9.6 has a tendency to be saturated with a larger value of $V_{T,max}/\Delta V_T$. It may be considered that the detergency is governed not only by the interfacial electrical action, but

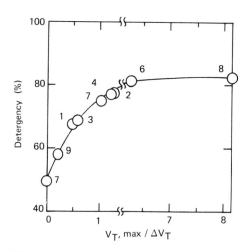

Figure 9.6 Relation between the detergency for soiled cloth and $V_{T,max}/\Delta V_T$ ratio. The numbers in the figure denote detergent solution numbers in Table 9.2.

also by complicated phenomena related to the thermal or mechanical energy of the system, and the various actions of surfactants.

9.3.3 Stability Constant for Heterocoagulation

In discussing the stability for the redeposition of soil onto fiber, Lange proposed the use of the geometrical mean of the ζ potentials of soil and fiber as a measure of stability when the potentials are nearly equal [24]. Durham remarked, on the other hand, that the electrical repulsive force between soil and fiber is proportional to the product of their ζ potentials, and discussed the stability in terms of these potentials [16,25]. Marshall et al. also proposed the following stability constant θ [26]:

$$\theta = \frac{\zeta_1 \zeta_2}{\kappa A_{123}} \tag{9.9}$$

where ζ_1 and ζ_2 are the ζ potentials of the soil and fiber, respectively. All of them pointed out the importance of ζ potentials or surface potentials of fiber and soil in discussing the stability of the system. However, it is also important in the discussion that a hetero-factor of soil and fiber potentials should be taken into consideration.

The following new constant θ_H, in which the difference in potential is taken into consideration, can better explain the actual system as described below [19]:

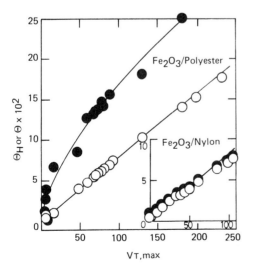

Figure 9.7 Relation between the stability constant, θ_H(○) or θ(●), and $V_{T,max}$ for Fe_2O_3 particle/fiber.

$$\theta_H = \frac{\zeta_1 \zeta_2}{\kappa A_{123}} f\left(\frac{\zeta_2}{\zeta_1}\right) \qquad (9.10)$$

where $f(\zeta_2/\zeta_1) = a \exp(b(\zeta_2/\zeta_1))$, a and b are constants, and $\zeta_2/\zeta_1 \geq 1$. (Details of this equation are described in Ref. 19.)

Figure 9.7 shows the value of θ_H for the system of Fe_2O_3/polyester plotted against the value of $V_{T,max}$, which is, as described above, closely correlated to the detergency for soiled cloth and the redeposition of soil onto fiber. The value of θ_H has a linear relation with $V_{T,max}$. For reference, the value of θ, in which the hetero-factor $f(\zeta_2/\zeta_1)$ is not taken into account, is also shown in Fig. 9.7, where θ is very different from θ_H and does not have a linear relation with $V_{T,max}$. This is a typical example of how the factor $f(\zeta_2/\zeta_1)$ exerts an effect on the stability constants θ and θ_H. Figure 9.7 includes the result for a Fe_2O_3/nylon system in which nylon fiber and Fe_2O_3 particles have ζ potentials of the same degree, and therefore both θ and θ_H are correlated with $V_{T,max}$.

There is also a clear correlation in the relation between θ_H and the amount of Fe_2O_3 redeposition on polyester or nylon fiber; the amount decreases with increasing θ_H [19]. As to the relation between θ_H and the detergency for soiled cloth made with carbon black or mud, the detergency also shows a good correlation with θ_H as shown in Fig. 9.8.

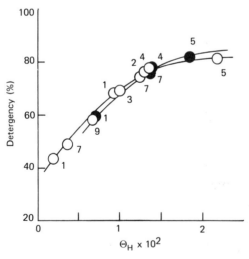

Figure 9.8 Relation between the stability constant θ_H and the detergency for carbon black-soiled cloth (○) [23] or mud-soiled cloth (●) [22]. The numbers in the figure denote detergent solution numbers in Table 9.2.

When θ_H is smaller than 1.5×10^{-2}, the detergency increases linearly with θ_H. However, the detergency versus θ_H curve leads to saturation when θ_H is larger than 2×10^{-2}.

REFERENCES

1. W. C. Powe, in *Detergency* (W. G. Cutler and R. C. Davis, eds.), Part 1, Marcel Dekker, New York, 1972, pp. 31-63.
2. T. Tachibana, *Hyomen*, 1(11), 35 (1963).
3. E. J. W. Verwey and J. Th. G. Overbeek, in *Theory of the Stability of Lyophobic Colloids*, Elsevier, Amsterdam, 1948.
4. A. Kitahara, *Kagaku-no-Ryoiki*, 24, 402 (1970).
5. F. Tokiwa, in *Interfacial Electrical Phenomena* (A. Kitahara and A. Watanabe, eds.), Kyoritsu, Tokyo, 1972, pp. 298-312.
6. E. Götte, W. Kling, and H. Mahl, *Melliand Textilber.*, 35, 1252 (1954).
7. H. Schott, in *Detergency* (W. G. Cutler and R. C. Davies, eds.), Part 1, Marcel Dekker, New York, 1972, pp. 153-235.
8. J. Visser, *J. Colloid Interface Sci.*, 34, 26 (1970).
9. W. Kling, H. Lange, G. Bohme, H. Krupp, G. Sandstede, and G. Walter, *Proc. 4th Int. Congr. Surf. Activ.*, 2, 439 (1964).
10. H. C. Hamaker, *Physica*, 4, 1058 (1937).
11. H. Lange, in *Solvent Properties of Surfactant Solutions* (K. Shinoda, ed.), Marcel Dekker, New York, 1967, pp. 117-188.
12. R. Hogg, T. W. Healy, and D. W. Fuerstenau, *Trans. Faraday Soc.*, 62, 1638 (1966).
13. B. V. Derjaguin, *Discuss. Faraday Soc.*, 18, 85 (1954).
14. T. Imamura and F. Tokiwa, *Nippon Kagaku Kaishi, 1973*, 648 (1973).
15. J. Gregory, *Adv. Colloid Interface Sci.*, 2, 396 (1969).
16. K. Durham, *J. Appl. Chem.*, 6, 153 (1956).
17. T. H. Shuttleworth and T. G. Jones, *Proc. 2nd Int. Congr. Surf. Activ.*, IV, 52 (1957).
18. H. Lange, *Fette, Seifen, Anstrichm.*, 65, 231 (1963).
19. T. Imamura and F. Tokiwa, *Nippon Kagaku Kaishi, 1974*, 405 (1974).
20. T. Imamura and F. Tokiwa, *Nippon Kagaku Kaishi, 1972*, 2177 (1972).
21. T. Imamura and F. Tokiwa, *Nippon Kagaku Kaishi, 1973*, 2051 (1973).
22. T. Imamura and F. Tokiwa, *Nippon Kagaku Kaishi, 1976*, 869 (1976).
23. T. Imamura, *Nippon Kagaku Kaishi, 1975*, 943 (1975).
24. H. Lange, *Am. Dyestuff Rep.*, 50, 433 (1961).
25. K. Durham, *Proc. 2nd Int. Congr. Surf. Activ.*, IV, 60 (1957).
26. J. K. Marshall and J. A. Kitchener, *J. Colloid Interface Sci.*, 22, 342 (1966).

10
Flotation

Shinnosuke Usui Research Institute of Mineral Dressing and
Metallurgy, Tohoku University, Sendai, Japan

10.1. Adsorption of Collectors 286
10.2. Activation and Depression 289
10.3. Interfacial Electrochemical Studies 290
10.4. Particle-Bubble Attachment 292
10.5. Slime Coating and Carrier Flotation 295
References 296

If we introduce gas bubbles into a mixed aqueous suspension of hydrophobic particles and hydrophilic particles, the former are captured by rising bubbles and separated from the latter as a froth. This is the principle of flotation by which useful minerals can be separated from useless gangue minerals. A collector is a reagent by which mineral surfaces are made hydrophobic. Alkyl xanthate is a typical collector for sulfide minerals to separate them from siliceous gangue minerals. In some cases a depressant is used to make the mineral surfaces hydrophilic or to prevent the adsorption of collectors. Although the free energy of collector adsorption contains chemical and electrical terms, as seen in Eq. (2.52), it is convenient to classify flotation from the viewpoint of collector adsorption. One is a system in which the chemisorption of collectors plays an important role, as in the case of sulfide minerals with mercapto-type collectors; and the other is a system in which the electrostatic interaction between collectors and mineral surfaces is responsible for the collecting action, as in the case of oxide or silicate minerals with ionic surfactants.

Interfacial electrical phenomena in relation to flotation will be described in what follows.

10.1 ADSORPTION OF COLLECTORS

A typical example showing the significance of the electrostatic conditions in flotation behavior is shown in Fig. 10.1, where the zeta (ζ) potentials and flotation recoveries of goethite as a function of pH are given. The isoelectric point (IEP) of goethite is seen to be located at pH 6.7. In acidic solutions goethite is positively charged and good flotation is obtained with an anionic collector (sodium dodecyl sulfate, SDS), while in alkaline solutions goethite is negatively charged and is floated by a cationic collector (dodecylammonium chloride, DAC). Similar behavior is reported in corundum flotation [2].

In Fig. 10.2 the ζ potentials of corundum as a function of the concentration of collector (SDS) and of inorganic electrolytes (NaCl, Na_2SO_4) are shown [3,4]. The IEP of corundum is at pH 9.4. The ζ

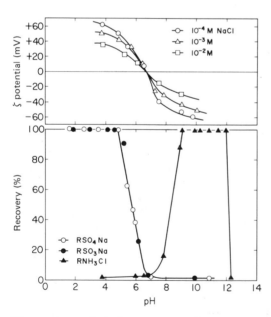

Figure 10.1 Relation between flotation recovery and ζ potential of goethite as a function of pH. Collector: Sodium dodecyl sulfate (RSO_4Na), sodium dodecyl sulsonate (RSO_3Na), and dodecylammonium chloride (RNH_3Cl). Collector concentration: 1×10^{-4} mol/liter. (From Ref. 1.)

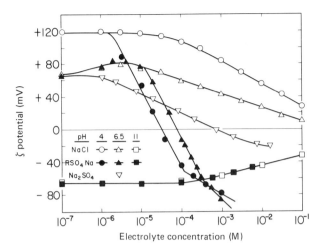

Figure 10.2 Zeta potential of corundum as a function of the concentration of SDS(RSO_4Na), NaCl, and Na_2SO_4 at various pH values. (From Refs. 3 and 4, reprinted by permission of the publisher, The Electrochemical Society, Inc.)

potential of corundum decreases in magnitude with increasing NaCl concentration at any pH and no sign reversal takes place, indicating that NaCl acts as an indifferent electrolyte. At pH 4 and 6.5, where corundum is positively charged, the ζ potential decreases abruptly at a particular collector concentration and sign reversal is seen, indicating the specific adsorption of the collector at the corundum surface. On the other hand, at pH 11, where corundum is negatively charged, there is no difference in the effect of electrolyte concentration on the ζ potential between RSO_4Na and NaCl, indicating that RSO_4^- behaves as an indifferent electrolyte for corundum at this pH value. As seen in the curve at pH 6.5, where corundum is positively charged, sign reversal occurs at around 10^{-3} mol/liter Na_2SO_4, indicating the specific adsorption of SO_4^{2-} on corundum surfaces. The abrupt change in ζ potential curves at low collector concentration at pH 4 and 6.5 was interpreted in terms of two-dimensional micelle formation of RSO_4^- at the corundum surface in such a way that long-chain ions are oriented with the charged head toward the surface and with the hydrocarbon chain sticking out into the solution (Fig. 10.3). By analogy with the micelle formation in bulk solution, Gaudin and Fuerstenau [5] called the micelle at the surface a "hemi-micelle." Hemi-micelle formation was also reported in negatively charged quartz and cationic collector (dodecylammonium acetate, DAA) systems [6]. Since collector ions serve as counterions in diffuse double layers, surface con-

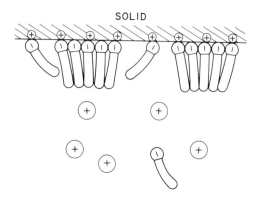

Figure 10.3 Schematic illustration of the hemi-micelle formation. (From Ref. 25, by permission of the American Institute of Mining, Metallurgical, and Petroleum Engineers, Inc.)

centration becomes high even when the bulk concentration of the collector is low [see Eq. (2.30)]. The critical concentration at which the hemi-micelle forms depends on the chain length of the collector concerned; the longer the chain length, the lower the critical concentration. Fuerstenau [6] reported that, in the quartz-amine system, hemi-micelle formation occurred at collector concentrations as low as about 1/100 times the critical micelle concentration (CMC) in the bulk. Since the collector ion concentration in a diffuse double layer also depends on the surface potential of the solid [see Eq. (2.30)], the pH of the solution has a significant effect on hemi-micelle formation on oxide and silicate surfaces. Figure 10.4 shows the interrelation among contact angle, surface coverage, ζ potential, and flotation recovery as a function of pH on a quartz-DAA (4×10^{-5} M) system [7]. The negative ζ potential of quartz increases with increase in pH until pH 8, at which point a sudden change in ζ potential toward the positive direction takes place. This breaking of the ζ potential curve reflects the onset of hemi-micelle formation of DA^+ on quartz surfaces as a result of the increase in the surface concentration of DA^+ counterions and is well correlated to such parameters as surface coverage, contact angle, and flotation recovery. The reduction in flotation in high pH (> 12) solutions has been explained in terms of the depletion of DA^+ concentration due to its hydrolysis to form free amine molecules, and in terms of the desorption of DA^+ as a result of competitive adsorption of alkali metal cations from the pH regulator. The competitive adsorption between DA^+ and K^+ was evidenced in differential capacity curves on mercury-aqueous solution interfaces [8].

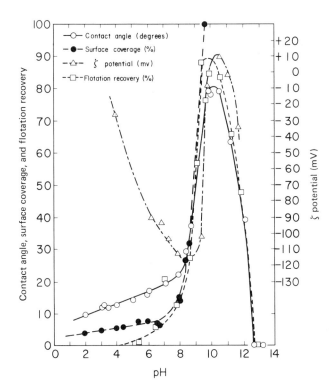

Figure 10.4 Correlation of adsorption density, contact angle, and ζ potential with flotation recovery of quartz as a function of pH. DAA: 4×10^{-5} mol/liter. (From Ref. 7, by permission of the American Institute of Mining, Metallurgical, and Petroleum Engineers, Inc.)

10.2 ACTIVATION AND DEPRESSION

Ion exchange in an electrical double layer is of importance in oxide flotation systems. The ratio of adsorption amount of collector ion (Γ_a) to that of competitive inorganic ions (Γ_b) of the same valency is given by Eq. (10.1), provided that the difference in their ionic sizes is ignored:

$$\frac{\Gamma_a}{\Gamma_b} = \frac{c_a}{c_b} \exp\left(\frac{\Delta \phi}{kT}\right) \tag{10.1}$$

where c is the bulk concentration and $\Delta \phi$ is the difference in the specific adsorption potential between the collector ion (a) and the inorganic ion (b). Fuerstenau [6] demonstrated that the negative ζ potential of quartz in aqueous media became zero at a DAA concentration of 3×10^{-4} mol/liter, but in the presence of 1×10^{-3} mol/liter of NaCl, the DAA concentration at which the zero ζ potential is brought about was found to be 1.4×10^{-3} mol/liter. This was explained by the competitive adsorption between Na^+ and DA^+; the higher the inorganic ion concentration, the higher the critical concentration of collector for hemi-micelle formation. This is quite different from micelle formation in bulk solutions, because CMC in the bulk solutions decreases with the addition of inorganic electrolytes, which is of great importance in understanding the depressing action in the flotation systems where interfacial electrical conditions predominate. In acid pH ranges where corundum is positively charged, the flotation of corundum with SDS is depressed by the addition of NaCl or Na_2SO_4, which results in the more pronounced effect of the latter. This is explained by the divalent SO_4^{2-} being specifically adsorbed on corundum surfaces, creating an excess negative charge in the Stern layer and decreasing the adsorption of collector anion. By the same principle, a depressing action with cations was reported in the cationic flotation of quartz [9].

In contrast to the above, SO_4^{2-} acts as an activator in flotation with cationic collectors. Corundum is positively charged at pH 6 and no flotation is obtained with the cationic collector (DAA). In this case, however, corundum flotation becomes possible by the addition of 5×10^{-3} mol/liter of Na_2SO_4 [2], with which the ζ potential of corundum acquires a negative sign (see Fig. 10.2). Nakatsuka, Shimoiizake, and co-workers [10,11] studied the activating action of polyvalent anions such as SO_4^{2-} and PO_4^{3-} in the cationic flotation of various oxide and silicate minerals and found the selectivity of the activating action. For instance, in an acidic media anion activation occurred for rutile, zircon, and illumenite, while no activation was found for magnetite, quartz, feldspar, and hypersthene. This selectivity of anion activation was utilized to separate minerals that were difficult to separate owing to the similarity of their points of zero charge (PZCs).

10.3 INTERFACIAL ELECTROCHEMICAL STUDIES

Electrokinetic studies give us useful information in understanding the relation between flotation and interfacial electrical conditions. On the other hand, the electrochemical titration technique offers information which is not available by electrokinetic measurements. Freyberger, Iwasaki, and de Bruyn [12,13] applied the potentiometric titration technique to silver sulfide-aqueous solution interfaces. This technique involves the potentiometric titration of potential-determining ions (Ag^+) of an aqueous Ag_2S suspension in the presence of a supporting electrolyte at constant pH to determine the adsorption density of potential-determining ions as a function of pAg. pAg can be converted to ψ_0 by the relation $\psi_0(v) = 0.059 (pAg^0 - pAg)$, where

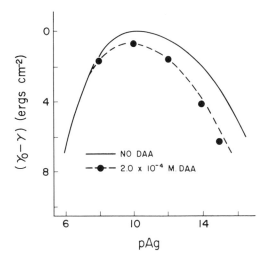

Figure 10.5 Electrocapillary curves at silver sulfide-aqueous solution interfaces in the absence and the presence of DAA (2.0×10^{-4} mol/liter). $NaNO_3$, 1×10^{-2} mol/liter, pH 4.7. (From Ref. 18.)

pAg^0 (= 10.2) refers to the PZC. Thus σ as a function of ψ_0 is obtained, from which the electrocapillary curve ($\gamma = -\int \sigma\, d\psi_0$) and differential capacity curve ($C = d\sigma/d\psi_0$) are derived. This procedure was applied to systems containing collectors. Figure 10.5 shows the electrocapillary curves relative to γ_0 (interfacial tension at PZC) for Ag_2S-aqueous solution interfaces in the absence and presence of DAA (2×10^{-4} mol/liter) at 1×10^{-2} mol/liter $NaNO_3$ and at pH 4.7 [18]. The interfacial tension is seen to decrease by several erg/cm^2 in the presence of DAA in the region where Ag_2S is negatively charged. Parks and de Bruyn [14] extended the potentiometric titration to a hematite-aqueous solution system where the adsorption density of potential-determining ions (H^+ and OH^-) was determined by the difference in pH before and after the addition of H^+ or OH^- ions. The general trend of differential capacity versus potential curves of hematite in contact with KNO_3 aqueous solution showed similar behavior to those of a mercury- or silver iodide-aqueous system, showing a minimum capacity at PZC in dilute electrolyte concentrations. Breeuwsma and Lyklema [15] demonstrated that the electrical double layer on hematite has properties between those of a nonporous substance (silver iodide or mercury) and those of porous silica.

Fuerstenau and Yamada [16] reported that the flotation with a long-chain ionic collector is enhanced by the addition of neutral molecules such as higher alcohols, probably due to their coadsorption at mineral surfaces. It was also suggested that the adsorption of undissociated amine molecules was responsible for amine flotation [17-21]. Somasundaran [22] mentioned that the complex formed between ionic and molecular forms plays an important role in flotation. Adsorption of organic substances in relation to double-layer structure has been best studied on mercury-aqueous solution interfaces, although no direct comparison with flotation behavior is available. By applying the Gibbs adsorption equation for a system of mercury in contact with aqueous solutions containing DAA and KOH in the presence of KF as a supporting electrolyte, Usui and Iwasaki [23] evaluated the adsorption density of dodecylammonium ion (Γ_{DA^+}) and that of undissociated dodecyl amine molecule (Γ_{DA}) separately at around the PZC of mercury, demonstrating that the ratio $\Gamma_{DA}/\Gamma_{DA^+}$ was considerably higher than that of c_{DA}/c_{DA^+} (c: concentration) in the bulk solution under corresponding pH.

As to the interfacial electrical phenomena in relation to flotation, comprehensive review articles by de Bruyn and Agar [24] and Aplan and Fuerstenau [25] are available.

10.4 PARTICLE-BUBBLE ATTACHMENT

The attachment of particles to an air bubble in liquid media depends on the rupture of intervening thin liquid films between air and particle surfaces, which has been described in Sec. 3.6. However, particle-bubble attachment in flotation is characterized by the process occurring under dynamic conditions in which not only surface forces but also hydrodynamic factors should be taken into consideration. The complexity is amplified by the interfacial electrical conditions around a moving bubble, which differ considerably from those of a moving solid [26]. Derjaguin and Dukhin [27] analyzed the particle-bubble attachment in which the surface layer around a bubble consisted of three layers (Fig. 10.6). In the first and outermost layer, particles move in accordance with the hydrodynamic law and no surface forces are involved. The second layer, the thickness of which is in the order of 10^{-4}-10^{-3} cm, is called the diffusio boundary layer, in which particles experience diffusiophoretic forces depending on their surface charge and electric field in this layer. In the third and innermost layer, particle motion is governed by surface forces, which consist of double-layer interaction forces and molecular interaction forces. It is necessary for particles to come into the third layer to achieve particle-bubble attachment.

Relatively large particles can enter the second layer due to the inertia force, but small particles travel along the streamline and cannot enter the second layer from the first layer; a critical size (b_c) for the entrance was given by

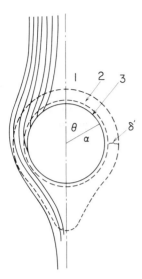

Figure 10.6 Schematic representation of a bubble rising in a liquid medium.

$$b_c = \frac{9}{(48)^{1/2}} \left(\frac{v\eta}{ga \, \Delta\rho} \right)^{1/2} \qquad (10.2)$$

where v is the rising velocity of a bubble, η the viscosity of the medium, g the gravitational constant, a the bubble radius, and $\Delta\rho$ the difference in density between particle and medium. The critical size b_c was estimated to be about 13 µm for a bubble of 1 mm radius ($\Delta\rho = 1$) in aqueous media. Particles that succeeded in passing into the second layer experience a special kind of force inherent to a moving bubble. When a bubble ascends in an aqueous surfactant solution, molecules or ions adsorbed at the front of bubble surfaces are stripped off backward along the surface and are accumulated at the bottom of the bubble. Surfactant molecules (ions) migrate from the interior of the bulk toward the depletion layer at the front of a bubble, while they migrate by diffusion toward the bulk from the accumulated part at the rear so as to establish an equilibrium state, thus forming a diffusion layer around a bubble in which the concentration gradient of surfactant molecules is established. The thickness (δ') of this diffusion layer was estimated to be about 10 µm for a bubble of radius 1 mm. In the case of the ionic surfactant, since the diffusion coefficients are different between anions and cations, an electric field (diffusion potential) is established in this diffusion layer. The normal component of the electric field is given by Eq. (10.3) in the case where surface activity of the surfactant is not very strong:

$$X_n = \frac{2RT}{F} \frac{(D_+ - D_-)U_\infty \Gamma_0}{D_+D_-(v_+ + v_-)ac_0} \cos\theta \qquad (10.3)$$

where $D_{+(-)}$ is the diffusion coefficient of cation (anion), U_∞ the rising velocity of a bubble, Γ_0 and c_0 the adsorption density and concentration of the surfactant at equilibrium, $v_{+(-)}$ the valency of cation (anion), and θ the polar angle. There are cases where the field strength becomes of the order of magnitude of several thousand volts. This layer is called the dynamic double layer or second double layer. When a charged particle smaller than δ' passes into the dynamic double layer, it suffers an electrophoretic force depending on the sign and magnitude of its charge (diffusiophoresis). In the case where the smooth and flat surface of a particle larger than δ' approaches the depletion layer of a bubble, the acting forces are quire complex, depending on two opposing factors. One is the repulsive component arising from the resistance against drainage of the intervening liquid layer; the other is electrostatic attraction between charges on mineral and bubble surfaces that are generated by the difference in diffusion rate between surfactant ions and their counterions migrating from mineral surfaces toward the depletion area of bubble surfaces. Surface forces operating in the third layer are analyzed in terms of the heterocoagulation theory.

Derjaguin and Schukakidse [28] studied the flotation of stibnite as a function of pH in the presence of frother (hexyl alcohol) only and analyzed the results by colloid stability theory by assuming that the surface charge of a bubble surface remains constant (surface potential being undefined) and the Hamaker constant A is 3×10^{-13}/erg. The criterion for flotation was obtained as

$$m = \frac{\varepsilon\delta\zeta^2}{A} < 3 \qquad (10.4)$$

where δ is the thickness of the diffuse double layer, ζ the ζ potential of stibnite, and ε the dielectric constant of water. The flotation recovery begins to decrease when m exceeds 3, in good agreement with the theoretical prediction. Bleier, Goddard, and Kulkarni [29] studied the amine flotation of quartz from the viewpoint of heterocoagulation between bubble and mineral particle. Blake and Kitchener [30] examined the stability of thin aqueous film between a methylated silica plate and a bubble in view of the disjoining pressure. Laskowski [31] reviewed the bubble-particle attachment in flotation from the view point of the stability of the disjoining film between a bubble and a particle.

In the theoretical analysis of bubble-particle attachment, it is essential to know the ζ potential of the bubble in aqueous media. Electrokinetic phenomena at gas-liquid interfaces have not been as fully explored as those at solid-liquid interfaces. The electrophoretic mobiliby measurements were made by several investigators [32-37], including the pioneering work of McTaggert [32]. Dibbs, Sirois, and Bredin [38] measured the electric current produced by a bubble rising in relation to the bubble-particle attachment. Usui, Sasaki, and Matsukawa [39,40] measured the Dorn potential to calculate the ζ potential of a bubble in aqueous surfactant solutions, where size dependency was indicated in contrast to the case of solid particles.

10.5 SLIME COATING AND CARRIER FLOTATION

Flotation becomes difficult if fine foreign particles cover the surface of the mineral to be floated. This is called slime coating. Ince [41] pointed out that the electrostatic interaction between slime and valuable minerals is of significance. Sun [42] studied the slime coating of quartz, calcite, and fluorite particles on sulfide minerals and reported that the slime coating was heavy when the ζ potentials of the slime and the sulfide particles were of opposite sign and large in magnitude, or when the ζ potential of the slime was low. Conversely, the slime coating was light when the ζ potential of sulfide mineral was low and that of the slime was high. Slime coating was avoided if the ζ potentials of both sulfide and slime were of the same sign and large in magnitude. The significance of electrostatic interaction in slime coating was also reported by Gaudin, Fuerstenau, and Miaw [43,44] on sulfide or oxide mineral flotation. In general, both oxide and silicate minerals acquire more negative charge in high-pH solutions, which provides favorable conditions for preventing slime coating. By using this principle, Iwasaki et al. [45] and Sato and Shimoiizaka [46] obtained good results in separating silicate minerals from goethite, where silicate minerals were floated in the former and goethite was floated in the latter.

Selective flocculation is a very promising method for obtaining successful metallurgical results, as evidenced by the "selective flocculation and flotation" method in iron ore beneficiation [47], in which iron oxide was selectively flocculated by starch and was obtained as a sink while silica was floated by a cationic collector. Studies on selective flocculation were carried out by Kitchener and others [48-51] on various systems.

Adhesion of fine particles to a large particle is not always harmful for flotation. This type of coagulation is utilized in recovering fine minerals. Bubble-particle attachment becomes difficult as particle size becomes small. In this situation carrier mineral particles of appropriate size are introduced so that fine particles are coagulated on the carrier particles and floated together. This is called carrier flotation or ultraflotation. Carrier flotation has been used in the purification of clay minerals, in which calcite is used as a carrier mineral on which colored titaniferrous impurity fine particles are coagulated [52]. Coagulation between particles of different sizes was discussed by Samygin and Barskii [53]. Warren [54] studied the effect of the hydrophobic nature of particle surfaces on shear flocculation between small and large particles. Recent information regarding fine-particle processing, including selective flocculation, is available in Refs. 55 and 56.

REFERENCES

1. I. Iwasaki, S. R. B. Cooke, and A. F. Colombo, *U. S. Bur. Mines, Rep. Invest. 5593*, 1960.
2. H. J. Modi and D. W. Fuerstenau, *Trans. AIME, 217*, 381 (1960).
3. D. W. Fuerstenau and H. J. Modi, *J. Electrochem. Soc., 106*, 336 (1959).
4. J. H. Modi and D. W. Fuerstenau, *J. Phys. Chem., 61*, 640 (1957).
5. A. M. Gaudin and D. W. Fuerstenau, *Trans. AIME, 202*, 958 (1955).
6. D. W. Fuerstenau, *J. Phys. Chem., 60*, 981 (1956).
7. D. W. Fuerstenau, *Trans. AIME, 208*, 1365 (1957).
8. S. Usui and I. Iwasaki, *Trans. AIME, 247*, 213 (1970).
9. D. M. Hopstock and G. E. Agar, *Trans. AIME, 241*, 466 (1968).
10. K. Nakatsuka, R. Nagai, and J. Shimoiizaka, *Nippon Kogyo Kaishi, 84*, 27 (1968).
11. K. Nakatsuka, I. Matsuoka, and J. Shimoiizaka, *Proc. 9th Int. Miner. Process. Congr.*, 1970, p. 251.
12. W. L. Freyberger and P. L. de Bruyn, *J. Phys. Chem., 61*, 586 (1957).
13. I. Iwasaki and P. L. de Bruyn, *J. Phys. Chem., 62*, 594 (1958).
14. G. A. Parks and P. L. de Bruyn, *J. Phys. Chem., 66*, 967 (1962).
15. A. Breeuwsma and J. Lyklema, *Discuss. Faraday Soc., 1971*(52), 324 (1971).
16. D. W. Fuerstenau and B. J. Yamada, *Trans. AIME, 223*, 50 (1962).
17. R. W. Smith, *Trans. AIME, 226*, 427 (1963).
18. I. Iwasaki and P. L. de Bruyn, *Surf. Sci., 3*, 299 (1965).
19. H. C. Li and P. L. de Bruyn, *Surf. Sci., 5*, 203 (1966).
20. P. Somasundaran and D. W. Fuerstenau, *Trans. AIME, 241*, 102 (1968).
21. S. K. Mishra, *Int. J. Miner. Process., 6*, 119 (1979).
22. P. Somasundaran, *Int. J. Miner. Process., 3*, 35 (1976).
23. S. Usui and I. Iwasaki, *Trans. AIME, 247*, 220 (1970).
24. P. L. de Bruyn and G. E. Agar, in *Froth Flotation*, 50th Anniversary Volume (D. W. Fuerstenau, ed.), AIME, New York, 1962, p. 91.

25. F. Aplan and D. W. Fuerstenau, in *Froth Flotation*, 50th Anniversary Volume (D. W. Fuerstenau, ed.), AIME, New York, 1962, p. 170.
26. S. S. Dukhin, in *Research in Surface Forces* (B. V. Derjaguin, ed.), Vol. 2, Consultants Bureau, New York, 1966, p. 54.
27. B. V. Derjaguin and S. S. Dukhin, *Bull. Inst. Min. Metall.*, 70, 221 (1961),
28. B. V. Derjaguin and N. D. Shukakidse, *Bull. Inst. Min. Metall.*, 70, 569 (1961).
29. A. Bleier, E. D. Goddard, and R. D. Kulkarni, *J. Colloid Interface Sci.*, 59, 490 (1977).
30. T. D. Blake and J. A. Kitchener, *J. Chem. Soc. Faraday Trans. I*, 68, 1435 (1972).
31. J. Laskowski, *Miner. Sci. Eng.*, 6, 223 (1974).
32. H. A. McTaggart, *Philos. Mag.*, 27, 297 (1914); 28, 367 (1914); 44, 386 (1922).
33. A. Alty, *Proc. R. Soc. Lond. A*, 106, 315 (1924); 112, 235 (1926).
34. S. Komagata, *Denki Kagaku*, 4, 380 (1936)
35. N. Bach and A. Gilman, *Acta Physicochim. URSS*, 9, 1 (1938).
36. G. L. Collins, M. Motarjemi, and G. J. Jameson, *J. Colloid Interface Sci.*, 63, 69 (1978).
37. C. Cichos, *Neue Bergbautech.*, 1, 941 (1971); 2, 928 (1972).
38. H. Dibbs, L. L. Sirois, and R. Bredin, *Can. Metall. Q.*, 13, 395 (1974).
39. S. Usui and H. Sasaki, *J. Colloid Interface Sci.*, 65, 36 (1978).
40. S. Usui, H. Sasaki, and H. Matsukawa, *J. Colloid Interface Sci.*, 81, 80 (1981),
41. C. R. Ince, *Trans. AIME*, 87, 261 (1930).
42. S. C. Sun, *Trans. AIME*, 153, 479 (1943).
43. D. W. Fuerstenau, A. M. Gaudin, and H. L. Miaw, *Trans. AIME*, 211, 792 (1958).
44. A. M. Gaudin, D. W. Fuerstenau, and H. L. Miaw, *Can. Min. Metall. Bull.*, 53, 960 (1960).
45. I. Iwasaki, S. R. B. Cooke, D. H. Harraway, and H. S. Choi, *Trans. AIME*, 223, 97 (1962).
46. T. Sato and J. Shimoiizaka, *Nippon Kogyo Kaishi*, 81, 25 (1965).
47. A. F. Colombo and D. W. Frommer, in *Flotation*, A. M. Gaudin Memorial Volume (M. C. Fuerstenau, ed.), Vol. 2, AIME, New York, 1976, p. 1285.
48. R. J. Pugh and J. A. Kitchener, *J. Colloid Interface Sci.*, 35, 656 (1970).
49. B. Yarar and J. A. Kitchener, *Trans. Inst. Min. Metall.*, 79, C23 (1970).

50. J. P. Friend and J. A. Kitchener, *Chem. Eng. Sci.*, 28, 1071 (1973).
51. A. D. Read and A. Whitehead, *Proc. 10th Int. Miner. Process. Congr.*, London, 1973, p. 949.
52. F. A. Seeton, *Deco Trefoil*, 27, 7 (1963).
53. V. D. Samygin, A. A. Barskii, and S. M. Angelova, *Colloid J.* (English Trans.), 30, 435 (1969).
54. L. J. Warren, *J. Colloid Interface Sci.*, 50, 307 (1975).
55. A. D. Read and C. T. Hollick, *Miner. Sci. Eng.*, 8, 202 (1976).
56. P. Somasundaran, ed., *Fine Particles Processing*, AIME, New York, 1980.

11
Fibers

Toshiro Suzawa Hiroshima University, Higashihiroshima, Hiroshima, Japan

11.1. Introduction 299
11.2. Electrokinetic Properties of Fiber Surface 300
11.3. Electrokinetic Properties of Fiber in Dyeing 303
    11.3.1. Fiber structure and surface dyeability 305
    11.3.2. Dye structure and surface dyeability 311
11.4. Electrokinetic Properties of Fiber and Adsorption of Surfactants 316
    11.4.1. Adsorption behavior of fibers in surfactant solution below the CMC 317
    11.4.2 Adsorption behavior of fibers in surfactant solution above the CMC 318
    References 320

11.1 INTRODUCTION

At the interface between charged fibrous materials such as fibers and papers and an aqueous solution containing various electrolyte ions, an electrical double layer is set up, because the molecules constituting the surface of fibrous substances dissociate into ions, or because ions and dipoles in solution are adsorbed onto the surface. The electrokinetic or zeta (ζ) potential is the difference in potential between a part of the double layer, which is fixed tightly so that it does not move when the solution streams along the surface, and the interior of the solution. Consequently, the ζ potential plays an important role in wet processing of fibrous materials, such as dyeing, detergency and papermaking, and so on. Usually, the ζ potentials for fibrous materials are measured using streaming potential, streaming current, and electroosmosis methods.

11.2 ELECTROKINETIC PROPERTIES OF FIBER SURFACE

The ζ potential for negative fibers such as cotton, rayon, poly(vinyl alcohol) (Vinylon, etc.), polyester (Terylen, Tetron, Dacron, etc.), polyacrylonitrile (Cashimilon, etc.), and polyethylene fibers in aqueous solution shows a negative value. This is because these fibers have an acidic dissociable group such as COOH group for cotton, rayon, polyvinyl alcohol), and polyester fibers and an SO_3H group for polyacrylonitrile fiber [1]. It was recognized that the pH corresponding to one-half of the saturated value of ζ determined from the ζ versus pH curves for these fibers agreed with the pK region of the acidic dissociable group (e.g., COOH, SO_3H) for the fiber [1].

The variation in potential with distance according to the Gouy-Chapman theory of the electrical double layer can be expressed by the Eversole and Boardman equation [2]:

$$\ln \tanh \frac{Ze\psi}{4kT} = \ln \tanh \frac{Ze\psi_\delta}{4kT} - \kappa x \qquad (11.1)$$

where ψ is the potential in a plane of distance x from the Stern layer where the potential is ψ_δ, Z the valency of the counterion, e the electronic charge in esu, κ the reciprocal of the double-layer thickness, k the Boltzmann constant, and T the absolute temperature. The ζ potential is often substituted for ψ. Using the surface potential ψ_δ calculated from a plot of log tanh $(Ze\zeta/4kT)$ against κ, the surface charge density onto a cellulose fiber such as cotton or rayon was estimated [3]. It agreed well with the theoretical surface charge density calculated from data of the carboxyl group content of the fiber and the BET nitrogen surface area for the water-swollen uncollapsed fiber.

In amphoteric fibers such as wool, silk, and polyamide nylon, the isoelectric point (IEP) is determined from the effect of pH on ζ of the fiber. In manufacturing nylon fibers, drawing and heat-setting processes result in changes in the fibrous structure, especially in the fine structure. The effect of these processes on the IEP of the fibers was determined by the ζ-potential method [4-7]. The effects of the degree of crystallinity, the birefringence, and the end group on the IEP of the spun, drawn, and heat-setted nylon 6 fibers, respectively, were investigated [7]. From these results it was found that the IEP of the fiber was almost unaffected by the spinning temperature and hardly changed by the drawing process, but was changed by the heat-setting process. In general, it was shown that the mole fraction of the amino end group or the carboxyl end group contributed mainly to the IEP of nylon fibers, and that the IEP was little affected by the degree of crystallinity or the birefringence.

It was recognized [8] that the ζ potential (absolute value) of a polymer solid such as fiber in water lowers with time approximately according to Eqs. (11.2) or (11.3):

$$\frac{-d\zeta}{dt} = k(\zeta - \zeta_\infty) \qquad (11.2)$$

or

$$-\ln \frac{\zeta - \zeta_\infty}{\zeta_0 - \zeta_\infty} = kT \qquad (11.3)$$

where $-d\zeta/dt$ corresponds to the hydration velocity and k is the rate constant. ζ_0 and ζ_∞ denote the initial ζ potential immediately after immersion in water and the equilibrium value of the ζ potential, respectively. $(\zeta_0 - \zeta_\infty)/\zeta_0$ and $(\zeta_0 - \zeta)/\zeta_0$ represent the hydration capacity for the dry sample and the degree of hydration in some time after the immersion in water, respectively.

Table 11.1 shows the relation between the V value and the hydration capacity mentioned above [8,9]. V is the total volume of water in liters per kilogram of dry fiber, which is estimated from the amount of moisture absorbed by the unit kilograms of fiber at 100% relative humidity. As listed, the hydration capacity of the higher hydrophilic fibers having a larger V value is larger than that of the higher hydrophobic fibers having a smaller V value. This result suggests the hydration capacity estimated from the degree of lowering in the ζ potential is used as a measure of the degree of swelling or the degree of moisture sorption. Using this method, a comparison of hydration and water sorption capacity of silk with those of silklike fibers was made [10]. From these facts, it was also recognized that the ζ potential of the higher hydrophobic fiber is larger than that of the higher hydrophilic fiber [8].

From the results of temperature dependence of the apparent rate constant k for the process of lowering the ζ potential of polymers such as poly(vinyl acetate), poly(vinyl formal), and nylon 6 in water with time, Eq. (11.2), it was concluded that the determining process is hydration in the temperature range $T < T_g$ (T_g: glass transition point) and diffusion in the range $T > T_g$, and that the temperature where two linear plots, log k_1 versus $1/T(°K)$ and log k_2 versus $1/T(°K)$ intersect, can be defined as the T_g of the polymer immersed in water [11]. Similar investigations were carried out with ethyl methacrylate-2-hydroxyethyl methacrylate copolymer/water, ethyl methacrylate-glycidyl methacrylate copolymer/water, methyl methacrylate-2-diethylaminoethyl methacrylate copolymer/water, and methyl methacrylate-N-(hydroxymethyl)-acrylamide copolymer/water systems [12].

Table 11.1 Relation Between Moisture Sorption Ability of the Fiber and Degree of Lowering in Its ζ potential

Fiber	V value[a] (liters/kg fiber)	ζ potential		Hydration capacity,[b] $(\zeta_0 - \zeta_\infty)/\zeta_0$
		ζ_0	ζ_∞	
Viscose rayon	0.40	28.30	8.50	0.70
Cotton	0.24	54.00	30.20	0.44
Mercerized cotton	0.33	74.00	24.40	0.67
Silk fibroin	0.28	59.90	23.46	0.61
Cellulose diacetate (CH_3COOH 50%)	0.14	71.00	50.10	0.42
Cellulose dinitrate (N = 10.2%)	0.12	45.10	29.80	0.33
Nylon	0.06	—	—	—
Dynel	0.02	51.15	35.00	0.31
Orlon	0.02	—	—	—
Tevilon	0.02 (0.005)	66.12	54.90	0.19
Glass fiber	0.005	41.10	35.19	0.14
Polyethylene tere-phthalate (Terylen)	0.005	81.62	74.20	0.10
Cellulose dinitrate (N = 13.7%)	0.03	55.05	49.40	0.09
Cellulose diacetate (CH_3COOH 62.5%)	0.01	108.10	100.06	0.07

[a]Estimated from the amount of moisture sorption (liters) per unit kilogram of fiber at 100% R.H.
[b]Estimated from the degree of lowering of ζ potential.
Source: Reproduced with permission from K. Kanamaru, Kagaku to Kogyo, 12, 89 (1958); copyright by the Chemical Society of Japan.

To determine whether the ζ potential of the fiber is attributable to the charge in the amorphous regions or to that in the surface of crystal regions, the ζ potential of poly(vinyl alcohol) fibers in water was measured [13]. The fibers were heat-setted at 220°C and then formalized by various degrees to the condition to be acetalized in the amorphous region alone. The ζ potentials (negative) of these fibers increased and their degree of swelling decreased with increasing degree of formalization. Also the ζ potentials of poly(vinyl chloroacetal) fibers with various degrees of chloroacetalization showed similar results [13]. Moreover, the ζ potentials (negative) for wet- and dry-spun poly(vinyl alcohol) Vinylon fibers, respectively, increased with increasing degree of formalization and decreasing degree of crystallinity [14]. On the other hand, the ζ potentials (negative) of poly(vinyl alcohol) fibers having almost the same degree of crystallinity increased and their degree of swelling decreased with increasing degree of formalization [15]. Such results suggest that the ζ potential of the fiber more or less results from the charge due to the COO^- group, and so on, in the amorphous region forming the fine structure of the fiber, and that, in general [8], the ζ potential of the higher hydrophobic fiber is larger than that of the higher hydrophilic fiber.

The ζ potentials of cotton treated with different cross-linking agents, such as dimethylol ethyleneurea (DMEU), dimethylol propyleneurea (DMPU), and dimethylol dihydroxyethyleneurea (DMDHEU), were measured in an aqueous solution of various kinds of inorganic salts [16,17]. The ζ potential of each treated fiber decreased, but the surface charge density increased with increasing electrolyte concentration. Also, the degree of increase in the surface charge density for cotton treated with different cross-linking agents was in the order DMDHEU > DMEU > DMPU. This is attributed to the contribution of the hydroxyl group and the polarity of carbonyl group in the cross-linking agent and the hydrogen-bond formation between the hydroxyl groups in cellulose and the cross-linking agent [17].

11.3 ELECTROKINETIC PROPERTIES OF FIBER IN DYEING

In the adsorption process of dyes onto a fiber surface, the ζ potential at the interface of the fiber and an aqueous solution of the dye plays an important role. An electrical double layer in the dyeing process is complex, because in addition to dye, the dyeing bath also contains acid, alkali, inorganic salt, and surfactant. Therefore, in order to investigate the relation between dyeability and electrokinetic property of a fiber surface, not only the change of ζ potential but also that of surface charge density, σ, should be considered in the dyeing system.

From the values of the ζ potential, the σ of the fiber in aqueous solutions of the dye was calculated using Eq. (11.4), which [18] was derived from Boltzmann's distribution law and from Gouy's theory:

$$\sigma = \pm \left(\frac{kT\varepsilon}{2\pi}\right)^{1/2} \left\{\sum_j n_j \left[\exp\left(-\frac{e_j\zeta}{kT}\right) - 1\right]\right\}^{1/2} \qquad (11.4)$$

where ε is the dielectric constant, n_j is the number of j cations or anions per unit volume in the bulk solution, and e_j is $Z_j e$. Here e is the electronic charge and Z_j is the valence of the j cation or anion.

Furthermore, the amounts adsorbed per unit area of the fiber surface, which is expressed as the surface dye adsorption in g/cm² of fiber, from the differences, $\Delta\sigma$, between the surface charge density of the fiber in an aqueous solution with dye and that without it were calculated [18,19].

The free energy of dyeing $\Delta\overline{G}$ was calculated using Eqs. (11.5)–(11.9) derived by Ottewill, Rastogi, and Watanabe [20].

$$\zeta = \left(\frac{C}{C + C_d}\right)^{1/2} \frac{\zeta^* + k_1 k_2 C_d}{1 + k_2 C_d} \qquad (11.5)$$

$$k_1 = \left(\frac{2\pi \times 1000 kT}{\varepsilon CN}\right)^{1/2} N_1 \frac{Z_d}{|Z|} \qquad (11.6)$$

$$k_2 = \frac{\exp(-\Delta\overline{G}/kT)}{55.6} \qquad (11.7)$$

$$\left(\frac{d\zeta}{d\log C_d}\right)_{\zeta=0} = -2.303 \left(\frac{C}{C + C^\circ_d}\right)^{1/2} \zeta^* \left(1 + \frac{\zeta^*}{k_1}\right) \qquad (11.8)$$

$$\frac{1}{C^\circ_d} = -k_2 \left(1 + \frac{k_1}{\zeta^*}\right) \qquad (11.9)$$

where C_d and C are the concentration of dye and indifferent salt; Z_d and Z the valence of dye and indifferent salt; ζ and ζ^* the ζ potentials of the fiber in the aqueous solution with and without dye, respectively; C°_d the concentration of dye at the isoelectric point of the ζ versus $\log C_d$ curve; N_1 the maximum number of available sites per unit area of the fiber surface; k the Boltzmann constant; N Avogadro's number; and $\Delta\overline{G}$ the electrochemical free energy of adsorption for dye. Thus $\Delta\overline{G}$ can be expressed as the sum of electrical and chemical terms; that is, $\Delta\overline{G} = \Delta G + Ze\psi_\delta$, where ΔG is the chemical free energy of adsorption for dye, and ψ_δ is the Stern potential and can be approximated by ζ. Provided that ψ_δ equals zero, $\Delta\overline{G}$ equals ΔG.

k_1 is calculated using Eq. (11.8) and k_2 by Eq. (11.9). The free energy of dyeing $\Delta \bar{G}$ is calculated by applying k_2 to Eq. (11.7). The heat of dyeing ΔH and the entropy of dyeing ΔS are evaluated in the usual manner.

11.3.1 Fiber Structure and Surface Dyeability

ζ Potential and Surface Dye Adsorption of Fiber in Dye Solution

The ζ potential of negative fibers (i.e., cotton, rayon, and Vinylon) in an aqueous solution of the direct dye, Congo Red, was measured [14,21]. With increasing concentration of dye, there was an increase of the negative value of the ζ potential of the fiber, which passed through a maximum value and thereafter decreased. This may be attributed to the adsorption of the dye anion and the compression of the thickness of an electrical double layer. Also, the amount of dye adsorbed per unit area of fiber surface (i.e., the surface dye adsorption) increased and then approached a constant value with increasing concentration of dye. These results may be due mainly to the formation of the hydrogen bond and to the van der Waals force between the fiber and the dye.

In the case of the dyeing of negative fibers such as poly(vinyl alcohol) and polyacrylonitrile fibers by the cationic dye [15,22-24] and that of amphoteric fiber such as nylon 6 by the acid dye [4,5], the sign of the ζ potential of fibers in the dye solution reversed and then its values approximated the saturated value with increasing concentration of dye. These facts suggest that these fibers combine with the dye mainly by an electrostatic bond.

Difference in ζ Potential and Surface Dyeability of Natural and Synthetic Fiber in Dye Solution

In general, it is believed that synthetic fiber is more difficult to dye than natural fiber. But the surface dye adsorption (expressed in g/cm^2 of fiber), which was calculated from the differences between the surface charge density of the fiber with dye and that without it, was greater for synthetic fiber than for natural fiber [25,26]. As shown in Table 11.2, the surface dye adsorption for synthetic fibers, (i.e., nylon 6 and polyurea Urylon) dyed with an acid dye, Orange II, was about 10 times that of natural fiber, (i.e., wool) dyed with the same dye. The same facts were recognized [25] for a natural fiber, cotton, and a synthetic fiber, Vinylon, dyed with a direct dye, Congo Red, and also for a natural fiber, wool, and a synthetic fiber, polyacrylonitrile Cashimilon, dyed with a basic dye, Malachite Green. From these results, in reference to surface dye adsorption, it may be concluded that synthetic fibers are easy to dye compared with natural fibers.

Table 11.2 Difference in Surface Dyeability of Natural and Synthetic Fiber in an Acid Dye Solution

(1) Wool/Orange II System, pH 3.75

Dye concentration (mol/liter) × 10^4	ζ (mV)	σ (esu/cm^2)	$\Delta\sigma \left(\text{electronic unit}/\text{cm}^2\right) \times 10^{-11}$	Surface dye adsorption (g/cm^2 fiber) × 10^{10}
0	−9.0	−140.0	0	0
1.0	−21.0	−333.4	4.03	2.4
2.0	−26.5	−461.1	6.69	3.9
3.0	−27.4	−487.7	7.24	4.2
4.0	−24.0	−480.6	7.09	4.1

(2) Nylon 6/Orange II System, pH 3.4

Dye concentration (mol/liter) × 10^4	ζ (mV)	σ (esu/cm^2)	$\Delta\sigma \left(\text{electronic unit}/\text{cm}^2\right) \times 10^{-12}$	Surface dye adsorption (g/cm^2 fiber) × 10^{10}
0	+64	+1360.2	0	0
1.0	−86.5	−2366.2	7.8	45

2.0	-107.1	-3948.9	11.1	65
3.0	-105.0	-3974.9	11.1	65
5.0	-113.8	-5132.1	13.5	79

(3) Urylon/Orange II System, pH 3.75

Dye concentration (mol/liter) × 10^4	ζ (mV)	σ (esu/cm^2)	$\Delta\sigma$ $\left(\text{electronic unit}/\text{cm}^2\right)$ × 10^{-12}	Surface dye adsorption (g/cm^2 fiber) × 10^{10}
0	+94.5	+393.0	0	0
0.5	-75.0	-275.5	6.8	38.9
1.0	-94.5	-452.4	8.5	43.1
2.0	-101.5	-576.8	9.7	56.4
3.0	-103.0	-648.7	10.4	60.6

Source: Reproduced with permission from T. Suzawa, *Kogyo Kagaku Zasshi*, *64*, 573 (1961); *65*, 127 (1962); copyright by the Chemical Society of Japan.

Improvement in Dyeability and the Change in its ζ Potential by the Modification of Fiber

As stated above, the determination of the isoelectric point of the amphoteric fiber is performed by estimating the effect of pH on ζ of the fiber. The improvement in dyeability by the electrostatic bond between the fiber and the dye may be achieved through the appearance of the isoelectric point on the negative fiber or the shift of isoelectric point of the fiber to a more alkaline region by the introduction of an electropositive group. Such attempts have been made for various synthetic fibers: for example, the polyacrylonitrile fiber produced by copolymerizing with various basic monomers, or the polypropylene fiber produced by graft copolymerizing with basic vinyl monomers. As an example for the measurement of the ζ potential of these fibers, the ζ for the poly(vinyl alcohol) Vinylon [27], the polyacrylonitrile Dralon [28], Exlan, Acrilan [27], and Vonnel [24,29] were measured.

The relation between the ζ potential and the dyeability of natural and man-made fibers (i.e., silk, viscose rayon, acetate rayon, nylon, Vinylon, Dynel, and Acrilan) was investigated by the introduction of an electropositive group or center on fibers [30,31]. In order to introduce the positive group or center, acetate rayon and Vinylon fibers were cross-linked by urea resins [30] and silk, viscose rayon, nylon, Dynel, and Acrilan fibers were treated in solutions of varying concentrations of copper acetate, chromium acetate of silver nitrate, potassium hydroxide, and dextrose in a fixed proportion (1:2:2) [31]. Figure 11.1 shows the effect of pH on ζ and dye exhaustion of Dynel fiber fixed withe a fixing solution containing copper acetate. With increasing concentration of copper acetate in the solution, the isoelectric point appeared at a concentration of 2% and then shifted to a more alkali region. Consequently, dye exhaustion increased. This suggests the contribution of an electrostatic bond between the positive center on the fiber and the dye anion.

As described above, hydrophobic synthetic fibers such as polyester fibers possess a high negative ζ-potential value. Also, poly(ethylene terephthalate) fibers are highly crystalline and do not contain chemically reactive groups. Therefore, this fiber is usually dyed with a disperse dye. Recently, polyester fiber was modified through the graft copolymerization of acrylic acid and acrylonitrile onto the fiber by both radiation and chemical methods, and their electrokinetic properties were reported [32,33]. The electrokinetic properties of these grafted polyester fibers in the presence of cationic dyes were also investigated [33]. As the amount of the graft monomers increased, the negative value of its ζ potential at a given pH decreased. This was attributed to an increase in the hydrophilicity of the fibers as a result of grafting. Also, a lowering of surface charge density and surface area of fibers as a result of grafting were observed.

In the case of an acrylic acid-grafted polyester fiber, the negative value of the ζ potential and the surface charge density of the fibers

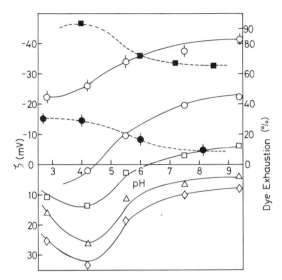

Figure 11.1 Effect of pH on ζ and dye exhaustion of Dynel fiber fixed with a fixing solution containing copper acetate. Open symbols and solid symbols represent ζ potential and dye exhaustion, respectively; the percentages of copper acetate in fixing solution are 0 ($\Diamond$), 2 ($\circ$), 5 ($\square$), 10 ($\triangle$), and 20 ($\Diamond$). [Redrawn with permission from Kogyo Kagaku Zasshi 60:611 (1957); copyright by the Chemical Society of Japan.]

decreased with increasing concentration of cationic dye in solution; the higher the amount of acrylic acid grafted on the fibers, the faster was the decrease in the ζ potential. This result suggests that by the electrostatic bond, cationic dye is adsorbed at the carboxyl group introduced to the fiber [33]. As shown in Fig. 11.2, the negative value of the ζ potential of acrylonitrile-grafted polyester fiber in an aqueous solution of the concentration (1×10^{-5} mol/liter) of the cationic dyes decreased linearly. Also, the ζ potential of the fibers for a constant graft content decreased progressively with increasing concentration of dye. This was attributed to the neutralization of negative charge as a result of an increase in adsorption of dye cations at the -CN group in the grafted fiber [33].

Comparison of the Surface Area Covered by Adsorbed Dye Molecules in a Dye Solution (ζ-Potential Method) with the BET Specific Surface Area of the Fiber

In many investigations on dyeing, the surface area of the fiber has been measured by a Brunauer-Emmett-Teller (BET) plot using nitrogen

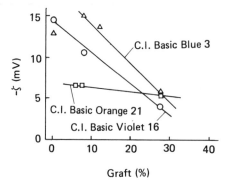

Figure 11.2 Zeta potential of acrylonitrile-grafted polyester fibers versus graft percent for dye concentration 1×10^{-5} mol/liter. [Redrawn with permission from J. Applied Polym. Sci. 23:2139 (1979); copyright by John Wiley & Sons, Inc.]

or water vapor. Also, when the measurement of the surface area covered by adsorbed dye molecules in dye solution is carried out directly, the occupied area and the orientation of dye molecules on the fiber should be known for the application of the BET adsorption equation. But it is actually difficult to know these parameters. Therefore, the surface area covered by adsorbed dye molecules on the fiber, which is determined from the slope of the surface dye adsorption versus total dye adsorption curves (the ζ-potential method), was investigated for various fibers [4-6,14,19,21,34,35]. Here, as stated above, the surface dye adsorption is calculated from the differences between the surface charge density of the fiber with dye and without it. The total dye adsorption is the amount of dye adsorbed by the fibers which is used to measure the ζ potential. The surface dye adsorption versus the total dye adsorption curves of the heat-set nylon 6 fibers in an acid dye solution were found to be linear, as shown in Fig. 11.3. From the reciprocal of the slope of these straight lines, the surface area covered by adsorbed dye molecules was calculated.

As shown in Table 11.3, the surface areas covered by adsorbed dye molecules for nylon 6 fibers treated under different conditions of spinning, drawing, and heat setting in acid dye solution were compared with the BET specific surface areas of the fibers determined by the adsorption of nitrogen and water vapor [35]. From these results it was suggested that both surface areas were independent of spinning temperature and dependent on draw ratio and heat-setting conditions, and that they were governed primarily by the change in the degree of crystallinity. Also, the surface area determined by the ζ-potential method was always larger than that determined by the adsorption of nitrogen (about 10^3 cm^2/g of fiber). Moreover, the former was of the

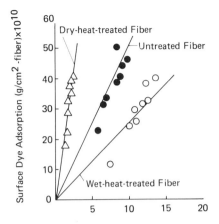

Figure 11.3 Relation between surface dye adsorption and total dye adsorption for nylon 6 fibers treated by different heat setting. [Redrawn with permission from Bull. Chem. Soc. Jpn. 41:539 (1968); copyright by the Chemical Society of Japan.]

same order as the surface area determined by the adsorption of water vapor (about 10^6 cm^2/g of fiber).

The surface area covered by adsorbed dye molecules for cellulose fibers such as cotton and rayon in a direct dye Congo Red solution (pH 10) [21] and that for wet- and dry-spun poly(vinyl alcohol) Vinylon fibers in the same dye solution [14] were determined. From the comparison of the surface area with the BET specific surface area, a result was obtained similar to that of the nylon 6 fiber mentioned above. Moreover, the surface area covered by adsorbed dye molecules was independent of the pH of the solution and chemical structure of the dye [4,25,34].

11.3.2 Dye Structure and Surface Dyeability

ζ Potential of Fiber in a Direct Dye Solution

With increasing concentration of dye, the negative surface potential of cotton increased with dibasic Bonzopurpurine 4B and decreased with tetrabasic Chlorazol Sky Blue FF [36]. The ζ potential (negative) of cotton fiber in aqueous solutions of direct dyes of benzidine derivative such as Congo Red and Benzopurpurine 4B and 10B having two SO_3Na groups in its dye molecule gave almost the same value, but the ζ potential in aqueous solution of Orange R of the same derivative having one SO_3Na group and one COONa group was less than that in aqueous solution of the former three dyes [19]. It was recently found

Table 11.3 Comparison of Surface Area Covered by Adsorbed Dye Molecules in an Acid Dye Solution with BET Specific Surface Area of Nylon 6 Fiber

Fiber			BET specific surface area		Surface area covered by adsorbed dye molecule on fiber $(cm^2/g\ fiber) \times 10^{-6}$
			Adsorption of nitrogen $(cm^2/g\ fiber) \times 10^{-3}$	Adsorption of water vapor $(cm^2/g\ fiber) \times 10^{-6}$	
Fiber spun under different spinning temperature	Spinning temperature (°C)	280	1.41	0.98	1.55
		290	1.38	0.98	
		300	1.39	1.01	
Drawn fiber	Draw ratio	1.0	4.53	1.44	4.40
		3.0	1.50	0.89	2.90
		4.0	1.28	0.69	1.80-2.00
Heat-set fiber	Untreated fiber		1.28	0.69	1.80-2.00
	Wet-heat-treated fiber		0.97	4.00	4.00
	Dry-heat-treated fiber		0.51	0.50	0.60

Source: Reproduced with permission from T. Suzawa and T. Yamaoka, Nippon Kagaku Kaishi, 1974, 2250 (1974); copyright by the Chemical Society of Japan.

that the ζ potential of the formaldehyde fiber in aqueous solutions of Congo Red and Chlorazol Sky Blue FF increased linearly with increasing concentration of dye and that the ζ potential of the fiber in the latter dye solution was larger than that in the former dye solution [37].

ζ Potential of Fiber in an Acid Dye Solution

The ζ potential and the surface dye adsorption for wool [38], nylon 6 [34], and polyacrylonitrile Vonnel [39] fibers in an acidic solution of monoazo acid dyes such as Orange I, II, and IV were determined. With increasing concentration of dye, the sign of the ζ potential for each fiber in these dye solutions changed from positive to negative because of the electrostatic bond between the basic groups of fiber and dye anion. Also, the surface dye adsorption increased in the same manner as the ζ with increasing concentration of dye. As shown in Fig. 11.4, the ζ potential for nylon 6 fiber became smaller in the order of Orange IV, I, and II [34]. This fact suggests the contribution of the intermolecular hydrogen bond between the amide group in the fiber-forming polymer molecule and the polar group in the dye molecule, such as the free OH group of Orange I and the NH group of Orange IV. It also suggests the contribution of the van der Waals force between the hydrophobic part in the dye molecule, such as the diphenylamine portion of Orange IV and the hydrocarbon chain portion of the fiber.

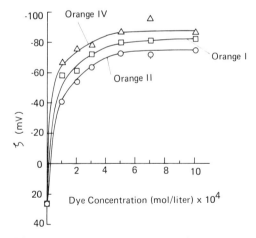

Figure 11.4 Zeta potential of nylon 6 fiber with increasing concentration of the acid dyes (pH 4, 25°C). [Redrawn with permission from Sen-i Gakkaishi (Journal of the Society of Fiber Science and Technology, Japan) 27:108 (1971); copyright by the Society of Fiber Science and Technology, Japan.]

Table 11.4 Temperature Dependence of Free Energy of Dye Adsorption and Maximum Number of Orange IV Molecule Adsorbed per Square Centimeter of Polyacrylonitrile Fiber Vonnel P

	Temperature			
	15°C	20°C	25°C	30°C
ζ^* (mV)	47.3	45.5	42.0	39.6
C_d° (mol/liter) $\times 10^5$	3.10	3.80	4.65	5.80
$(d\zeta/d \log C_d)_{\zeta=0}$ (mV)	−39.10	−39.0	−39.0	−39.0
N_1 (cm^{-2} fiber)	24.8×10^{11}	24.1×10^{11}	23.1×10^{11}	22.5×10^{11}
$\Delta \overline{G}$ (kcal mol^{-1})	−8.55	−8.54	−8.48	−8.41

Source: Reproduced with permission from T. Suzawa and K. Kawakami, *Nippon Kagaku Kaishi*, 1975, 1134 (1975); copyright by the Chemical Society of Japan.

In the case of the dyeing of polyacrylonitrile fiber Vonnel P in acid dye solution [39], a linear relation was found between ζ and log C_d (C_d is the dye concentration) and the slope was independent of temperature. As shown in Table 11.4, the free energy of dyeing $\Delta \bar{G}$ calculated using Eqs. (11.5)-(11.9) from the slope of the ζ versus log C_d curve decreased with increasing temperature. This corresponds to the decrease in the surface dye adsorption with rise of temperature. Also, the occupied areas of the dye molecule on the fiber at the maximum dye adsorption are calculated from the maximum number, N_1, of available sites per unit area of fiber surface [refer to Eq. (11.6)]. From these occupied areas for Orange I and IV on the fiber, it was suggested that these dyes were coarsely adsorbed on Vonnel P fiber.

ζ Potential of Fiber in a Cationic Dye Solution

The sign of the ζ potential (negative) for poly(vinyl alcohol) Vinylon [22], polyacrylonitrile Dralon [28], and Vonnel P [24] fibers in a cationic dye solution changed from negative to positive because of the electrostatic bond between the acidic end group, such as the COO^- group for poly(vinyl alcohol) fiber or the SO_3^- group for polyacrylonitrile fiber, and the dye cation. Then the ζ potential (positive) approximated the saturated value.

In the case of these fiber-dye systems, the free energy of dyeing $\Delta \bar{G}$ was also calculated from the slope of the ζ versus log C_d curve, as mentioned in the preceding section [22,24]. The $\Delta \bar{G}$ of the poly(vinyl alcohol) fibers having a different degree of formalization and the same degree of crystallinity in a cationic dye Methylene Blue solution increased with increasing degree of formalization [15]. This corresponds to the increase in the surface dye adsorption with increasing degree of formalization. The heat of dyeing $\Delta \bar{H}$ and the entropy of dyeing $\Delta \bar{S}$ were evaluated from the temperature dependence of $\Delta \bar{G}$. With increasing degree of formalization, the $\Delta \bar{H}$ (negative) increased and the $\Delta \bar{S}$ (positive) decreased. The positive value of $\Delta \bar{S}$ for each fiber may be attributed to the formation of a hydrophobic bond as a driving force, in addition to the electrostatic bond.

ζ Potential of Fiber Dyed with Reactive Dyes

As described above, the electrokinetic behavior of dyeing has been studied extensively, but very little work has been done on fibers dyed with reactive dyes. The ζ potential of cellulosic fibers dyed with reactive dyes was investigated by using reactive dyes having monochlorotriazine, vinylsulfone (both monofunctional and bifunctional), and trichloropyrimidine reactive systems [40]. In order to investigate the effect of the reaction between the reactive dye molecule and the cellulosic fiber, two sets of experiments were performed. In one, the dye was allowed to react to form a covalent bond with cellulose. In the other, the dye was present in an unreacted form. Equal quantities

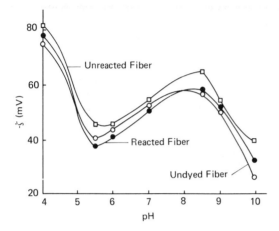

Figure 11.5 The ζ versus pH curves for cellulose fiber dyed with reactive dye Cibacron Brilliant Blue BR at 7.76 mmol/kg of the fiber. [Redrawn with permission from Colloid and Polymer Sci. 254:1030 (1976); copyright by Dr. Dietrich Steinkopff Verlag GmbH & Co. KG.]

of dye were dyed in reacted and unreacted form for individual dyes. Figure 11.5 shows the ζ versus pH curves for the fiber dyed with Cibacron Brillaint Blue BR. At pH 7, the value of the ζ potential for the fiber containing a dye in the unreacted form is greater than the corresponding value for the blank undyed sample. This suggests the effect of SO_3H groups present in the adsorbed dye molecule on the ζ potential of the fiber. However, when the same amount of dye is present in the reacted form on the fibers, the ζ potential drops considerably. This may be attributed to the difference between the decrease in ζ potential as a result of blocking the OH groups in the fiber due to reaction with the dye molecule and the increase in ζ potential as a result of SO_3H groups present in the dye molecule. Also, the effect of the nature and function of the reactive systems of a dye molecule on the ζ potential and the surface charge density of fibers dyed with reactive dyes were investigated [40,41].

11.4 ELECTROKINETIC PROPERTIES OF FIBER AND ADSORPTION OF SURFACTANTS

In connection with detergency or dyeing, the relation between adsorbability of surfactants and electrokinetic property of the fiber surface has been investigated. It has been carried out in aqueous solutions of surfactant concentration below or above the critical micelle concentration (CMC) [42].

11.4.1 Adsorption Behavior of Fibers in a Surfactant Solution Below the CMC

In aqueous solutions of surfactant concentration below the CMC, the ζ potential of the following fibers was measured: cotton, silk, and wool as natural fibers [43,44]; viscose rayon as regenerated fiber [44]; acetate as semisynthetic fiber [44]; polyamide-nylon 6 and 66 [43,44], poly(vinyl alcohol)-Vinylon [46,47], poly(vinyl chloride)-Teviron [29], polyacrylonitrile-Cashimilon [29,45,46], Kanekalon [29] and Vonnel [29], polyester-Tetron [46,47], polyethylene-Hizex [46], and polypropylene-Pylen [46,47] as synthetic fibers.

In the adsorption between the fiber and the surfactant having opposite electrical signs, such as an anionic surfactant sodium dodecyl sulfate (SDS) and amphoteric fibers—wool, nylon 6 and 66 (in more acidic pH than its isoelectric point) [43,44,46,47] or a cationic surfactant dodecyl pyridinium bromide (DPB)—and negative fibers—poly(vinyl alcohol), polyacrylonitrile, polyethylene, and polypropylene fibers [46,47]—with the increase in surfactant concentration the ζ potential of the fiber in the surfactant solution reversed its sign and then increased in value mainly by the electrostatic bond between the fiber and the surfactant. In the adsorption between the fiber and the surfactant having the same electrical sign, such as a negative fiber and SDS or amphoteric fiber (in more acidic pH than its isoelectric point) and DPB, by increasing the surfactant concentration, the ζ potential of the fiber increased by the van der Waals force between the hydrophobic groups of the fiber and that of the surfactant, or the ζ value remained almost constant by the electrostatic repulsion force between them [43,44,46,47].

Also, from the values of ζ potential, the surface charge density σ of the fiber in aqueous solutions of the surfactant was calculated by means of Eq. (11.4). Moreover, the amounts of ionic surfactant adsorbed per unit area of the fiber surface, which is expressed as the surface adsorption in g/cm^2 of fiber, were calculated from the difference $\Delta \sigma$ between the surface charge density of the fiber in aqueous solution with surfactant and that without it [43,45-47]. The surface adsorption of the surfactant increased with increasing surfactant concentration, and the increase was greater the more hydrophobic the fiber became (and the smaller the degree of swelling of the fiber) [46,47], regardless of pH.

In aqueous acidic solutions (pH 3.4) of a nonionic surfactant, poly(ethylene glycol nonyl phenyl ether), the negative ζ of the hydrophobic fibers such as polyacrylonitrile, polyethylene, and polypropylene fibers decreased at first and then approached constant values with increasing surfactant concentration [46]. This suggests that oxonium cation is produced from the nonionic agent.

The ζ potential of amphoteric fibers such as nylon 6 and polyacrylonitrile Vonnel P in aqueous acidic solutions (pH 3.4) of sodium alkyl

Table 11.5 Heat of Adsorption and Entropy of Adsorption for Sodium Alkyl Sulfates on Vonnel P Fiber

	SDS	STS
$\Delta \overline{H}$ (kcal mol^{-1})	−6.8	−6.1
$\Delta \overline{S}$ [cal (mol·deg)$^{-1}$]	5.3	8.7

Source: Reproduced with permission from T. Suzawa and T. Takahashi, Yukagaku, 25, 798 (1976); copyright by the Japan Oil Chemists' Society.

sulfates increased by the van der Waals force in addition to the electrostatic bond with increasing chain length of the surfactant [45,48,49].

The free energy of adsorption $\Delta \overline{G}$ for sodium alkyl sulfates on Vonnel P fiber or sodium 2-sulfonated fatty acid methyl esters on nylon 6 fiber, respectively, was calculated using Eqs. (11.5)-(11.9) from the slope of the ζ versus log C_s (C_s: the concentration of surfactant) curve, and so on [48,49]. The negative values of $\Delta \overline{G}$ increased with rise of temperature and with increasing chain length of the surfactant. The heat of adsorption $\Delta \overline{H}$ and the entropy of adsroption $\Delta \overline{S}$ for SDS and sodium tetradecyl sulfate (STS) on Vonnel fiber, which are evaluated from the temperature dependence of the $\Delta \overline{G}$, are shown in Table 11.5. The positive values of $\Delta \overline{S}$ indicate the importance of entropic interaction and the formation of hydrophobic bonds as a driving force for adsorption of the surfactant on the fiber surface.

11.4.2 Adsorption Behavior of Fibers in a Surfactant Solution Above the CMC

In the adsorption isotherms of anionic or cationic surfactants on solids such as carbons, metals, and fibers [50-57], a discontinuity appears in the concentration, which approximately coincides with the CMC of the surfactant, and the maximum appears at a higher concentration than the CMC. Also, the concentration of maximum adsorption corresponds to the critical washing concentration (CWC) in detergency [58]. In contrast to these facts, several studies [59-64] did not produce such maximums in the isotherms. In connection with these phenomena, the measurement of the electrokinetic potential of fibers in surfactant solution in the vicinity of the CMC and above its concentration is of importance.

The following ζ potential values of fibers were measured in aqueous solutions of surfactant concentration above the CMC. The ζ potentials

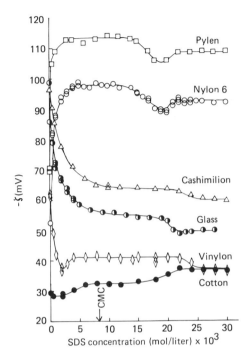

Figure 11.6 Zeta potentials for fibers with SDS concentration (25°C). [Redrawn with permission from Bull. Chem. Soc. Jpn. 43:2326 (1970); copyright by the Chemical Society of Japan.]

of cotton and wool fibers became constant in aqueous sodium cetyl sulfate (SCS) solutions above the CMC (5×10^{-4} mol/liter, at pH 11 and 55°C) [43]. Also, in aqueous cetyl pyridinium chloride solutions, the ζ value of cotton became nearly constant for some distance above the CMC (6.8×10^{-4} mol/liter, at pH 6.4 and 25°C) and diminished slightly at still higher concentrations [43]. Further, in aqueous sodium alkyl sulfate solutions in the vicinity of the CMC, the maximum ζ potential of fibers such as cotton, wool, silk, viscose rayon, and nylon fibers either appeared or did not appear with increasing surfactant concentration [44]. Moreover, a change in the ζ value for such fibers as cotton, poly(vinyl alcohol) Vinylon, nylon 6, polyacrylonitrile Cashimilon, polypropylene Pylen, and glass in aqueous sodium alkyl sulfates, SDS, STS, and SCS, solutions was investigated with increase in surfactant concentration [65,66]. As shown in Fig. 11.6, the negative ζ potentials of fibers in aqueous surfactant solutions change rapidly with increasing surfactant concentrations, but they remain nearly constant in the vicinity of the CMC. At much higher concentrations above the CMC, in the vicinity of the concentration of 2×10^{-2} mol/liter for SDS, the ζ values again gradually change; thereafter, passing a break or a minimum, they become approximately constant again. The concentration at the break or the minimum observed nearly corresponded to that at the break of the η_{spm}/C_m (C_m is the micellar concentration), which is reduced viscosity for the micelle [65,66] versus C_m curve. From this fact it may be suggested that the change in ζ potentials at high concentrations is connected with that in the structure of the micelle [65,66]. Also, the negative ζ potentials of fibers in aqueous solutions of mixed sodium alkyl sulfates changed sharply with increasing concentration of the mixtures, and thereafter, passing a maximum in the vicinity of the CMC, changed completely as the total concentrations and the mixing ratios of the surfactants changed [66].

REFERENCES

1. T. Suzawa, *Kogyo Kagaku Zasshi*, 62, 232, 1231 (1959); *63*, 1069, 2205 (1960); *64*, 573 (1961); *Dyestuffs Chem.*, *8*, 543 (1963).
2. W. G. Eversole and W. W. Boardman, *J. Chem. Phys.*, 9, 798 (1941).
3. S. R. S. Iyer and R. Jayaram, *J. Soc. Dyers Colour.*, 87, 338 (1971).
4. T. Suzawa, T. Saito, and H. Shinohara, *Bull. Chem. oc. Jpn.*, 40, 1596 (1967).
5. T. Suzawa and T. Saito, *Bull. Chem. Soc. Jpn.*, 41, 539 (1968).
6. T. Suzawa and H. Takahashi, *Kogyo Kagaku Zasshi*, 72, 906 (1969).
7. T. Suzawa, *Kogyo Kagaku Zasshi*, 74, 2146 (1971).
8. K. Kanamaru, *Kagaku to Kogyo*, 12, 89 (1958); T. Suzawa, *Kogyo Kagaku Zasshi*, 64, 573 (1961).
9. K. Kanamaru, *Kolloid-Z.*, 168, 115 (1960).
10. A. Kitamura and A. Shibamoto, *J. Sericult. Sci. Jpn.*, 44, 307 (1975).
11. K. Kanamaru and M. Hirata, *Kolloid Z. Z. Polym.*, 230, 206 (1969).
12. M. Hirata and S. Iwai, *Nippon Kagaku Kaishi*, *1972*, 2397 (1972); *1973*, 2202 (1973); M. Hirata, Y. Okano, and S. Iwai, *Nippon Kagaku Kaishi*, *1974*, 589, 1758 (1974).
13. O. Yoshizaki, *Kobunshi Kagaku*, 12, 420 (1955).
14. T. Suzawa and N. Kitagawa, *Kogyo Kagaku Zasshi*, 73, 1858 (1970).
15. T. Suzawa and Y. Yamashita, *Colloid Polym. Sci.*, 259, 973 (1981).

16. G. L. Madan, S. K. Shrivastava, and N. T. Baddi, *Kolloid Z. Z. Polym.*, *251*, 483 (1973).
17. G. L. Madan and S. K. Shrivastava, *Colloid Polym. Sci.*, *255*, 269 (1977).
18. T. Suzawa, *Kogyo Kagaku Zasshi*, *63*, 148 (1960).
19. T. Suzawa, *Kogyo Kagaku Zasshi*, *64*, 573 (1961).
20. R. H. Ottewill, M. C. Rastogi, and A. Watanabe, *Trans. Faraday Soc.*, *56*, 854 (1960); R. H. Ottewill and A. Watanabe, *Kolloid-Z.*, *170*, 38; *171*, 33; *173*, 122 (1960).
21. T. Suzawa and T. Mukai, *Kogyo Kagaku Zasshi*, *73*, 2451 (1970).
22. T. Suzawa and K. Touma, *Nippon Kagaku Kaishi*, *1974*, 855 (1974).
23. T. Suzawa, *Kogyo Kagaku Zasshi*, *65*, 124 (1962).
24. T. Suzawa and M. Takaoka, *Nippon Kagaku Kaishi*, *1972*, 2424 (1972).
25. T. Suzawa, *Kogyo Kagaku Zasshi*, *64*, 573 (1961).
26. T. Suzawa, *Kogyo Kagaku Zasshi*, *65*, 127 1962).
27. O. Yoshizaki, *Kobunshi Kagaku*, *15*, 761 (1958).
28. O. Glenz and W. Beckmann, *Melliand Textilber.*, *38*, 296 (1957).
29. K. Yamaki, *Kogyo Kagaku Zasshi*, *65*, 2036 (1962).
30. K. Kanamaru and S. Ota, *Kogyo Kagaku Zasshi*, *60*, 452 (1957).
31. K. Kanamaru and S. Ota, *Kogyo Kagaku Zasshi*, *60*, 611 (1957).
32. P. D. Kale and H. T. Lokhande, *J. Appl. Polym. Sci.*, *19*, 461 (1975).
33. H. T. Lokhande and N. R. Mody, *J. Appl. Polym. Sci.*, *23*, 2139 (1979).
34. T. Suzawa, K. Nagareda, and Y. Mizuhiro, *Sen-i Gakkaishi*, *27*, 108 (1971).
35. T. Suzawa and T. Yamaoka, *Nippon Kagaku Kaishi*, *1974*, 2250 (1974).
36. S. M. Neale and R. H. Peter, *Trans. Faraday Soc.*, *42*, 478 (1946).
37. H. T. Lokhande and A. S. Salvi, *J. Appl. Polym. Sci.*, *21*, 277 (1977).
38. T. Suzawa, *Kogyo Kagaku Zasshi*, *63*, 1066 (1960).
39. T. Suzawa and K. Kawakami, *Nippon Kagaku Kaishi*, *1975*, 1134 (1975).
40. H. T. Lokhande and A. S. Salvi, *Colloid Polym. Sci.*, *254*, 1030 (1976).
41. H. T. Lokhande and A. S. Salvi, *Colloid Polym. Sic.*, *256*, 1021 (1978).
42. T. Suzawa, *Yukagaku*, *11*, 619 (1962).
43. J. S. Stanley, *J. Phys. Chem.*, *58*, 536 (1954).

44. M. von Stackelberg, W. Kling, W. Benzel, and F. Wilke, Kolloid-Z., 133, 67 (1954).
45. T. Suzawa, Kogyo Kagaku Zasshi, 65, 2042 (1962).
46. T. Suzawa, Kogyo Kagaku Zasshi, 66, 1002 (1963).
47. T. Suzawa and M. Yuzawa, Yukagaku, 15, 20 (1966).
48. T. Suzawa and T. Takahashi, Yukagaku, 25, 796 (1976).
49. T. Suzawa and T. Hakozaki, Yukagaku, 29, 260 (1980).
50. M. L. Corrin, E. L. Lind, A. Roginsky, and W. D. Harkins, J. Colloid Sci., 4, 485 (1949).
51. A. L. Meader and B. A. Fries, Ind. Eng. Chem., 44, 1636 (1952).
52. R. D. Vold and A. K. Phansalkar, Recl. Trav. Chim. Pays-Bas, 74, 41 (1955).
53. A. Fava and H. Eyring, J. Phys. Chem., 60, 890 (1956).
54. R. D. Vold and N. H. Sivaramakrishnan, J. Phys. Chem., 62, 984 (1958).
55. H. C. Evans, J. Colloid Sci., 13, 537 (1958).
56. H. J. White, Y. Gotshal, L. Robenfeld, and F. H. Sexsmith, J. Colloid Sci., 14, 598 (1959).
57. F. Z. Saleeb and J. A. Kitchener, J. Chem. Soc., 1965, 911.
58. T. Tachibana, A. Yabe, and M. Tsubomura, J. Colloid Sci., 15, 278 (1960).
59. A. S. Weatherburn and C. H. Bayley, Text. Res. J., 22, 797 (1952).
60. M. Hayashi and Y. Nakazawa, Radioisotopes, 8, 149 (1959).
61. M. Hayashi, Bull. Chem. Soc. Jpn., 33, 1184 (1960).
62. C. Sakai and S. Komori, Yukagaku, 14, 66 (1965).
63. J. A. Kitchener, J. Photogr. Sci., 13, 152 (1965).
64. B. E. Gordon, G. A. Gillies, W. T. Shebs, G. M. Hartwig, and G. R. Edwards, J. Am. Oil Chem. Soc., 43, 232 (1966).
65. Y. Iwadare and T. Suzawa, Bull. Chem. Soc. Jpn., 43, 2326 (1970).
66. Y. Iwadare, Bull. Chem. Soc. Jpn., 43, 3364 (1970).

12
Paper

Hiroshi Yamada* Sanyo-Kokusaku Pulp Co., Ltd., Tokyo, Japan

12.1. Introduction 323
12.2. Electrokinetic Properties of Cellulose Fibers 324
12.3. Interactions Between Cellulose and Polyelectrolytes 324
    12.3.1. Electrokinetic effect of polyelectrolytes 324
    12.3.2. Adsorption of polyelectrolytes onto cellulose fibers 326
    12.3.3. Drainage, retention of fines, and flocculation mechanism 327
12.4. Electrokinetic Properties and Retention of Fillers 330
12.5. Sizing of Paper 332
12.6. Miscellaneous 333
12.7. Conclusion 334
    References 334

12.1 INTRODUCTION

Suspensions of paper stocks are generally regarded as colloidal systems. Recently, studies on destabilization of these systems have become of great importance in the paper industry for the improvement of machine production, efficient utilization of resources, and pollution abatement. In particular, attention has been focused on retention and drainage at the wet end of a paper machine, and extensive studies have been carried out of fundamental and practical aspects in the last decade. In this chapter the present state of the development of electrokinetics in papermaking is reviewed.

---

*Dr. Yamada is now retired.

12.2 ELECTROKINETIC PROPERTIES OF CELLULOSE FIBERS

The zeta (ζ) potential of cellulose fibers is affected by the type of wood, the pulping process, the degree of cooking, and the presence of residual substances [1]. Table 12.1 shows the ζ potential of different pulps obtained from mills [2]. Bleaching has less effect than the wood species or pulping process used. Absolute values obtained by the three measuring methods do not agree with one another, even though correlations showing the same trend might exist.

Whether or not the ζ potential of fine fibers can represent the whole system has intrigued many investigators [3]. Although the fines fraction has been confirmed to carry higher electric charges compared with long fibers [4,5], different results are obtained using dry-milled pulp [6,7]. It is also reported that the ζ potential of long fibers as obtained by the streaming method is close to that of the fines of the same pulp measured by a microelectrophoresis technique [8,9].

Electrophoretic mobility increases with carboxyl content of the pulp [6,10]. The ozonization of thermomechanical pulp is said to enhance the ζ potential for the same reason [11].

The ζ potential increases with the degree of beating or refining of cellulose fibers. This is believed to be due to higher adsorption of substances of negative charge and increase in sites having negative charge on the surface [8,12]. However, some reports differ from this [13,14].

The ζ potential of fibers decreases with decreasing pH, gradually from 9 to 4 and rapidly from 4 to 3 toward the isoelectric point of about 2 [8,9,12]. This may be caused by the suppression of dissociation of functional groups.

Inorganic salts lower the ζ potential of fibers due to the compression or destruction of the double layer surrounding them [12,15]. The effect of salts is shown to become more pronounced with valency of cation [15-17]. The addition of small amounts of aluminum sulfate (alum) causes a marked effect on the ζ potential [8,15,18]. The sign of the ζ potential may be reversed under certain circumstances [19-21]. There is such a case where the ζ potential is constant at -3 and $+5$ mV regardless of the incremental addition of alum [22].

12.3 INTERACTIONS BETWEEN CELLULOSE AND POLYELECTROLYTES

12.3.1 Electrokinetic Effect of Polyelectrolytes

Many kinds of polyelectrolytes, such as polyacrylamide (PAM), polyethylenimine (PEI), polyamine (PA), polyamine-polyamide-epichlorohydrin (PPE), urea- or melamine-formaldehyde (UF or MF), and modified starch, have been employed in electrokinetic studies.

Table 12.1 Zeta Potential of Various Pulps Measured by Three Methods (Unwashed, 22 ± 1°C)

Type of pulp	Streaming potential/streaming current method		Repap instrument (−mV)	Laser-Zee meter (−mV)
	−mV	−B[a]		
1. Spruce sulfite, bleached	14.2 ± 0.6	14.2	23.5 ± 3	32
2. Spruce sulfite, unbleached	14.9 ± 0.7	21.6	22.5 ± 3.5	38
3. Spruce, pine sulfate, unbleached	19.4 ± 0.3	11.6	21.5 ± 3	36
4. Spruce magnefite, bleached	20.4 ± 0.3	18.2	23.5 ± 5	36
5. Beech sulfite, bleached	18.6 ± 0.3	11.7	37 ± 5	38
6. Beech magnefite, bleached	20.8 ± 0.3	10.3	25.5 ± 2	40
7. Spruce, pine sulfate, bleached	21.9 ± 0.4	15.8	35 ± 5	42
8. Birch sulfite, unbleached (high-yield pulp)	15.1 ± 0.2	21.7	27.5 ± 6.5	44
9. Spruce, fir thermo-mechanically ground, bleached	31.0 ± 0.2	13.0	35 ± 3	55
10. Ground, unbleached	—	—	28 ± 5	44

[a] −B refers to a measure of plug density. The higher the plug density, the higher the value of −B.

It was found that the higher the molecular weight or the degree of substitution, the less the cationic demand for PAM [23,24]. This is because high molecular weight and high cationic density make the configuration of the molecule elongate in solution, so that the polymer can be adsorbed efficiently and covers more negative sites on the surface. Also, the cationic demand increases with increase in surface area, indicating that the polymer may not come into the cellulose, but reacts with the outer hydrodynamic surfaces.

Comparing the effectiveness of polymers, the cationic demand for PAM was found to be two and half times as much as for PEI at pH 5 [9]. By means of electrostatic titration, the cationic demand was shown to increase in the order cationic PAM, PPE, UF resin, and cationic starch [20,21]. A plateau was observed at a ζ potential of -3 mV when wet-strength resin or cationic starch was used [22]. The effect of polymers is also described in other publications [5,12,18,22,25,26].

The cationic demand may not always increase in proportion to the carboxyl content of cellulose because of the difference of penetrability of polymers [10]. Pulp beating enhances the cationic demand [9,12,20,21]. The region of the plateau was shown to become more extensive as the beating proceeds, while specific conductivity continues to rise steadily with increasing addition of polyamide resin [14].

The cationic demand increases with increasing pH due to an increase in the negative charge of the pulp [9,12,23,24]. Salt concentrations diminish the cationic demand, the extent of reduction being larger with calcium salt than with sodium salt [10,20,21,23,24].

The effectiveness of cationic polymers is interfered with by the various organic components dissolved out of pulp or added deliberately. Among those components are humic acid, lignosulfonic acid, xylan, oxidized starch, and others [9,23,24,27,28].

12.3.2 Adsorption of Polyelectrolytes onto Cellulose Fibers

Many extensive investigations have been made based on adsorption kinetics and flocculation of fibers.

When using cationic PAM, adsorption generally increases with decrease in molecular weight or degree of substitution of polymer, and with increase in amount of addition or temperature [23,29-31]. On the other hand, flocculation increases with increasing molecular weight, sol concentration, and stirring speed [32]. Adsorption of cationic PAM is much higher than that of anionic or nonionic PAM [33]. Adsorption is also affected by pH and ionic strength [30,32].

Adsorption of PEI depends on the molecular weight, the amount of addition, and the pH [34,35]. In contrast to the low adsorption and the high charge-reversing ability of high-molecular-weight PEI, adsorption of low-molecular-weight PEI seems to occur at the inner surfaces of pore structure not effected electrokinetically. The maximum flocculation was observed to occur in a different mobility range in high- and in low-molecular-weight PEI [35].

Cationic polymers can be completely adsorbed at a fixed pH, which varies with the structure of the polymer and shifts to the alkaline side with increasing basicity of the amino group [36]. Adsorption of modified starches of different ionic characters is also reported [37-39].

Adsorption increases with lapse of time and speed of agitation because of the exposure of fresh surfaces and the diffusion of polymer into inaccessible regions. In a system involving fine fibers alone, the change of mobility is much faster than that of adsorption and the equilibrium is realized within 1 min. When long fibers are present, mobility decreases gradually toward the negative side with time. From these results it can be seen that adsorption of the polymer occurs more rapidly onto fines than onto long fibers [10,12,20,23,29,30,40-42].

The structure of the adsorbed layer varies with the affinity of cellulosic substrates to polymer. Whether the adsorption proceeds to form monolayers or multilayers is determined by the combined effects of specific surface area of fibers, carboxyl content of fibers, molecular area of one adsorbed segment, and cationic charge of the polymer [43].

The fact that adsorption increases with increasing amount of carboxyl group suggests the ionic bonds or the cation exchange for the adsorption mechanism [44]. However, the adsorption appears to proceed in two steps through ionic bonds and van der Waals forces or hydrogen bonds before and after neutralization [10,45]. Physical entrapment of the polymer in the fine pore structure of fibers and the necessity of taking the accessible surface area of fibers into account have also been proposed [46,47].

12.3.3 Drainage, Retention of Fines, and Flocculation Mechanism

Paper stocks contain varying amounts of fine particles, including filler or pigment. The term "fines fraction" is used to denote all materials that pass through the 75-μm screen [48]. If fine particles are agglomerated, they will attach to the fiber and drainage and total retention will be improved. Although the paper stock tends to flocculate due to the presence of long fibers, it is necessary to use polymeric aids for keeping a tolerable degree of drainage and retention.

Although optimum drainage and retention are found to occur at a ζ potential of 0 mV in some instances [9,10,27,49], they do not always correlate with charge neutralization and may occur before neutralization, depending on the paper stock and the polymer type used [3,5,14].

The effect of aluminum salts on the dewatering and retentive activity of cationic PAM has been studied, and both aluminum hydroxide and sulfate ion are found to take part in the whole mechanism [50]. The drainage and fines retention are affected by the ionic domain in water

[17,20,21]. The effects of organic substances are also given [9, 28, 51].

Cationic PAMs containing various molar ratios of amino group have been studied to evaluate the filtration rate [52]. High-molecular-weight PEIs having a narrow distribution are most effective for filtration [53]. Cationic guar gum is said to be a unique dryness aid [54]. High-molecular-weight polyoxyethylene was found to be the most effective retention aid for newsprint furnish [55,56]. The synergistic effect on flocculation can be obtained by the combination of low-molecular-weight cationic polymer and high-molecular-weight anionic polymer [6,57-63]. This is called the two-step or dual system.

From measurements of the flocculation degree of pulp using a light probe, it was assumed that the cause of flocculation by the addition of polymer is due to bridging between fibers or fiber fragments, and irreversible deflocculation takes place after the flocculation is destroyed by shear forces [64]. Although fibers are redispersed above a fixed dosage of polymer, fines remain attached to fiber at the same concentration [65]. This is due to the decreased collapse of flocs as well as the increased flocculation with fibers.

A dynamic retention/drainage jar can be used for the study on fines retention at different degrees of turbulence. In a purely electrokinetic system as obtained by the use of a metallic cation, retention decreases under strong turbulence, but flocculation is reformed rapidly after the turbulence comes to a standstill. Flocculation close to this mode also occurs when using low-molecular-weight PEI. The mechanism for these soft flocs appears to be electrostatic attraction and is characterized by ease of redispersion and a reversible degree of flocculation.

On the other hand, flocculation obtained using a high-molecular-weight cationic polymer or two-step polymer dosage exhibits high resistance and tenacity to redispersion at severe turbulence and is believed to be formed through bridging of the polymer. These hard flocs show resistance to short-time strong turbulence, but extended exposure as in the case of repeated redispersion causes reduction to a primary retention level similar to soft flocs having little tendency to reform again [6,61-63,66].

When bridges are formed between particles, neutralization of the negative charge is unnecessary and the adsorbed polymer joins the neighboring surfaces even in the state of insufficient neutralization. This was also recognized in the experiment, which measured the relative amount of floc through the decrease of scattered light from the sedimentation path [57].

Insufficient neutralization at the point of maximum flocculation has led to the concept of a patch model [42]. As a result of local neutralization by the trains of polymer segments, domains of both positive and negative charge may coexist mosaiclike on the surface and the polymer extends from the surface forming loops. Using this

(a)

(b)
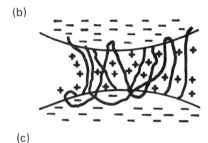

(c)

(d)
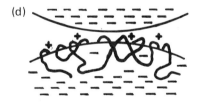

Figure 12.1 Models for (a) initial polymer adsorption, (b) initial flocculation via a patch bridge, (c) reconformation of polymer on surface during redispersion, and (d) reflocculation via electrostatic attraction. (From Ref. 66.)

model, the behavior of the adsorbed polymer on collision with particles can be shown as illustrated in Fig. 12.1.

From the fact that polymer adsorption is faster onto fines than fibers and that the adsorption process is reversible, the flocculation mechanism, including heteroflocculation between cationized fines and fibers followed by the bridging between particles, has also been proposed [30].

It should be mentioned that strong flocculation is accompanied by undesirable effects on formation uniformity of the sheet. This poses a dilemma of how to combine retention, drainage, and formation.

It was observed that if a highly flocked system is subjected to controlled turbulence, the coarse flocs can be smoothed out with relatively little loss in retention and drainage [48]. Based on this result, the concepts of macroflocculation (involving fiber) and microflocculation (deposition of fines on fiber) have been derived. The latter is desirable because of the favorable effects on retention and drainage as well as formation. It is important, however, to control the turbulence carefully in connection with the headbox system design.

If the degrees of retention are measured under the influence of different degrees of turbulence, it is possible to give an indication of the effectiveness of polymers for a certain paper stock system. This flocculation index also enables one to know the agitation speed to be used in the laboratory if one-pass retention of fines on a machine is known [66]. Although the dynamic jar may be used with confidence to forecast the fines retention and drainage, modified techniques capable of simulating machine conditions closer have been proposed [67,68]. The Minidrinier Retention Tester, which enables one to know the retention characteristics just behind the headbox, has been described [69].

12.4 ELECTROKINETIC PROPERTIES AND RETENTION OF FILLERS

The ζ potential of natural and synthetic fillers is given in Fig. 12.2 as a function of pH. Similar data are also presented by other authors [20,22,71]. The ζ potential increases with decrease in particle size [70,72]. The relationship between the mobility of fillers and the concentration of salts, especially that of aluminum sulfate, has been studied extensively [19,20,22,73-76]. The isoelectric points of alumina, anatase, and rutile shift to a lower pH in the presence of fine fibers due to heteroflocculation [74,77]. The effect of cationic polymers on the ζ potential of clay, titanium dioxide, and calcium carbonate has also been reported [20,78-80].

Filler retention is usually evaluated as either one-pass (first-pass) retention or overall retention. The former, which refers to the ratio of amount of filler retained in the sheet to that present in the paper stock at the headbox, is most frequently used for simplicity. It is desirable to enhance the one-pass retention for maintaining cleanliness in the machine and eliminating overload in the saveall system [81].

Filler retention is generally governed by mechanical filtration and chemical adsorption. It varies significantly with shape, size, and surface structure of filler particles [73,82-84].

Filler retention increases with the degree of pulp beating due to the improvement in mutual configuration between fibers and fillers as well

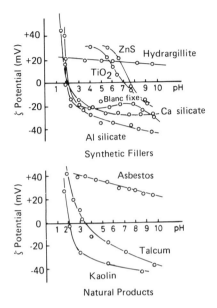

Figure 12.2 Zeta potential of synthetic and natural fillers. (From Ref. 70.)

as the filtration behavior [82,85]. It is also affected by pH, addition of alum, coexisting cation and anion, and the presence of fine fibers [71,73,75-77,83,85-87]. The effect of the order of addition has also been studied in systems involving fiber, titanium dioxide, and alum with or without polymer [78].

Maximum filler retention can be obtained by lowering the ζ potential of the system using cationic or anionic PAM or PPE resin [23,26,27]. Cationic PAM has proved to be effective for retaining China clay in a conventional acidic medium, whereas PEI is useful for chalk in a neutral medium [88]. Cationic PAM with high molecular weight was most effective for retention of titanium dioxide [33,89]. The effect of PEI has been evaluated using filtrated and fractionated pulps [90]. A combination of alum and PAM or PEI enhances the efficiency of the polymer to some extent [88,91,92]. This is true in hard water as well as in soft water [93].

In some cases, however, the filler retention was shown not to be maximum at a ζ potential of 0 mV [94,95]. It was pointed out that any level of retention can be obtained using a high-molecular-weight anionic polymer coupled with a low-molecular-weight cationic polymer, while the ζ potential can be varied arbitrarily from -10 to $+20$ mV.

The effectiveness of cationic starch is affected by the amount of alum not being added, but present in the system. Thus its use becomes unsuitable in systems involving high levels of alum [96,97]. The effectiveness is also lost by the anionic soluble substances. The degree of substitution and the distinction between quaternary and tertiary are important in selecting the starch for best performance [98].

The synergistic effect obtained from a combination of MF resin or PEI and Polyox has been described [99]. Because the ζ potential of the system is -11 to -12 mV in this synergistic flocculation, good formation is obtained compared to the usual flocculation.

Filler retention is influenced significantly by the shear forces during drainage, and the control of ζ potential affects only weak shear-sensitive interparticle forces [95]. Retention loss caused by agitation introduced prior to the slice is rapidly and reversibly recovered when the polymer is used [100]. Also, filler retention decreases with increasing agitation speed and clay concentration and increases with the addition of fine fibers [94]. From these results, a sufficient number of active sites to retain the fine particles and sufficiently strong links between fibers and the retained particles seem to be two major factors. Polymers usually increase the number of sites available for fines retention. Although cationic polymers improve the dynamic retention to 35-53%, the two-step process shows a retention of over 90% [81].

12.5 SIZING OF PAPER

Sizing is another field which has been studied in the light of the electrokinetic process [101-103]. Factors such as type of pulp, type of sizing agent, pH, concentration of aluminum salts, water hardness, presence of foreign contaminants, and use of cationic polymers have been investigated extensively [4,26,83,86,104-112]. Particular emphasis has been placed on the interaction of rosin size with aluminum salts in relation to the mechanism of sizing.

It has previously been suggested that size retention may be treated as a special case of sol coalescence between the size precipitate particles and cellulose surface, where the electrokinetic potential determines whether the precipitate particles and the fibers can approach each other closely enough to exert van der Waals forces or hydrogen bonds as short-range attractive forces.

Following this view, the retention of rosin and aluminum is caused by physical adsorption, and the superimposed electrostatic forces accelerate or arrest the flocculation of particles [105]. For the best sizing, the net surface charge of the sized fibers should be controlled or moderated and the formed precipitates should be well distributed and properly oriented by, for example, sulfate ions.

In addition, the chemical combination of size precipitates and uronic acid groups in pulp was proposed as a modification [107]. These chemical forces combined with van der Waals forces are the essential factors in holding the size precipitates to the fiber surface. The electrostatic attraction is a secondary factor important for pulling the particles to the surface in the first place.

The composition of size precipitates has been reported to be $Al(Ab)_{1.9}(OH)_{1.1}(H_2O)_{1.7} + 1.2HAb + 0.6n$ (Ab: combined abietic acid, HAb: free abietic acid, n: neutral component) in standard precipitate [112].

Sizing with a dispersed-type sizing agent is substantially the same as a soap type, but the formation of aluminum resinate is carried over to the dryer section, contrary to the immediate reaction of the soap type [105]. This is because the aluminum resinate is formed only on the surface of the particles, leaving most of the unreacted resin in the inner part.

In sizing with a synthetic sizing agent such as alkylketene dimer or stearic anhydride, the cationic polymer promotes size retention [25]. The reaction mechanism of aliphatic sizing agents has been reported [113]. The importance of ζ-potential measurement in surface sizing has also been described [114].

12.6 MISCELLANEOUS

The ζ potentials of coating-grade pigments such as kaolin, calcium carbonate, aluminum hydroxide, titanium dioxide, and satin white have been measured with and without dispersants [71,72,115-121]. Rheological properties of slurries have also been studied together with mobility of the system [122]. The ζ potential of pigments in the presence of adhesives is reported as a function of pH in China clay-casein, China clay-starch, satin white-casein, and satin white-starch systems [117]. Mobility and apparent viscosity in the clay-starch system are also given [116]. Other studies have been made using poly(vinyl alcohol) as the adhesive [123,124].

Electrokinetic considerations are successful in the utilization of secondary fibers [5,23,24,27,57,58,125,126], the recovery of white water [5,23,24,83,125,127], the treatment of effluent [28,128-130], the dyeing of paper [27,28,63,131], and in studies on colloidal pitch [132,133], on abrasions of wire [134] and so on.

Electrokinetic charge is also a factor in the deposition of anionic acrylic latex [135] or of cationic styrene or styrene-butadiene latex onto the pulp fibers [136,137] and in the treatment of fillers with cationic polymers for the enhancement of ash content in paper [138,139]. A ζ-potential measurement was also successfully carried out in mill trials [22,51,82,128,140,141] and in the development of specialty papers [5,13,40,57]. Care is needed in the correct use of polymers [142] and carrying forward a program for process optimization [143]. Continuous monitoring or automatic control of the ζ potential can be achieved by the use of equipment such as the Laser Zee System 3000.

12.7 CONCLUSION

As described above, the ζ potential is not a decisive parameter that characterizes the colloidal state in most papermaking systems. It is only one parameter of many and when used in conjunction with others, it is useful for indicating the state of the system [3]. It is likely, however, that the measurement of the ζ potential has a distinct advantage over information available from other sources. It exhibits an unexpected strength in clearing the points previously unnoticed and offers new views for interpretation or theoretical support for problems relating to the colloidal phenomena.

REFERENCES

1. A. W. McKenzie, *Appita*, 22, 82 (1968).
2. A. Schausberger and J. Schurz, *Papier*, 33, 148 (1979).
3. R. A. Stratton and J. W. Swanson, *Tappi*, 64(1), 79 (1981).
4. D. Eklund, *Nor. Skogind.*, 21, 140 (1967).
5. F. Hoffmann, F. Müller, E. Rohloff, and H. Tretter, *Papier*, 29, 529 (1975).
6. K. W. Britt and J. E. Unbehend, *Tappi*, 57(12), 81 (1974).
7. R. Branion and M. Arcelus, *Trans. Tech. Sect.* (Can. Pulp Pap. Assoc.), 5, TR1 (1979).
8. M. J. Jaycock and J. L. Pearson, *Sven. Papperstidn.*, 78, 167 (1975).
9. D. Horn and J. Melzer, *Papier*, 29, 534 (1975).
10. E. Strazdins, *Tappi*, 57(12), 76 (1974).
11. V. Hornof, G. H. Neale, and D. Leblanc, *Cellul. Chem. Technol.*, 14, 191 (1980).
12. E. Strazdins, *Tappi*, 55(12), 1691 (1972).
13. J. F. McKague, D. O. Etter, J. O. Pilgrim, and W. H. Griggs, *Tappi*, 57(12), 101 (1974).
14. R. Anderson and J. Penniman, *Pap. Trade J.*, 158(38), 22 (1974).
15. G. Mugler and P. Mugler, *Zellst. Pap.*, 25, 137 (1976).
16. F. Onabe and J. Nakano, *Kami Pa Gikyoshi* (Japan), 24, 193 (1970).
17. J. C. Walkush and D. G. Williams, *Tappi*, 57(1), 112 (1974).
18. V. Balodis, *Appita*, 21, 96 (1967).
19. A. W. McKenzie, V. Balodis, and A. Milgrom, *Appita*, 23, 40 (1969).
20. R. W. Davison, *Tappi*, 57(12), 85 (1974).
21. R. W. Davison and R. E. Cates, *Pap. Technol. Ind.*, 16, 107 (1975).
22. R. G. Anderson and J. G. Penniman, *Pap. Trade J.*, 158(2), 56 (1974).

23. T. Lindström and C. Söremark, *Papier*, *29*, 519 (1975).
24. T. Lindström, C. Söremark, C. Heinegård, and S. Martin-Löf, *Tappi*, *57*(12), 94 (1974).
25. E. Strazdins, *Tappi*, *53*(1), 80 (1970).
26. E. Poppel and I. Bicu, *Papier*, *23*, 672 (1969).
27. J. Melzer, *Papier*, *28*(10A), V33 (1974).
28. W. Günder and W. Auhorn, *Wochenbl. Papierfabr.*, *103*, 581 (1975); *Paper* (World R. and D. No.), 20 (1975).
29. T. Lindström and C. Söremark, *J. Colloid Interface Sci.*, *55*, 305 (1976).
30. T. Lindström, C. Söremark, and L. Eklund, *Trans. Tech. Sect.* (Can. Pulp Pap. Assoc.), *3*, TR114 (1977).
31. G. Carlsson, T. Lindström, and C. Söremark, *Sven. Papperstidn.*, *80*, 173 (1977).
32. J. Böhm and P. Luner, in *Fiber-Water Interactions in Papermaking*, Vol. 1, Technical Division, British Paper and Board Industry Federation, London, 1978, p. 251.
33. G. J. Howard, F. L. Hudson, and J. West, *J. Appl. Polym. Sci.*, *21*, 1, 29 (1977).
34. B. S. Das and H. Lomas, *Pulp Pap. Mag. Can.*, *74*, T281 (1973).
35. D. Horn and J. Melzer, in *Fiber-Water Interactions in Papermaking*, Vol. 1, Technical Division, British Paper and Board Industry Federation, London, 1978, p. 135.
36. H. Tanaka, K. Tachiki, and M. Sumimoto, *Tappi*, *62*(1), 41 (1979).
37. J. Marton and T. Marton, *Tappi*, *59*(12), 121 (1976).
38. J. Marton, *Tappi*, *63*(4), 87 (1980).
39. S. Hernádi and P. Völgyi, *Wochenbl. Papierfabr.*, *106*, 355 (1978).
40. E. Strazdins, *Tappi*, *60*(7), 113 (1977).
41. W. F. Linke, *Tappi*, *51*(11), 59A (1968).
42. J. W. S. Goossens and P. Luner, *Tappi*, *59*(2), 89 (1976).
43. F. Onabe, *J. Appl. Polym. Sci.*, *22*, 3495 (1978); *23*, 2909, 2999 (1979); *24*, 1629 (1979).
44. N. A. Bates, *Tappi*, *52*(6), 1157 (1969).
45. D. L. Kenaga, W. A. Kindler, and F. J. Meyer, *Tappi*, *50*(7), 381 (1967).
46. G. G. Allan and W. M. Reif, *Sven. Papperstidn.*, *74*, 25 (1971).
47. B. Alinče, *Cellul. Chem. Technol.*, *8*, 573 (1974).
48. K. Britt, *Pulp Pap. Can.*, *80*, T152 (1979).
49. E. Poppel. I. Bicu, and V. Dobronauteanu, *Papier*, *29*, 93 (1975).
50. E. E. Moore, *Tappi*, *56*(3), 71 (1973); *58*(1), 99 (1975).
51. J. G. Penniman, *Pap. Trade J.*, *163*(15), 35 (1979).

52. K. Suzuki and H. Tanaka, *Mokuzai Gakkaishi* (Japan), *23*, 204 (1977).
53. V. V. Lapin, *Zellst. Pap.*, *24*, 260 (1975).
54. J. G. Penniman, *Pap. Trade J.*, *163*(5), 62 (1979).
55. R. H. Pelton, L. H. Allen, and H. M. Nugent, *Pulp Pap. Can.*, *81*, T9 (1980).
56. C. H. Tay, *Tappi*, *63*(6), 63 (1980).
57. U. Beck, F. Müller, J. W. S. Goossens, E. Rohloff, and H. Tretter, *Wochenbl. Papierfabr.*, *105*, 391 (1977).
58. F. Müller and U. Beck, *Papier*, *32*(10A), V25 (1978).
59. J. H. Klungness and M. P. Exner, *Tappi*, *63*(6), 73 (1980).
60. E. E. Moore, *Tappi*, *59*(6), 120 (1976).
61. K. W. Britt, *Tappi*, *56*(10), 46 (1973).
62. K. W. Britt and J. E. Unbehend, *Tappi*, *59*(2), 67 (1976).
63. K. W. Britt, A. G. Dillon, and L. A. Evans, *Tappi*, *60*(7), 102 (1977).
64. J. M. Muhonen and D. G. Williams, *Tappi*, *56*(10), 117 (1973).
65. C. A. King and D. G. Williams, *Tappi*, *58*(9), 138 (1975).
66. J. E. Unbehend, *Tappi*, *59*(10), 74 (1976); *60*(7), 110 (1977).
67. K. W. Britt and J. E. Unbehend, *Tappi*, *63*(4), 67 (1980).
68. D. Abson, R. M. Bailey, C. D. Lenderman, J. A. Nelson, and P. B. Simons, *Tappi*, *63*(6), 55 (1980).
69. F. MK Werdouschegg, *Tappi*, *60*(7), 105 (1977).
70. O. Huber and J. Weigl, *Wochenbl. Papierfabr.*, *97*, 359 433 (1969).
71. M. J. Jaycock and J. L. Pearson, *J. Appl. Chem. Biotechnol.*, *25*, 815 (1975).
72. J. Weigl, G. Waltner, and E. Weyh, *Wochenbl. Papierfabr.*, *104*, 439 (1976).
73. S. N. Iwanow and W. Ljadowan, *Zellst. Pap.*, *19*, 101 (1970).
74. M. J. Jaycock and J. L. Pearson, *Sven. Papperstidn.*, *78*, 289 (1975).
75. M. J. Jaycock and J. L. Pearson, *J. Appl. Chem. Biotechnol.*, *25*, 827 (1975).
76. M. J. Jaycock and J. L. Pearson, *J. Colloid Interface Sci.*, *55*, 181 (1976).
77. M. J. Jaycock, J. L. Pearson, R. Counter, and F. W. Husband, *J. Appl. Chem. Biotechnol.*, *26*, 370 (1976).
78. A. W. McKenzie and G. W. Davies, *Appita*, *25*, 32 (1971).
79. H. R. Bryson, *Pap. Technol. Ind.*, *16*, 372 (1975).
80. J. Weigl and L. Huggenberger, *Wochenbl. Papierfabr.*, *102*, 886 (1974).
81. K. W. Britt, *Tappi*, *56*(3), 83 (1973).
82. W. E. Frankle and J. G. Penniman, *Pap. Trade J.*, *157*(32), 30 (1973).

83. K. Nishikawa, Y. Inoue, Y. Fujioka, and S. Takahashi, *Kami Pa Gikyoshi* (Japan), 32, 76, 419 (1978).
84. J. Weigl, *Papier*, 33(10A), V105 (1979).
85. D. G. Williams, *Tappi*, 56(12), 144 (1973).
86. R. Counter, M. J. Jaycock, and J. L. Pearson, *Sven. Papperstidn.*, 79, 193 (1976).
87. L. Lason and E. Szwarcsztajn, *Cellul. Chem. Technol.*, 13, 523 (1979).
88. B. Blanchin, G. Gervason, P. Vallette, and G. Sauret, in *Fiber-Water Interactions in Papermaking*, Vol. 1, Technical Division, British Paper and Board Industry Federation, London, 1978, p. 151.
89. R. H. Schiesser, *Tappi*, 59(10), 71 (1976).
90. B. Borchers, H.-J. Hartmann, R. Nicke, and M. Tappe, *Zellst. Pap.*, 28, 14 (1979).
91. R. Nicke and H.-J. Hartmann, *Zellst. Pap.*, 25, 360 (1976).
92. B. Lipič, *Wochenbl. Papierfabr.*, 105, 269 (1977).
93. W. E. Frankle and J. L. Sheridan, *Tappi*, 59(2), 84 (1976).
94. D. F. Rutland, A. Y. Jones, P. M. Shallhorn, J. Tichy, and A. Karnis, *Pulp Pap. Can.*, 78, T99 (1977).
95. J. N. Arno, W. E. Frankle, and J. L. Sheridan, *Tappi*, 57(12), 97 (1974).
96. D. D. Halabisky, *Tappi*, 60(12), 125 (1977).
97. L. P. Avery, *Tappi*, 62(2), 43 (1979).
98. P. G. Stoutjesdijk and G. Smit, *Wochenbl. Papierfabr.*, 103, 897 (1975).
99. H. F. Arledter and A. Mayer, *Papier*, 29(10A), V32 (1975).
100. P. Arvela, J. W. Swanson, and R. A. Stratton, *Tappi*, 58(11), 86 (1975).
101. R. W. Davison, *Tappi*, 58(3), 48 (1975).
102. W. H. Griggs and B. W. Crouse, *Tappi*, 63(6), 49 (1980).
103. E. Strazdins, *Tappi*, 64(1), 31 (1981).
104. E. Poppel, *Zellst. Pap.*, 23, 79 (1974).
105. E. Strazdins, *Tappi*, 48(3), 157 (1965); 60(10), 102 (1977).
106. R. J. Kulick, *Tappi*, 60(10), 74 (1977).
107. E. J. Vandenberg and H. M. Spurlin, *Tappi*, 50(5), 209 (1967).
108. R. Counter, M. J. Jaycock, and J. L. Pearson, *Sven. Papperstidn.*, 78, 333 (1975).
109. L. J. Stryker, B. D. Thomas, and E. Matijević, *J. Colloid Interface Sci.*, 43, 319 (1973).
110. K. Roberts, J. Kowalewska, and S. Friberg, *Sven. Papperstidn.*, 77, 239 (1974).
111. T. Lindström and C. Söremark, *Sven. Papperstidn.*, 80, 22 (1977).
112. R. W. Davison, *Tappi*, 47(10), 609 (1964).

113. S. Hernádi, J. Deme, and M. Sümegi, Zellst. Pap., 28, 225 (1979).
114. O. Huber and J. Weigl, Wochenbl. Papierfabr., 101, 702 (1973).
115. W. Schempp and J. Schurz, Papier, 27, 257 (1973).
116. P. Sennett, J. P. Olivier, and H. H. Morris, Tappi, 52(6), 1153 (1969).
117. H. Benninga, A. Harsveldt, and A. A. De Sturler, Tappi, 50(12), 577 (1967).
118. O. Huber, Papier, 29, 525 (1975).
119. O. Huber, Wochenbl. Papierfabr., 101, 657 (1973).
120. B. W. Greene and A. S. Reder, Tappi, 57(5), 101 (1974).
121. H. Hentschel, Wochenbl. Papierfabr., 103, 888 (1975).
122. S. K. Nicol and R. J. Hunter, Aust. J. Chem., 23, 2177 (1970).
123. W. Schempp, W. Friesen, and J. Schurz, Wochenbl. Papierfabr., 101, 369 (1973).
124. W. Schempp, W. Friesen, and J. Schurz, Tappi, 58(1), 116 (1975).
125. J. G. Penniman, Pap. Trade J., 62(7), 36 (1978); 62(10), 47 (1978).
126. E. Strazdins, Papier, 34(10A), V49 (1980).
127. K. Daucik, W. Frankle, and J. G. Penniman, Paper, 183, 567 (1975).
128. W. J. J. Laseur, Wochenbl. Papierfabr., 101, 473 (1973).
129. I. P. G. Stoutjesdijk and G. Smit, Nor. Skogind., 29, 120 (1975); Paper (World R. and D. No.), 42 (1975).
130. R. H. Windhager, Pulp Pap. Can., 76, T340 (1975).
131. J. Marton, Tappi, 63(2), 121 (1980).
132. L. H. Allen, Pulp Pap. Can., 76, T139 (1975).
133. L. H. Allen, Trans. Tech. Sect. (Can. Pulp Pap. Assoc.), 3, TR32 (1977); 6, TR8 (1980).
134. J. Weigl, G. Waltner, and E. Weyh, Wochenbl. Papierfabr., 105, 500 (1977); 107, 317 (1979).
135. J. J. Latimer and R. A. Gill, Tappi, 56(4), 66 (1973).
136. B. Alince, Pap. Trade J., 163(5), 52 (1979).
137. B. Alince and A. A. Robertson, Tappi, 61(11), 111 (1978).
138. A. Breunig, Wochenbl. Papierfabr., 106, 655, 901 (1978).
139. G. Schweizer and J. Weigl, Wochenbl. Papierfabr., 106, 895 (1978).
140. A. Duff, M. Bergatt, and J. Rabideau, Pulp Pap. Can., 76(3), 58 (1975).
141. E. R. Sandstrom, Pap. Trade J., 163(2), 47 (1979).
142. R. J. Dobbins, Pap. Trade J., 162(21), 42 (1978).
143. J. G. Penniman and W. E. Frankle, Pulp Pap., 50(12), 66 (1976).

13
Electrocapillary Emulsification

Kazuo Nishizawa Laboratory of Interfacial Technique, Kyoto, Japan

13.1. Introduction 339
13.2. Electrocapillary Emulsification 340
13.3. Apparatus 342
13.4. Applications 345
     13.4.1. Solvent dyeing 345
     13.4.2. Cosmetic creams 348
     13.4.3. Microcapsules 348
     13.4.4. Electrocapillary spinning 350
     13.4.5. A practical emulsifying machine 354
     References 354

13.1 INTRODUCTION

The interfacial tension between oil and water decreases by applying a potential difference to the interface, and hence emulsification takes place when the potential difference is sufficiently high. This is called electrocapillary emulsification. This process was applied to the formation of the bath of solvent dyeing, cosmetic creams, and microcapsules, as well as to the spinning of gelatin fiber. The process was found to be useful, since very fine particles could be obtained in the absence or presence of very small amounts of surfactants.

The electrocapillary phenomenon is the change in interfacial tension created by the application of a potential difference between two phases in contact with each other. Although this has been studied extensively on mercury in contact with aqueous solutions, it can be extended to cover other systems consisting of two immiscible liquid phases [1-7]. It is important to notice that another variable, the interfacial potential difference, is introduced here in addition to the usual variables, such as the solution composition and temperature, to the interfacial systems.

An application of this phenomenon is in the formation of water/oil or oil/water emulsion by applying a fairly low voltage compared to electric dispersion [8,9]. This method is based on the decrease to almost zero in interfacial tension by the application of a potential difference to the interface, spontaneous emulsification taking place by interfacial fluctuation [10,11]. The emulsions formed are in general highly monodispersed and stable, and hence of great practical value, since physical properties of emulsions are in general strongly dependent on the size distribution.

In the case of ordinary emulsification processes by the mechanical or spontaneous emulsification methods, properties of emulsions formed, such as the particle size distribution, stability, and so on, are usually controlled by using ordinary surfactants at high concentrations. In many cases this causes a considerable disadvantage, and in addition the use of synthetic surfactants often gives rise to serious problems. The present technique has in this sense a great advantage, since emulsions are formed in the absence of, or at least in the presence of very small amounts of surface-active materials [8,10].

13.2 ELECTROCAPILLARY EMULSIFICATION

A schematic diagram of the simple apparatus for studies on electrocapillary emulsification is given in Fig. 13.1. The aqueous phase in a syringe (B) is introduced into the oil phase through a Teflon capillary (G; inner diameter about 0.5 mm, and outer diameter about 1 mm) at the tip of a glass syringe, by driving the motor (A). The platinum electrodes (F and I) in the aqueous (E) and oil (H) phases, respectively, are connected to the variable DC supply (K). The sign of the applied voltage E is conveniently taken as that of the aqueous phase with reference to that of the oil phase.

When the applied voltage E is increased, the drop size decreases at first, due to the decrease in interfacial tension, and then a shower of fine droplets is formed for E higher than the critical voltage of emulsification, E_c. In the case of oil/water emulsification, the oil phase is introduced from the syringe into the aqueous phase.

Although, in addition to the microscopic emulsion droplets, comparatively large drops are also formed at the critical voltage of emulsification, the value of E_c is considered to be the minimum voltage necessary to give rise to continuous emulsification. This critical voltage is strongly influenced by the composition of the system.

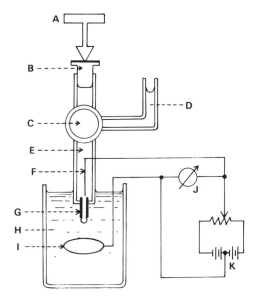

Figure 13.1 Schematic diagram of electrocapillary emulsification. A, driving motor; B, syringe; C, three-way cock; D, water reservoir; E, water phase; F, platinum electrode; G, Teflon capillary; H, oil phase; I, platinum electrode; J, voltmeter; K, DC power supply.

Figure 13.2 shows the relation between E_c and the specific conductivity of the oil phase λ_0 for the water/oil emulsification. Here the aqueous phase is 10^{-4} mol/liter KCl and the oil phase, MIBK (methyl isobutyl ketone), contains increasing amounts of sodium dodecyl sulfate (SDS). It was noticed that at first E_c decreases in magnitude with increasing λ_0; emulsification takes place more easily when the SDS concentration in the oil phase is higher. This can be ascribed to the decrease in interfacial tension caused by the adsorption of SDS and also to the decrease in the ohmic drop in the oil phase. However, when λ_0 approaches the specific conductivity of the aqueous phase λ_W, E_c increases in magnitude indefinitely. It was found that no emulsification takes place when λ_0 becomes higher than λ_W (shadowed range in Fig. 13.2).

Figure 13.3 shows the relation between E_c and λ_W for oil/water emulsification; the oil phase is 10^{-2} mol/liter tetrabutylammonium chloride (TBAC, surface-inactive organic electrolyte) in MIBK and the aqueous phase contains increasing amounts of KCl. A tendency

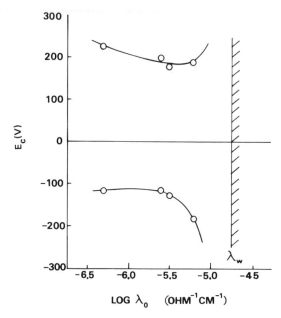

Figure 13.2 Influence of specific conductivity of oil phase on the critical voltage of emulsification. Water phase: 10^{-4} mol/liter KCl. Oil phase: SDS in MIBK.

similar to that in the above case is noted; E_c increases in magnitude indefinitely as λ_W approaches λ_0, no emulsification taking place for λ_W higher than λ_0 (shaded range in Fig. 13.3).

Figure 13.4 shows the summary of results obtained under various conditions [9,10]. The oil/water emulsion is formed when the ionic strength of the oil phase is higher than that of water, while the water/oil emulsion is formed under the opposite condition. Thus we can conclude that the ionic strength of the discontinuous (inner) phase must be higher than that of the continuous (outer) phase in order for electrocapillary emulsification to take place. This condition can be treated theoretically, based on the electrostatic free energy of the system.

13.3 APPARATUS

It is necessary in practice to emulsify the dispersed phase as much as possible in a given time interval. Instead of a single capillary as shown in Fig. 13.1, a bundle of capillaries can be employed. Although it is possible to fix a number of parallel capillaries by using an adhesive agent, it is more convenient to use a nozzle with numerous

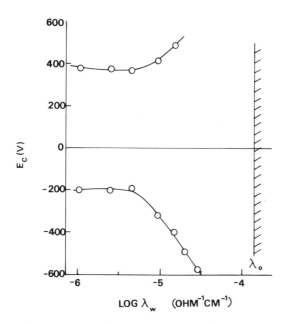

Figure 13.3 Influences of specific conductivity of water phase on the critical voltage of emulsification. Water phase: KCl. Oil phase: 10^{-2} mol/liter TBAC in MIBK.

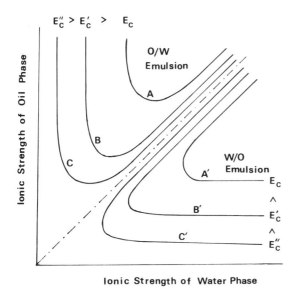

Figure 13.4 Condition of electrocapillary emulsification. Increase in E_c: oil/water type; A → B → C → •••; water/oil type; A' → B' → C' → •••.

pores. Since the system contains organic liquid, a Teflon nozzle with fluorine rubber packings is recommended. It is important to make a tight packing to prevent leakage of liquid, especially when emulsifying at high temperatures.

Figure 13.5 shows an example of a nozzle for the emulsification at a capacity lower than 500 ml. The bottom disk of the cap (B), of thickness 0.3-1.0 mm and diameter 20 mm, is perforated with 800 holes of diameter about 50 μm each. The top tube of the holder (A) is connected with the water feeding tube by means of a Teflon joint. The nozzle system is placed directly in the emulsification vessel.

The material of the nozzle plate and the holder can be selected according to the chemicals, temperature, and pressure used for emulsification. For the preparation of the bath for solvent dying, in which perchloroethylene is used, the bath, nozzle plate, and holder are made of Teflon. However, for the preparation of cosmetic cream, polyethylene may be used instead.

A simple way to prepare a nozzle plate with many pores is to drill many holes in the Teflon or polyethylene disk with a separation of 7-10 mm, into which poly(vinyl chloride) tubes of inner diameter 0.1-0.5 mm and length about 10 mm are inserted. The whole disk is sliced into thin disks of thickness 1-2 mm.

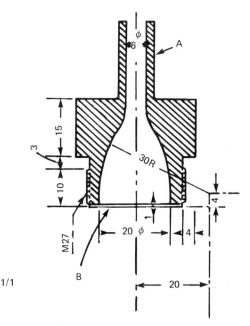

Figure 13.5 Nozzle system of electrocapillary emulsification (cap type).

In some cases a Teflon membrane filter can be used instead of the nozzle plate, the most convenient one being that of porosity 50%, pore radius about 5 μm, and thickness 0.5-1.0 mm. This gives a liquid flow rate of 3 ml/min at a pressure difference 0.5 kg/cm^2. The membrane filter nozzle will, however, clog in concentrated systems (e.g., saturated dye solution, polymer solutions, and suspensions of disperse dyes).

13.4 APPLICATIONS

13.4.1 Solvent Dyeing

The aqueous dye bath, which is usually used for dyeing, has various shortcomings. For instance, although the amount of water theoretically needed for dyeing is very small, in practice it is necessary to use a lot of water, and hence the loss of dye is enormous. Various proposals are needed to overcome these difficulties, so that only a small amount of water is used for dye adsorption. One possibility consists of the use of the emulsion of the water phase, containing dye dispersed in organic liquid, or the use of dyes that are soluble in organic solvent. For an emulsion type a highly dispersed and monodisperse dispersion is necessary to prevent uneven dyeing. This can be achieved only by the use of a large amount of surfactants, if the procedure of mechanical or spontaneous emulsification is employed. However, this also creates various difficulties (e.g., the interaction between surfactants and dyes taking place in the bath). On the other hand, electrocapillary emulsification is a useful procedure, since it can be achieved in the presence of a very small amount of surfactant. In fact, satisfactory results were obtained by using an emulsion-type dye bath prepared by the present method.

Preliminary dyeing tests were carried out in a beaker for various fibers at 90°C for 60 min. It was found that the nozzle plate with pores smaller than 60 μm in diameter could not be used because the pores became seriously clogged. It was also found that the emulsion became polydisperse when the fluctuation of flow rate of the water phase was higher than 10%.

Hence, a water/oil emulsion of dye bath was prepared by using a cap type of perforated nozzle plate with 650 pores of diameter 60 μm. An aliquot of aqueous dye solutions was emulsified in tetrachloroethylene, containing 0.2-1.0% Sorgen 40, at the flow rate 0.6 liter/min by applying −1000 to −4000 V. In dyeing cotton and rayon, the water phase contained sodium chloride or sodium sulfate as the builder.

Figures 13.6 and 13.7 show examples of the effect of the bath ratio and water content of dye absorption. The dye absorption reaches a maximum at a certain bath ratio and water content. Table 13.1 gives the optimum bath ratio and water content for wool, nylon 6, and other fibers. Table 13.2 shows the results of the solvent dyeing experiments

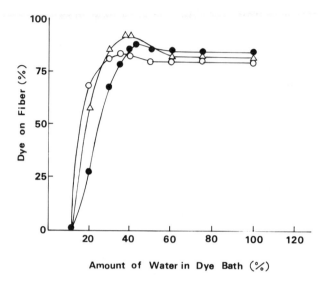

Figure 13.6 Influence of water content on dye absorption by wool. Acid dye (2% owf). 90°C, 60 min. Bath ratio: ○, 1:20, △, 1:30; ●, 1:50.

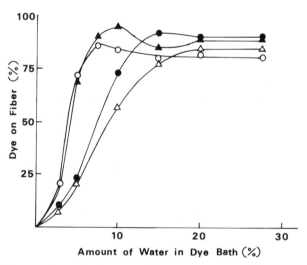

Figure 13.7 Influence of water content on dye absorption by nylon 6. Acid dye (2% owf). 90°C, 60 min. Bath ratio: ○, 1:10; ▲, 1:20 ●, 1:50; △, 1:75.

Table 13.1 Optimum Bath Ratio and Water Content for Dyeing

Fiber	Dye	Optimum bath ratio	Optimum water content (% owf)
Wool	Acidic	1:30	38
Cotton	Direct	1:15	30
Rayon	Direct	1:15	30
Polyacrylate	Basic	1:20	12
Nylon 6	Acidic and basic	1:20	10
Polyester	Disperse	1:20	8

Table 13.2 Dyeing of Wool by Solvent Dyeing Using Electrocapillary Emulsification at 2000 V[a]

Number	Dye absorption	Light fastness	Fastness to rubbing	Wash fastness
1	B	S	S	B
2	B	S	S	B
3	S	S	S	B
4	B	S	B	B
5	S	S	S	S
6	S	S	B	B
7	W	S	B	B
8	B	S	S	S

[a]B, better than; S, the same degree as; W, worse than the dyeing in aqueous dye bath.

compared with aqueous dyeing experiments carried out under the same conditions.

13.4.2 Cosmetic Creams

Electrocapillary emulsification can be used to prepare cosmetic creams containing minute amounts of surfactant. Table 13.3 gives the formulas tried. They do not contain extra additives (e.g., pigments, perfumeries, or antioxidants).

13.4.3 Microcapsules

Kondo and Arakawa [12] succeeded in preparing poly($\underline{N}^{\alpha},\underline{N}^{\varepsilon}$-L-lysine diethylterephthaloyl) (PPL) nanocapsules, of mean diameter 38 nm, containing sheep erythrocyte hemolysate, by using electrocapillary emulsification together with interfacial polymerization.

Table 13.3 Cosmetic Creams

Materials	Composition[a] (%)			
	A	B	C	D
Liquid paraffin	25	35.3	29	15
Squalane	10	–	10	30
Lanoline	5	3.7	5	5
Vaseline	6	1.5	2	10
Cholesterin	–	1.5	–	–
Lanoyl alcohol	1.6	–	–	–
Beeswax	11	–	–	10
Cerecin	8	14.7	–	12
Solid paraffin (m.p. 73°C)	–	–	10	–
S-90[b]	0.8	0.6	0.6	0.5
IPM[b]	–	16.3	–	–
DK-S-20[b]	–	–	6	–
Water	35	30	38.5	10

[a]A, a glossy, spreadable cream; B, an elastic cream (the degree of dispersion is not very high); C, a fine-textured cream; D, a soft, glossy, high quality cream.
[b]Emulsion stabilizer.

The sheep hemolysate, containing both 0.4 mol/liter L-lysine and 0.63 mol/liter sodium carbonate, was emulsified by the electrocapillary method in a cyclohexane-chloroform mixture (3:1 by volume), containing 0.04 mol/liter terephthaloyd dichloride, 5% tetra ethylammonium chloride, and 1.5×10^{-4} mol/liter sorbitan trioleate. The oil phase was cooled by ice and gently stirred during emulsification. When the applied voltage became higher than a critical value, emulsification started to take place and a shower of very fine emulsion droplets were formed. The applied voltage was kept at 85 V and flow rate at 0.042 ml/min. The polycondensation reaction then took place to form a PPL membrane at the droplet/oil interface.

The capsules thus formed were fractionated several times by centrifugation and redispersion in water by using polyoxyethylene sorbitan monolaurate as the dispersing agent. The fraction with a mean diameter of about 400 nm was washed with and then redispersed in isotonic phosphate buffer at pH 7.4.

The microcapsules formed by this method were of a very high degree of dispersion and were shown to give a myoglobinlike oxygen absorption curve with a resistance against flow higher than that of hymolystate-loaded PPL microcapsules of the same volume fraction.

The polydispersity of the capsules originally formed by this method was ascribed by Kondo and Arakawa to the slow rate of polycondensation taking place at the interface. This may be avoided by increasing reactant concentration or by using oil-soluble reactants.

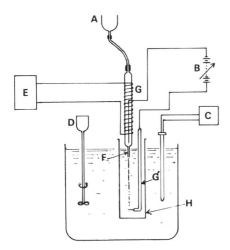

Figure 13.8 Apparatus for electrocapillary spinning of gelatin. A, reservoir of aqueous gelatin; B, high-tension DC supply; C, thermometer; D, stirrer; E, heating circuit; F, Teflon capillary; G, G', platinum electrodes; H, alcoholic dehydration bath.

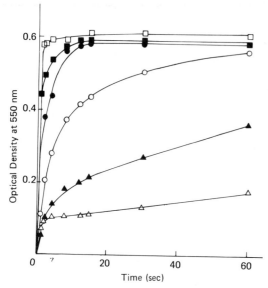

Figure 13.9 Change in optical density of water with time due to the dissolution of gelatin. Applied voltage: ○, 0; ●, 500; □, 1000 V. ▲, Untreated; △, Difco gelatin.

13.4.4 Electrocapillary Spinning

The principle of electrocapillary emulsification can be applied to the spinning of gelatin and other polymers, since the interfacial tension, and hence the radius of the jet from the capillary tip, decreases by the application of potential. The apparatus shown in Fig. 13.8 is in principle the same as that of emulsification. An aqueous solution of gelatin in the reservoir (A) was extruded into the alcoholic dehydration

Table 13.4 Relative Initial Dissolution Rate of Gelatin in Water

Type of gelatin with voltage applied	Relative initial dissolution rate
0	2.2
500	3.9
1000	66.7
Untreated	1.0
Difco	1.3

Table 13.5 Influence of Applied Voltage on the Diameter of Gelatin Thread

Applied voltage (V)	Diameter (μm)	Applied voltage (V)	Diameter (μm)
0	200		
+100	150	−100	112
+200	100	−200	85
+300	50	−300	62
+500	32	−500	37
+750	25	−750	25
+1000	20	−1000	17

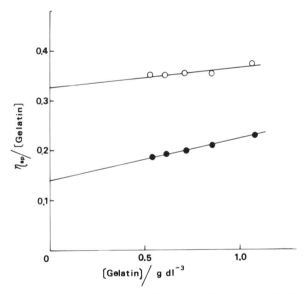

Figure 13.10 Reduced viscosity of gelatin solution as a function of gelatin concentration. ●, Untreated, [η] = 0.137; ○, 1000 V, [η] = 0.328.

bath (H) through the Teflon capillary (F) with an inner radius 0.5 mm. The gelatin content of the aqueous phase was 20% throughout and both aqueous and oil phases were kept at 50°C, using a constant-temperature bath to avoid coagulation.

The two platinum electrodes (G and G'), in the aqueous and oil phases, respectively, are connected to the variable DC supply (B). The distance of the electrodes was kept at 120 mm and the pressure head of aqueous phase at 100 mm. It was important to keep the water content of the ethanol bath lower than 35% during the spinning process.

If the spun thread of gelatin was not reeled, the gelatin aqueous solution extruded from the capillary tip made a vigorous irregular motion due to rapid dehydration, thus forming a lacelike clot. The clot thus obtained had a loose structure with a very high specific surface area, and hence showed a high dissolution rate in water. Figure 13.9 shows the dissoluation curve, at room temperature, of the gelatin sample prepared at various applied voltages. The figure contains for comparison the results obtained using both standard Difco gelatin and untreated gelatin. The vertical axis shows the optical density at 550 nm, which is considered to give the relative gelatin concentration, and the horizontal axis the time of dipping in water.

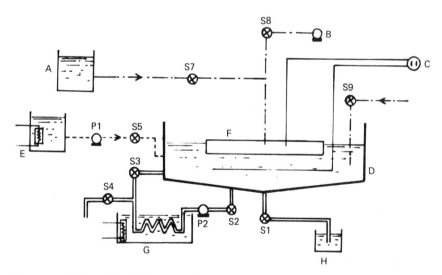

Figure 13.11 Block diagram of electrocapillary emulsification machine. A, water supply; B, vacuum pump; C, high-tension DC supply; D, emulsifying bath; E, oil supply; F, nozzle system; G, heat exchanger; H, emulsion outlet. —, High-tension circuit; – · – · –, water-phase path; ---- oil-phase path.

Figure 13.12 Emulsifying bath of the electrocapillary emulsification machine.

The amount of gelatin dissolved in water after 60 min of dipping increases with increasing applied voltage. Table 13.4 shows the relative values of the initial rate of dissolution. The initial rate of dissolution was found to increase by 60-fold with the application of 1000 V.

Table 13.5 shows the effect of the applied voltage on the thread diameter. The diameter of the thread decreases with increasing applied voltage.

It was found by a separate study, by using disk electrophoresis, that the separation of gelatin into hydrophobic and hydrophilic components took place by electrocapillary spinning [13].

Figure 13.10 shows viscosity, at 40°C, versus concentration curves of both gelatin treated at 1000 V and untreated gelatin, obtained by using the Ubbelohde viscometer. The increase in intrinsic viscosity also indicates that low-molecular-weight fractions have been removed by electrocapillary spinning.

From the results thus obtained, the present method can be used for the purification of water-soluble polymers in general. In fact, materials of poly(vinyl alcohol), gum arabic, and so on, with higher solubility than the original materials could be prepared.

13.4.5 A Practical Emulsifying Machine

Converting the pilot stage of the laboratory to the industrial scale of an emulsifying machine was manufactured in which 30% water/oil emulsions could be prepared at the rate 40 liters/hr. Figure 13.22 shows the block diagram and Fig. 13.12 the emulsifying bath. The bath is made of polyacrylate resin with Teflon spray coating to fascilitate the observation of detailed structure.

REFERENCES

1. J. Guastalla, *J. Chim. Phys.*, *53*, 470 1956); also in *Proc. 2nd Int. Cong. Surf. Activ.*, *3*, 112 (1957).
2. M. Blank and S. Feig, *Science*, *141*, 1117 (1963).
3. A. Watanabe, M. Matsumoto, and R. Gotoh, *Nippon Kagaku Kaishi*, *87*, 941 (1966).
4. A. Watanabe, M. Matsumoto, H. Tamai, and R. Gotoh, *Kolloid Z. Z. Polym.*, *220*, 152 (1967).
5. A. Watanabe, *Nippon Kagaku Kaishi*, *92*, 575 (1971).
6. M. Dupeyrat and J. Michel, *J. Colloid Interface Sci.*, *29*, 605 (1969).
7. E. Nakache, Thesis, University of Paris, 1972.
8. A. Watanabe, K. Higashitsuji, and K. Nishizawa, in *Colloidal Dispersions and Micellar Behavior* (K. L. Mittal, ed.), ACS Symposium Series No. 9, American Chemical Society, Washington, D.C., 1975, p. 97.
9. K. Higashitsuji and A. Watanabe, *Nippon Kagaku Kaishi*, *1979*, 974 (1979).
10. A. Watanabe, K. Higashitsuji and K. Nishizawa, *J. Colloid Interface Sci.*, *64*, 278 (1978).
11. K. Higashitsuzi and A. Watanabe, *Nippon Kagaku Kaishi*, *1979*, 1287 (1979).
12. T. Kondo and M. Arakawa, private communication.
13. Y. Nakamura and A. Watanabe, unpublished data.

14
Pigments and Paints

Isao Kumano Toyo Ink Mfg. Co., Ltd., Itabashi, Tokyo, Japan

14.1. General 355
14.2. The Surface Charge of Pigments in Water and Solvents 356
14.3. The Charge and Dispersion of Pigments in Paints 363
References 366

14.1 GENERAL

A pigment consists of fine, colored or colorless solid particles and is insoluble in water, solvents, varnishes, plastics, and other media in which it is used. Pigments are generally used to generate color and to hide the substrates in suspensions of paints, printing inks, pigment resin colors, and so on. In order to perform their function fully, pigments must be dispersed readily and maintain a stable dispersion as long as possible. However, pigments have a strong tendency to flocculate since the dispersion of fine particles is thermodynamically unstable. The ease and the stability of the dispersion of a pigment in a medium is dependent on several factors: for instance, the specific gravity, the size and its distribution, the shape, and the surface properties of the pigment particles. The surface area, the polarity, the functional groups, and the electrical properties of the pigment surface are important surface characteristics of a pigment. The surface charge is one of two important factors contributing to the stabilization of dispersion of the pigment in a medium, the theoretical explanation of which is described in Sec. 1.3.

When pigment particles are dispersed in water, solvents, and varnishes, they might have a surface charge, the origins of which can be attributed to

1. The dissociation of the surface functional groups
2. The isomorphic substitution of the lattice ions
3. The selective adsorption of certain ions from solution

The magnitude of the surface charge of the pigment can be measured in terms of the zeta (ζ) potential by electrophoresis or streaming potential measurement.

14.2 THE SURFACE CHARGE OF PIGMENTS IN WATER AND SOLVENTS

Titanium dioxide pitment is one of the most important and widely used pigments and a large number of articles have been published on its physical properties. In order to study the surface electrical properties of the pigment, a commercial titanium dioxide pigment is suspended in water having various pH values, and the pH of suspensions is measured [1]. Although no change is found in pH before and after the addition of the titanium dioxide pigment when the pH of original water is 7.2, the pH shifts to higher values with the addition of pigment when the original pH is below 7.2. This indicates a decrease in H^+ ions through adsorption onto the surface of pigment particles, and conversely the pH shifts to lower values when the original pH is above 7.2, indicating a decrease in OH^- ions in water. The titanium dioxide pigment is found to deposit on the cathode in an acidic suspension and on the anode in an alkaline suspension in electrodeposition, indicating that the pigment has positive and negative charges, respectively [1]. When the pigment is suspended in water of pH 7.2, no electrodeposition takes place, suggesting that there is no surface charge on the pigment particles, and thus the isoelectric point (IEP) of the titanium dioxide pigment is 7.2.

Almost all titanium dioxide pigments in commercial use generally have some kind of modification of the surface of particles through treatments and coatings of various sorts, mostly in the form of oxides such as silica and alumina, in order to improve their weather resistance and dispersion. It would be useful, therefore, to review and summarize the surface charge of these metal oxide pigments, referring to Table 14.1.

The charges of metal oxides can be attributed to

1. The amphoteric dissociation of surface OH groups
2. The adsorption of H^+ and OH^- ions from solution

The IEP of silica has been reported to be in the pH range 1-2, and silica has a negative charge at pH above 4 due to the dissociated sur-

Table 14.1 Isoelectric Point of Metal Oxide Pigments

Pigment	Description	Isoelectric point
α-Al_2O_3	Ignited (600-1000°C), aged (1 day)	6.4-6.7
	Aging of the above (water, 7 days)	9.1-9.5
α-$Al(OH)_3$	Hydrolysis of $Al(OC_2H_5)_3$	9.2
MgO	Optical grade	12.1-12.7
SiO_2	Synthesized sol	1.0-2.0
TiO_2	Hydrolysis of $TiCl_4$	6.0
	Ignition of the above	4.7
ZnO	Synthesized	9.3
α-Fe_2O_3	Mineral	6.6-6.8
	Synthesized from $Fe(NH_4)(SO_4)_2$	8.0
	Ignition of the above	6.5
Fe_3O_4	Reduction of natural Fe_2O_3	6.3-6.7
α-FeOOH	Synthesized from $FeCl_3$	6.7
Cr_2O_3	Synthesized from $CrCl_3$	7.0

Source: Ref. 2.

face SiOH groups or to OH^- ions adsorbed onto the surface from the solution [3].

Heat treatment of silica, however, would cause the condensation of free SiOH groups, resulting in a decrease in the surface charge together with some loss in the surface area probably due to sintering. The surface charge density of precipitated silica is shown to decrease from 40 $\mu C/cm^2$ to 8.8 $\mu C/cm^2$ at pH 9.0 by heat treatment at 800°C for 36 hr in air [4]. Metal oxides dehydrated by heat treatment can become hydrated, on the other hand, on exposure to water, resulting in a shift in IEP [2]. For example, the IEP of dehydrated alumina shifts from 6.7 to 9.2 after hydration, which is the same value as that of aluminum hydroxide. In iron oxide, prepared by hydrolysis and not subsequently dried at high temperature, the IEP shifts from 8.0 to 6.5 by heat treatment at 850°C, and returns to the original value by hydration. The IEP of titanium dioxide shifts from 4.7 to 6.2 by hydration in a similar manner.

The impurities contained in pigments have a remarkable effect on their surface charge [2]. The presence of 7.4 mol% of SO_4^{2-} ions in anhydrous alumina as a structural impurity is found to shift the IEP to a more acidic pH by 0.5 unit, and the presence of 3×10^{-4} M of SO_4^{2-} ions adsorbed onto the surface of alumina shifts its IEP to a more acidic pH by 2.9 units. Cl^- and $H_2PO_4^-$ ions adsorbed on the surface of alumina produce a similar effect to SO_4^{2-} ions, whereas the adsorption of Ca^{2+} and La^{2+} ions has the reverse effect of shifting the IEP to a more alkaline pH.

Pigment particles dispersed in water tend to flocculate rapidly in the pH range near the IEP of the pigments. The magnitude of the ζ potential of alumina, the IEP of which is 8.9 in KNO_3 solution, is observed to be less than 14 mV at pH 8.5-9.5. Its stability ratio, expressing the rate of flocculation, is unity, indicating rapid flocculation of pigment. However, at the pH region below 8.0, where the ζ potential would be larger than 15 mV, it has large stability ratios and very small sedimentation volumes, showing stable dispersion [5].

As shown in Table 14.1, the IEP of titanium dioxide rutile of high purity is 4.7 for the dehydrated one and 6.0 for the hydrated, whereas that of the commercial titanium dioxide pigment described previously is 7.2, caused by oxides such as silica and alumina coated on the surface of pigment particles.

Titanium dioxide pigments, coated with 5% silica and alumina in varying ratios, are electrodeposited in the pH range 7-12 [1]. As the ratio of silica in coatings becomes larger, the amount of deposited titanium dioxide pigment is found to increase and this could be explained only by a larger surface charge of pigment particles.

In a series of titanium dioxide pigments, coated with different amounts and proportions of silica and alumina, the IEP of the pigments is found to be directly related to the amount of alumina in the coatings [6]. With unvarying amounts of silica in the coatings, as the amount of alumina decreases the IEP of the pigments takes on lower pH values. The extrapolation of these data to zero alumina gives a value of about pH 1.5 for coatings of silica only, which corresponds to the IEP of pure silica. When the amount of alumina remains constant and the amount of silica is decreased, however, the IEP of the pigments is not that of pure alumina but remains constant at pH 4.4, which is similar to that of pure titanium dioxide of the base pigment. It has been shown by adsorption measurement of stearic acid and stearylamine onto the surface of pigment particles that aluminum atoms form a mixed oxide with silicon atoms in the coatings, and this is thought to be the reason why the pigment does not take the IEP of pure alumina.

It was also found that the IEP changes with the state of distribution of silica and alumina [7]. Pigments having additional silica coatings on top of the alumina coating exhibit the IEP of silica instead of that of alumina. When silica and alumina and manganese phosphate are coated on the surface of titanium dioxide pigment under alkaline condition, its negative ζ potential is larger than that of pigments coated with the same amount of these materials under acidic condition [8]. The differences in ζ potentials of the different coatings are thought to arise from the differences in distribution of both oxides in the coating. Under alkaline condition, for instance, aluminum hydroxide deposits first at a pH in the neighborhood of its IEP to distribute in inner layer of coatings, followed by the deposition of silica on top of the coatings. Thus the surface layer of coatings would be rich in silica, resulting in a more negative charge than expected. The ζ potentials of pigments, coated with silica, alumina, and manganese phosphate in different ratios, are found to vary linearly with changes in ratio under alkaline and acidic conditions, respectively, the differences being 20 to 60 mV larger for pigments coated under alkaline condition. The pigments covered by coatings rich in silica and consequently of a lower IEP generally have a better dispersion stability in water, as expected from the larger negative charge in the pH region above its IEP [6].

Carbon blacks are manufactured by thermal cracking and partial combustion of hydrocarbons, and are made up of nearly spherical particles in which carbon atoms are arranged in a graphitic layer structure together with very small amounts of combined hydrogen and oxygen on the surface. One of the characteristics of the surface of carbon blacks is the presence of functional groups such as carboxyl, phenolic hydroxyl, lactone, and quinone groups, which have close relations with the surface charge of carbon blacks.

The surface charges of several carbon blacks, including surface oxidized ones, are negative in electrolyte solutions [9]. The origins of the surface charge of carbon blacks are in both the dissociative acidic functional groups and the adsorbed charge-determining ions, of which the former is thought to give a predominating effect. The negative ζ potentials of oxidized carbon blacks become larger with the increase of surface acidic groups such as carboxyl and phenolic hydroxyl groups. In aqueous sodium hydroxide solution, where the charge-determining ions are OH^- ions, the contribution of the adsorbed OH^- ions is found to be greater than that of the dissociated functional groups. Quinone groups, present mainly on the surface of rubber-grade carbon blacks, render the surface charge positive by the adsorption of cations in solution [10,11].

Surface-active agents are often used to improve the dispersion of pigment in water associated with changes in surface charge. Titanium dioxide pigments coated with alumina may change their ζ potential depending on the concentration of the dispersing agent, sodium hexametaphosphate [12]. As the concentration of dispersing agent increases, the sign of the ζ potentials of the pigments changes from positive to negative, giving a maximum in magnitude at the characteristic concentration of 100 ppm. The negative ζ potentials of pigments then decrease gradually to zero with further increase in concentration of dispersing agent. It is assumed from these results that metaphosphate ions are adsorbed onto the positively charged surface of the pigment, decreasing and reversing its positive charge, while sodium ions are fixed on the negatively charged surface at higher concentrations of dispersing agent. It was also found that a larger amount of sodium metaphosphate is needed to make the surface charge negative and to keep the dispersion stable for titanium dioxide pigments having coatings richer in alumina than silica [1].

The negative charges of titanium dioxide and iron oxide pigments in water in the presence of sodium chloride and sodium hydroxide increase with the addition of 0.5-1.0 mM sodium dodecylpoly(oxyethylene)sulfate [$R_{12}O(EO)_nSO_3Na$] [13]. A further addition of a small amount of Ca^{2+} ions remarkably increases the negative ζ potential by cooperative adsorption of $R_{12}O(EO)_nSO_3^-$ ions and Ca^{2+} ions on the surface of pigment particles. The zeta potential reaches a maximum when the amount of adsorbed Ca^{2+} ions is equivalent to half of the adsorbed $R_{12}O(EO)_nSO_3^-$ ions and it decreases markedly at higher concentrations of Ca^{2+} ions. No increase of negative ζ potential is observed, of course, with the addition of Ca^{2+} ions only in the absence of $R_{12}O(EO)_nSO_3^-$ ions.

Pigments have a surface charge in organic solvents as well as in water. Inorganic pigments such as iron oxide, chrome yellow, and chrome oxide green are found to flocculate, to move up and down and to form bridges between the electrodes during electrophoresis in nonpolar organic solvents [14]. Because these pigments, however, are found to contain small amounts of water, they are first dried until the moisture content is less than 0.05-0.1%. Then they are tested to determine that they no longer flocculate in an electric field, exhibit a small sedimentation volume, and keep good dispersion in nonpolar organic solvents. Chrome yellow and chrome oxide green are observed to deposit on the anode in electrodeposition in a nonpolar media, indicating that they have a negative charge.

The surface charge of titanium dioxide pigments changes variously according to the kinds of solvents and coatings of the pigments. Titanium dioxide pigments exhibit a positive charge in benzene, methyl ethyl ketone [15], n-butanol [16], and a negative charge in ethyl acetate [15] and n-butylamine [16]. The surface charge of titanium dioxide pigments in alcohols becomes more negative as the number of carbon atoms of the alcohols decreases [17]. Titanium dioxide pigments with coatings richer in alumina than silica have a positive charge in methanol, ethanol, and n-butanol, whereas those coated mainly with silica have a negative charge in these same alcohols [1]. Although these pigments, having a definite positive or negative charge, exhibit good dispersion stability in alcohols, the pigments coated with equimolar amounts of silica and alumina exhibit almost zero charge, which is associated with poor dispersion.

Zinc oxide, calcium carbonate, and precipitated barium sulfate are observed to have a positive charge by electrophoresis in lower alcohols, and iron oxide, lithopone, clay, and carbon black have a negative charge [18].

In the case of organic pigments, fast yellow 10G has a negative charge in nonpolar solvents such as benzene and toluene and a positive charge in alcohols such as methanol, ethanol, and n-butanol, whereas phthalocyanine blue has a positive charge in both of these solvents [14]. The surface charge of organic pitments is thought to depend on their chemical structure and manufacturing processes. Substituents of azo pigments considerably affect the charge of the pigments and those pigments substituted with a chlorine atom and a nitro group such as fast yellow 10G and toluidine red have a negative charge. However, the pigments substituted with a sulfonic acid group such as red lake C and lithol red have a positive charge, owing to Ba^{2+} and Ca^{2+} ions which are taken into the pigment as precipitating agents and act as charge-determining ions. Phthalocyanine blue is basic in nature and tends to adsorb H^+ ions exhibiting positive charges, while phthalocyanine green, which is manufactured by substituting 15-16 atoms of chlorine in phthalocyanine blue, becomes acidic, resulting in a negative charge by the adsorption of anions in organic solvents [19].

Surface-active agents have a remarkable influence on the surface charge of pigments dispersed in organic solvents. Titanium dioxide pigments, which exhibit no charge in nonpolar organic solvetns, obtain a positive charge with the addition of the anionic surface-active agent sodium alkylnaphthalenesulfonate, and a negative charge of fast yellow 10G is reversed to positive with the addition of this surface-active agent associated with improved dispersion stability [14]. This is thought to be based on the formation of electrical double layers by dissolution of sodium ions of the surface-active agent into a trace amount of water around the pigment particles to give a positive charge, and by dissolution of organic sulfonate ions in the bulk of organic solvents.

Titanium dioxide pigments have a positive charge in nonpolar organic solvents such as n-hexane and p-xylene containing Aerosol OT (sodium di-2-ethylhexyl sulfosuccinate) [20]. As the concentration of Aerosol OT increases, the ζ potentials of pigments decrease gradually keeping the dispersion stable as long as they are larger than 50 mV at Aerosol OT concentrations of approximately 0.5-1.0 mM. A further decrease in ζ potentials at higher concentrations of Aerosol OT results in rapid separation of pigments. The experimental results on the relation between ζ potentials and dispersion stability are in close agreement with predictions based on the DLVO theory for stability ratios. α-Alumina and precipitated barium sulfate have a positive charge in those solvents containing Aerosol OT, whereas carbon black, which exhibits no charge in the absence of the surface-active agent [21], and phthalocyanine green have a negative charge in the same media.

As the IEP of metal oxide pigments is shifted by hydration, water influences the charge of pigment particles in organic solvents. Although completely dried titanium dioxide has a negative charge in p-xylene containing Aerosol OT 30 min after dispersion, the charge reverses to positive by hydration after 15 hr [20]. The negative charge of α-alumina in alcohols reverses to positive with the addition of water, whereas the charge of aluminum hydroxide in alcohols is always positive with any addition of water [22].

Carbon black has a negative charge in nonpolar solvents such as n-hexane, cyclohexane, and benzene containing Aerosol OT. As water is added into the dispersion, the negative ζ potential increases to a maximum, followed by a gradual decrease on further addition of water [23]. To explain the effect of water, it is proposed that in the presence of small amounts of water, Na^+ ions, which are formed due to the slight dissolution of Aerosol OT, dissolve into water solubilized in the micelles of Aerosol OT, and organic sulfonate ions of the surface-active agent are adsorbed onto the surface of carbon black to give a negative charge. When the amount of water is increased, however, water dissolving Na^+ ions becomes adsorbed onto the pigment particles by partition between the bulk solution and the pigment surface, resulting in a decrease in the negative charge of carbon black particles.

When two kinds of pigments having a different IEP are mixed in water, they have opposite signs from each other in the pH range between them associated with heavy flocculation. For example, an equimolar mixture of silica and alumina in an electrolyte solution, with IEP values of 2.0 and 9.0, respectively, has a stability ratio unity and a large sedimentation volume at pH 4.0-8.5, indicating heavy heteroflocculation [24]. The pH dependence of the viscosity of 2% w/v dispersions containing both oxides in various ratios in an electrolyte solution also reflects the dominant effect of coflocculation over the pH range 5.0-8.0.

A mixture of titanium dioxide and alumina, with IEP values of 5.9 and 8.9, respectively, flocculates within the pH range between them [25]. The flocculation, however, is observed to change dependently with time, and the lower limit of the pH of flocculation shifts to higher values with every redispersion. After standing 24 hr, rapid flocculation can occur only at a pH value near 8.9, which corresponds to the pH of poor dispersion of alumina. This is explained as Al^{3+} ions dissolving out of alumina and being adsorbed onto the surface of titanium dioxide, followed by the deposition of metal hydroxide to cover the whole surface of titanium dioxide particles at equilibrium [26]. Titanium dioxide pigments then act as if they were aluminum hydroxide.

The surface charge of clay particles is obtained by a deficiency in positive charge in crystal lattice by the isomorphic substitution of high-valence cations in the crystals, (i.e., Si^{4+} and Al^{3+}), by lower-valence cations (i.e., Mg^{2+}, Na^+, and Li^+). It has been indicated, on the other hand, that positive sites exist on the edges of crystals owing to exposed lattice aluminum atoms. Consequently, particles of clay might have two oppositely charged double layers, one associated with the face of clay plates and the other associated with the edge of the plates. The particles of clay, therefore, attract each other edge to face by the electrostatic attraction of opposite charges to form card house flocculates. With addition of cationic polymer, consisting of a linear chain of recurring N-substituted piperidinium chloride units alternating with methylene units along the chain, the negative charge of the faces is reversed to positive, resulting in the stabilization of dispersion [27].

Pigments charged with opposite signs in organic solvents behave similarly. Although titanium dioxide anatase dried at 110 and 250°C is observed to move to the cathode in electrophoresis in n-butanol, the pigment dried at 500°C moves to both electrodes, indicating the existence of pigment particles of both positive and negative charge [28]. The sedimentation velocity of the latter pigment is more than 100 times faster than that of the former pigment, suggesting an enhanced flocculation due to the mutual attraction of particles charged with opposite signs.

14.3 THE CHARGE AND DISPERSION OF PIGMENTS IN PAINTS

When pigments are used in paints and printing inks, they are dispersed in varnishes consisting of binder resins, solvents, and additives such as driers, plasticizers, dispersing agents, and others. The charge of pigments in these products plays an important roll in maintaining a good dispersion, but it might change considerably in the presence of varying amounts of these components. The stabilization by steric hindrance of the adsorbed layers is also effective, and it depends on the nature of the dispersion systems which of these two mechanisms might play the predominant roll.

Titanium dioxide pigments, coated with alumina with and without silica, have various ζ potentials in water-based paints, depending on the coatings of the pigment surface, the types and the concentrations of binder resins, the degree of milling, and the presence of neutralizing organic amines [29]. Even low concentrations of binder resins can increase the negative charge of pigments, and milling for a long time, on the other hand, can decrease the ζ potentials and reduce the pH values of the dispersions. The scattering power of paints is found to increase, independently of the types of resins, as the negative ζ potentials decrease.

Three kinds of phtalocyanine blue pigments (i.e., toner, resinated and polar) have negative ζ potentials of 45-50 mV in water, decreasing to 21-24 mV with the addition of alcohol used in many cases in practical water-based paints and printing inks [30]. The ζ potentials of pigments decrease further to 11-18 mV in varnishes containing acrylic, polyester, and alkyd resins, making stabilization by electrostatic repulsion difficult. Several stable dispersions, showing no color drift on finished panels, obtained by suitable combinations of pigments and resins are thought to be due to polymer layers adsorbed onto the surface, or to molecules of resin polymer penetrating into the electrical double layers of pigment particles.

The presence of binder resins can have considerable effect on the surface charges and the adsorbed layers of pigment particles in organic solvent systems as well as in water-based systems, resulting in complex phenomena of pigment dispersion. Inorganic pigments such as titanium dioxide, zinc oxide, chrome yellow, and iron oxide have no charge in benzene dissolving polystyrene, have a negative charge in resins of low acid value such as polymerized oil from linseed oil, and have a positive charge in resins of higher acid value such as alkyd, alkyd phenol, maleic, and acrylic resins [31]. Sedimentation of pigment particles is fast, and flow properties such as viscosity and yield value are worst in the dispersions in polystyrene varnish, better in acrylic resin varnish, and best in polymerized oil from linseed oil varnish associated with little flocculation in paints, excellent gloss, and excellent hiding power of paint film.

Titanium dioxide pigments, with and without coatings of alumina, have positive and negative ζ potentials in n-butanol and n-butylamine, respectively, and their dispersions are stabilized by an electric force when the repulsive potential exceeds the critical value 15 kT, which is the characteristic of stable dispersions. When these titanium dioxide pigments are dispersed in melamine resin or linseed oil dissolved in xylene, they have small negative or zero ζ potentials, but the stability of the dispersions is excellent, which is attributed to the steric hindrance of adsorbed polymers on the surface of the particles [16]. The ζ potential of titanium dioxide pigments, prepared by the sulfate process and coated with both silica and alumina, in pentaerythritol drying oil modified alkyd resins becomes more positive as the proportion of alumina in the coatings increases. The hiding power of these pigments also increases as the proportion of alumina increases, but detailed examination of results shows that the main factor controlling the stability of dispersion and hence the hiding power is not the ζ potential but the adsorption of resin molecules on the pigment surface [32].

Titanium dioxide rutile pigments have a positive charge in alkyd resin dissolved in xylene, and a negative charge in melamine resin dissolved in n-butanol and xylene [33]. The ζ potentials decrease exponentially as the concentration of resin increases, except at a low concentration range. The charge of the pigments in varnishes containing both alkyd and melamine resins varies with the order of contact of pigment with resin. The ζ potential of titanium dioxide pigments decreases gradually with the addition of melamine varnish to the paints prepared previously by dispersing the pigment in alkyd varnish. However, it differs from that of paints prepared in reverse order, suggesting that the added resin cannot completely replace the resins previously adsorbed onto the surface of the pigment particles.

The charge of the commercial phthalocyanine blue and green pigments, differing in crystal form, crystal stability, and resistance to flocculation, in various alkyd and amino resin varnishes differs depending on their nature and the type of resin [34]. The stability of dispersion is estimated from the attractive and the repulsive forces calculated by the DLVO theory and it is concluded that the dispersion is stabilized by an electrostatic repulsion for some of phthalocyanine blue pigments of larger particle size in alkyd melamine varnish and phthalocyanine green pigment in alkyd varnish. In the case of phthalocyanine blue pigments of finer particle size, the dispersion is stabilized in both alkyd and alkyd melamine varnishes by steric hindrance between layers of resin adsorbed on the particles.

In the electrophoresis of phthalocyanine blue pigments prepared in the form of isomeric and acycular particles and dispersed in rotogravure printing inks consisted of modified phenolic resin in toluene, the mobility and the amount of deposition of the acycular particles are found to be 40 times larger than that of the isomeric particles, suggesting that the contribution of the positive charge on the prismatic planes of particles is greater than that of the negative charge on the basal planes [35]. The yield values of the printing inks are larger for acycular particles due to the formation of networks between positively charged prismatic planes and negatively charged basal planes, similar to the card house flocculation of clay. The ζ potentials of pigments change with the addition of certain basic derivatives of copper phthalocyanine, which reduce the flocculation of phthalocyanine pigments by promoting the adsorption of resin onto the surface of the pigment particles [36]. As the amount of the basic derivative increases, the ζ potentials tend to decrease gradually together with a decrease in yield values, approaching the Newtonian flow in the neighborhood of zero ζ potential.

Phthalocyanine green pigment exhibits a negative charge in the printing ink varnish above, and when mixed with printing ink containing phthalocyanine blue pigment of positive charge, heavy coflocculation of blue and green pigments is observed. The coflocculation of blue and green pigments can be reduced to a large extent, and hence smooth mixture of both pigments is obtained with the addition of the basic derivative of copper phthalocyanine, owing to the steric hindrance of adsorbed polymer layers associated with the decrease in ζ potentials [35].

REFERENCES

1. H. Rechmann, *Farbe Lack*, 70, 861 (1964).
2. G. A. Parks, *Chem. Rev.*, 65, 177 (1965).
3. G. R. Joppien and K. Hamann, *J. Oil Colour Chem. Assoc.*, 60, 412 (1977).
4. D. E. Yates and T. W. Healy, *J. Colloid Interface Sci.*, 55, 9 (1976).
5. G. R. Wiese and T. W. Healy, *J. Colloid Interface Sci.*, 51, 427, 434 (1975).
6. G. D. Parfitt and J. Ramsbotham, *J. Oil Colour Chem. Assoc.*, 54, 356 (1971).
7. G. D. Parfitt, *Croat. Chem. Acta*, 45, 189 (1973).
8. W. Tang, S. Zhang, Z. Sany, and X. Wang, *Nan-cing Ta Hsueh Hsueh Pao, Tzu Jan K'o Hsueh*, 1979, 99 (1979).
9. T. Murata, Y. Matsuda, and H. Imagawa, *Denki Kagaku oyobi Kogyo Butsuri Kagaku*, 46, 42 (1978); 47, 334 (1979).
10. T. Murata, Y. Matsuda, and H. Imagawa, *Tanso*, 1979, 15 (1979).
11. T. Murata and Y. Matsuda, *Tanso*, 1979, 125 (1979).
12. S. Noro, A. Takamura, and M. Koishi, *Chem. Pharm. Bull.*, 27, 301 (1979).
13. M. Murata, *Nippon Kagaku Kaishi*, 1979, 984, 1294 (1979).
14. H. Brintzinger, R. Haug, and G. Sachs, *Farbe Lack*, 58, 5, 143 (1952); 59, 15, 59 (1953).
15. K. Tamaribuchi and M. L. Smith, *J. Colloid Interface Sci.*, 22, 404 (1966).
16. L. A. Romo, *J. Phys. Chem.*, 67, 386 (1963).
17. S. Yoshizawa, N. Watanabe, and I. Tari, *Denki Kagaku oyobi Kogyo Butsuri Kagaku*, 37, 521 (1969).
18. N. Shinada and T. Tomiyama, *Shikizai Kyokaishi*, 39, 347 (1966).
19. A. Vinther and P. Sørensen, *Farbe Lack*, 77, 317 (1971).
20. D. N. L. McGown, G. D. Parfitt, and E. Willis, *J. Colloid Sci.*, 20, 650 (1965).
21. K. E. Lewis and G. D. Parfitt, *Trans. Faraday Soc.*, 62, 1652 (1966).
22. L. A. Romo, *Discuss. Faraday Soc.*, 42, 232 (1966).
23. A. Kitahara, S. Karasawa, and H. Yamada, *J. Colloid Interface Sci.*, 25, 490 (1967).
24. R. D. Harding, *J. Colloid Interface Sci.*, 40, 164 (1972).
25. T. W. Healy, G. R. Wiese, D. E. Yates, and B. V. Kavanagh, *J. Colloid Interface Sci.*, 42, 647 (1973).
26. G. R. Wiese and T. W. Healy, *J. Colloid Interface Sci.*, 52, 452, 458 (1975).
27. A. P. Black, F. B. Birkner, and J. J. Morgan, *J. Colloid Interface Sci.*, 21, 262 (1966).

28. S. Yoshizawa, N. Watanabe, and I. Tari, *Denki Kagaku oyobi Kogyo Butsuri Kagaku*, *30*, 439 (1962).
29. M. Cremer, *J. Oil Colour Chem. Assoc.*, *60*, 385 (1977).
30. P. G. Schmidt, *J. Coat. Technol.*, *52*, 37 (1980).
31. G. Florus and K. Hamann, *Farbe Lack*, *62*, 260, 323 (1956).
32. M. J. B. Franklin, K. Goldsbrough, G. D. Parfitt, and J Peacock, *J. Coat. Technol.*, *42*, 740 (1970).
33. Y. Oyabu, H. Kawai, and Y. Nakanishi, *Shikizai Kyokaishi*, *35*, 98 (1961).
34. V. T. Crowl, *J. Oil Colour Chem. Assoc.*, *50*, 1023 (1967).
35. W. Ditter and D. Horn, *Proc. 4th Int. Conf. Org. Coat. Sci. Technol.*, 1978, p. 251.
36. W. Black, F. T. Hesselink, and A. Topham, *Kolloid Z. Z. Polym.*, *213*, 150 (1966).

15
Cosmetics

Shoji Fukushima Shiseido Co., Ltd., Yokohama, Japan

15.1. Introduction 369
15.2. Emulsion 370
    15.2.1. Cream 370
    15.2.2. ζ Potential of emulsion droplets 371
    15.2.3. Effect of surface potential on emulsion stability 376
    15.2.4. Gel in O/W cream 378
15.3. Suspension 378
    15.3.1. ζ Potential of solid particles 378
    15.3.2. Gelling agent 379
    References 383

15.1 INTRODUCTION

Although briefly and simply termed "cosmetics" in the literature, this simple word encompasses a broad range of studied and their intended purpose of use, related to each other by the fact that the systems are, in most cases, dispersions.

    A face powder is a mixture of several kinds of powder. The addition of a small amount of binder, oil, or wax to the above makes a compact-type face powder, and dispersion into an oily base results in a makeup foundation.

    Emulsion-type cosmetics are called creams. They can be classified into two types: an oil-in-water cream and a water-in-oil cream. Sometimes powders are added to improve the texture, such as the softness or lubricity. There is also a cream-type makeup foundation which is an emulsion containing powders.

A toothpaste is a dispersion of powders such as calcium carbonate or calcium diphosphate as polishing materials in an aqueous solution of water-soluble polymers such as carboxymethyl cellulose or sodium alginate, and a nail lacquer is a dispersion of pigments in a nonaqueous solution of nitrocellulose.

A shampoo consists of surface-active agents with additives to improve the performance and appearance, the additives not always being in a dissolved state. For example, an opacifying agent which is added to give a pearly lustrous appearance, or zinc pyrithione, which is used as an antidandruff agent, are in a dispersed state.

Since many cosmetics are dispersions, prevention of instability such as flocculation, settling, creaming, or coalescence of the dispersed materials is an essential factor to be considered in the formulation. Therefore, the electrical phenomena at the interface which is related to colloid stability, are of great importance in cosmetics.

However, almost no literature is to be found on cosmetics, perhaps because of the complexity of the formulation. This chapter undertakes to review some papers on the electrical phenomena of emulsion and suspension, and introduces technologies developed to stabilize dispersion-type cosmetics.

15.2 EMULSION

15.2.1 Cream

A cosmetic cream comes within the category of emulsion. According to the definition that an emulsion is a system consisting of one liquid phase dispersed as droplets in another liquid phase [1], creams are not always emulsions. Waxes, which are frequently incorporated in creams, may still be in a solid state. In some cases, inorganic powders such as titanium dioxide or talcum powder are included in the formulation.

Cetostearyl alcohol used for stabilizing an oil-in-water (O/W) cream forms lyotropic lamellar liquid crystals in the cream [2-5]. This phase is thought to be a semisolid. In spite of all these circumstances a cream is usually considered as an emulsion.

In order to discuss the stability of an emulsion, we must distinguish between coalescence, flocculation, creaming, and droplet growth by molecular diffusion [6-8]. Coalescence is the joining of small droplets to form larger ones, leading ultimately to two liquid layers. Flocculation is the sticking together of droplets to form clusters. In creaming the droplets rise or fall under the action of gravity and may be accelerated by coalescence and flocculation. Finally, droplet growth by molecular diffusion is the growth of larger droplets at the expense of smaller ones due to the difference in solubility between different-sized droplets. Any or all of these may occur after the emulsion has been prepared, but flocculation is mostly associated with the electrical phenomena at the droplet surface.

15.2.2 ζ Potential of Emulsion Droplets

Among several methods, the microelectrophoretic mobility measurement is considered to be the most appropriate method for determining the zeta (ζ) potential of emulsion droplets. However, the measurement of droplets of cosmetic creams is hardly possible due to the generally high volume fraction of internal phase. To overcome this difficulty, the cream is diluted with the dispersion medium when measuring the mobility. Riddick recommends a way of redispersing a small amount of the cream into the dispersion medium that was removed by centrifuging [9]. This method may cause coalescence of droplets, should the centrifugal force be too strong. Riddick says that centrifuging should be performed such that the separation occurs within 10-30 min.

Although several equations have been proposed for the ζ-potential calculation from the microelectrophoretic mobility [10-13], the following two are the most commonly used in the study of oil-in-water emulsion.

One is the Helmholtz-Smoluchowski equation,

$$\zeta = \frac{4\pi\eta v}{\varepsilon E} \tag{15.1}$$

where v is the microelectrophoretic mobility of droplets in the field strength E, ε the dielectric constant, and η the viscosity of the dispersion medium.

The second is the Henry equation, in which the effects of the droplet radius a and the thickness of the electric double layer $1/\kappa$ on mobility have been taken into consideration by the introduction of the function $f(\kappa a)$ into Eq. (15.1). It is as follows:

$$\zeta = \frac{4\pi\eta v}{\varepsilon E} \frac{1}{f(\kappa a)} \tag{15.2}$$

For κa above 300, $f(\kappa a) = 1$ should be accurate to within 1%; while for smaller values, $f(\kappa a)$ approaches 2/3, which is valid to within 1% when $\kappa a = 0.5$ [13].

In the experimental study, 1 is given for $f(\kappa a)$ when the droplet radius is expected to be much greater than the double-layer thickness [14-17]. In the reverse case, 2/3 is given. Elworthy and co-workers have used other values by estimating them precisely [18,19].

Sherman used Eq. (15.3) instead of Eq. (15.2),

$$\zeta = \frac{6\pi\eta v}{\varepsilon E} \frac{1}{f(\kappa a)} \tag{15.3}$$

but this is equivalent to Eq. (15.2), since 1.0 or 1.5 was given for the respective case [20].

Table 15.1 Zeta Potentials of Emulsion Particles Described in the Literature

Oil	System component		ζ Potential (mV)	Reference
	Surfactant	Electrolyte		
Nujol	—	H_2SO_4	−48	9
	—	NaOH	−74	9
IPM	—	H_2SO_4	−42	9
	—	NaOH	−84	9
Nujol	—	$AlCl_3$	−40	9
	—	$CaCl_2$	−43	9
	—	NaCl	−64	9
	—	$Na_4P_2O_7$	−93	9
IPM	—	$Na_2B_4O_7$	−80	9
Nujol	K-oleate	—	−122	9
	Natrosol	—	−40	9
IPM	Natrosol	—	−36	9
Light mineral oil	Span 80 + Tween 80	—	−45	14
Cottonseed oil	Span 80 + Tween 80	—	−42	14
Xylene	Oxytetracycline HCl	—	139.1	15

Oxytetracycline	HCl K-oleate	115.6	15
Oxytetracycline	HCl K-oleate	−118.4	15
Oxytetracycline	HCl K-citrate	145.8	15
Oxytetracycline	HCl K-citrate	−113.0	15
Oxytetracycline	HCl $K_3[Fe(CN)_6]$	102.8	15
Oxytetracycline HCl	$K_3[Fe(CN)_6]$	−56.7	15
Oxytetracycline HCl	$K_4[Fe(CN)_6]$	91.5	15
Oxytetracycline HCl	$K_4[Fe(CN)_6]$	−133.6	15
Ca-valerate	$NaNO_3$	−70.3	21
Ca-valerate	$Pb(NO_3)_2$	−62.5	21
Ca-valerate	$Cr(NO_3)_3$	−69.5	21
Ca-valerate	$Cr(NO_3)_3$	22.4	21
Ca-valerate	$Zr(NO_3)_4$	−76.9	21
Ca-valerate	$Zr(NO_3)_4$	34.5	21
Ca-valerate	$Th(NO_3)_4$	−68.5	21
Ca-valerate	$Th(NO_3)_4$	103.3	21
Ca-valerate +(EO)-octylphenyl ether	—	90	21
Ca-Caproate + (EO)-octylphenyl ether	—	100	21

Table 15.1 (continued)

Oil	System component		Electrolyte	ζ Potential (mV)	Reference
	Surfactant				
Xylene	Ca-Caprylate + (EO)-octylphenyl ether		—	60	21
Light mineral oil	Span 80 + Tween 80		NaCl	−42	16
	Span 80 + Tween 80		$CaCl_2$	40	16
Cottonseed oil	Span 80 + Tween 80		NaCl	−45	16
Liquid paraffin	PVA		—	−67	17
	PVA + cetyltrimethyl ammonium bromide		—	90	17
	PVA + Na-dodecyl benzene sulfonate		—	−85	17
Chlorobenzene	Hexadecanol		—	−59.4	18
	Hexadecanol + $(EO)_6$-hexadecyl ether		—	−60.1	18
Anisole	$(EO)_3$-hexadecyl ether		—	−76.5	19
	$(EO)_6$-hexadecyl ether		—	−69.0	19

	(EO)$_9$-hexadecyl ether	—	−73.3	19
	(EO)$_{25}$-hexadecyl ether	—	−37.3	19
Chlorobenzene	(EO)$_3$-hexadecyl ether	—	−78.4	19
	(EO)$_6$-hexadecyl ether	—	−74.6	19
	(EO)$_9$-hexadecyl ether	—	−74.6	19
	(EO)$_{25}$-hexadecyl ether	—	−31.7	19
Arachis oil	Na-palmitate	—	100	22
	Cetyltrimethyl ammonium bromide	—	100	22
Soybean oil	Egg lecithin	NaCl	−42	23
	Egg lecithin	KCl	−35	23
	Egg lecithin	LiCl	−33	23
	Egg lecithin	CaCl$_2$	−10	23
	Egg lecithin	CaCl$_2$	19	23
	Egg lecithin	MgCl$_2$	−14	23
	Egg lecithin	MgCl$_2$	15	23
	Egg lecithin	AlCl$_3$	45	23

Table 15.1 summarizes the ζ potentials of emulsion droplets as stated in literature [9,14-19,21-23]. Since ζ potentials vary with electrolyte concentration, only the maximum absolute value in each study was selected. We can see examples of emulsions having both positive and negative ζ potentials in spite of the fact that they consist of the same components. They are those which invert the sign with increase in the electrolyte concentration.

15.2.3 Effect of Surface Potential on Emulsion Stability

The author recommends that readers refer to the specialized textbooks by Kitchener and Musselwhite [6] and Becher [24] on the theory of the effect of surface potential on emulsion stability. Problems in practical systems are discussed below.

For an oil-water interface in an emulsion to be electrically charged, the emulsion should contain some ions, such as (a) ionic surface-active agents, (b) preservatives such as sodium benzoate, and (c) chelating agents. However, there are some cases where emulsion droplets bear electrical charge on the surface although the emulsion does not contain any electrolytes. Sherman explains that the relatively high electrophoretic mobility of the oil droplets may be due to the soap impurity in the emulsifier (sorbitane monolaurate) [25]. Becher believes that it arises as a result of hydrogen bonding at the ether oxygen of the polyoxyethylene chain and subsequent oxonium ion formation [14].

The DLVO theory discusses whether two particles can approach and contact each other as a result of the summation of attraction potential energy based on the London-van der Waals force and the repulsion potential energy due to the electrical surface charge [26]. When the total potential energy curve does not attain a maximum value, the particles flocculate strongly. If it reaches a very high maximum value at a particle separation, in other words, if the electrical repulsion force is strong enough to overcome the London-van der Waals attraction force at that distance, the particles are unable to approach close enough to flocculate. In such cases, however, since the potential curve reaches a shallow minimum which is referred to as a secondary minimum at a longer separation, the particles weakly flocculate at that distance [25,26].

Calculation of potential energy curves suggests that the secondary minimum is significant with spheres of 10 μm radius but insignificant at 0.1 μm radius. Whether it is important at 1.0 μm will depend on the magnitude of the Hamaker constant [27] and on the effect of the correction for retardation [6].

> The flocculation in the secondary minimum is weak but does not stop at pairs to ultimately form larger flocs. The structural viscosity of concentrated emulsions is considered to result from the flocculation in the secondary minimum [6,24-26].

Even though the surface charge is high enough to prevent flocculation, creaming cannot be avoided due to the action of gravity if the droplet size is larger than about 1 µm and the density is different from that of the dispersion medium. In that case, the droplets will rise or fall according to Stokes' law. Below 1 µm, thermal disturbance (i.e., Brownian motion) against the creaming should be considered.

Surface separation between the nearest droplets, H, is related to the volume fraction of interval phase, ϕ, the droplet diameter, D, and the packing state of the droplets. If spherical droplets of a uniform diameter are placed at the lattice point of the closest packing structure, H is as follows:

$$H = \left(3\sqrt{\frac{0.740}{\phi}} - 1\right)D \qquad (15.4)$$

where 0.740 is the value corresponding to the volume fraction of whole spheres against whole volume in the state of closest packing. Such a state is possible if all equidiameter droplets interact with one another under the same interaction force. The surface separation calculated by using Eq. (15.4) at various values of ϕ and D are tabulated in Table 15.2.

The calculation suggests that under practical conditions (e.g., at ϕ = 0.5 and D = 1.0 µm) the surface separation between the nearest neighbors is as high as 1400 Å. A more dilute emulsion gives a greater value. In contrast, published data of the surface separation at which the potential energy reaches the secondary minimum, H_{min}, was 100-200 Å [15,21].

One paper reports that H_{min} is 2000-3000 Å [14], but in that case, since $\phi \approx 0.15$ and D = 0.75, H is expected to be about 53,000 Å. Such discrepancies between H_{min} and H suggest that the secondary minimum

Table 15.2 Intersurface Distance H (µm) Calculated Using Eq. (15.4)

	D(µm)			
ϕ	0.1	0.5	1.0	5.0
0.1	0.0949	0.4746	0.9491	4.746
0.2	0.0547	0.2735	0.5470	2.735
0.3	0.0351	0.1757	0.3514	1.757
0.4	0.0228	0.1139	0.2279	1.139
0.5	0.0140	0.0699	0.1399	0.699
0.6	0.0073	0.0363	0.0726	0.363
0.7	0.0019	0.0095	0.0189	0.095

or the repulsion force by the surface charge cannot prevent ultimate creaming. Creaming may be avoided, however, by increasing the volume fraction of internal phase up to a corresponding value.

15.2.4 Gel in O/W Cream

Another method of preventing droplet creaming is to give the emulsion thixotropic properties. Cetostearyl alcohol is a material widely used for this purpose in the cosmetic industry. It acts as a consistency modifier as well as a stabilizer of the cream. It builds platelike lamellar liquid crystals by associating with the emulsifier and water, and the liquid crystals are considered to make the continuous phase gelatinous and act as a stabilizer in the cream by immobilizing the emulsion droplets [2-5].

15.3 SUSPENSION

15.3.1 ζ Potential of Solid Particles

Surfactants are frequently used in suspension-type cosmetics for dispersing powdery materials. Fukushima and Kumagai [28,29] studied the difference among four types of nonionic ethoxylated surface-active agents of the ether type, ester type, amine type, and amide type, as a dispersing agent for aqueous suspension of titanium dioxide pigments. They found that tri- to heptaethoxylauryl esters have a good dispersion ability in the suspensions of rutiles, and mono- to tetraethoxyamines have a good dispersion ability in those of rutiles and anatases. Esters and amines having more ethoxy numbers and all the ethers and amides were ineffective in stabilizing these suspensions (Table 15.3).

They also found that dispersibility was in close correlation with adsorption of the surface-active agent on the titanium oxide, and that the adsorption was electrostatic or attributed to ionic properties of the surface-active agent molecules and to the surface charge or the acidity of the oxide (Table 15.4). Ethers were not adsorbed, probably due to the nonpolarity of the ether oxygen. Esters with lower ethoxy numbers appear to have been adsorbed by the carbonyl oxygens that have a negative atmosphere. The infrared adsorption at 1738 cm^{-1} arising from the stretching vibration of an ester group shifted to 1723 cm^{-1} after adsorption. The nonadsorption of those of higher ethoxy numbers is possibly due to the weaker bonding force between the carbonyl group and the oxide surface compared with the solubilization force into the bulk water [29].

Amines of lower ethoxy numbers were adsorbed regardless of the surface charge of titanium oxides, and the saturation adsorption amount of tetraethoxylaurylamine was in the same order as the surface acidity of the oxides (Fig. 15.1). Amines often behave like a base, so they are considered to be adsorbed at acidic sites of the oxide surface.

Table 15.3 Turbidity of Supernatant Liquid of Titanium Dioxide Dispersions[a]

Ethoxy number	A-100				R-820			
	LE	LES	LAN	LAD	LE	LES	LAN	LAD
1			++				++	
2	−	−	++	−	−	−	++	+
3	−	−	++	−	−	++	++	−
4	−	−	++	−	−	++	++	−
5	−	−	−	−	−	++	−	−
6	−	−	−	−	−	++	−	−
7	−	−	−	−	−	++	−	−
8	−	−	−	−	−	+	−	−
9		−				−		
10		−	−			−	−	

[a]A-100, anatase (Ishihara-Sangyo Co., Ltd.); R-820, rutile (Ishihara-Sangyo Co., Ltd.); LE, ethylene oxide derivative of lauryl alcohol; LES, ethylene oxide derivative of lauric acid; LAN, ethylene oxide derivative of laurylamine; LAD, ethylene oxdie derivative of laurylamide; ++, turbid; +, fairly turbid; −, clear.
Source: Refs. 28 and 29.

Table 15.5 illustrates the electrophoretic mobilities of the titanium dioxides in aqueous solutions of hexaethoxylauryl ester (LES-6) and tetraethoxylaurylamine (LAN-4). The oxides dispersed in the LES-6 solution are negative in the surface charge, and those dispersed in the LAN-4 solution are positive, supporting the foregoing theory.

15.3.2 Gelling Agent

There are many cosmetics in which dispersed particles are stabilized against sedimentation by thixotropic property. A higher viscosity of dispersion medium will retard sedimentation of the particle based on Stokes' law, but in a medium that lacks thixotropy the particles will ultimately settle on a long-term storage. According to Stokes' law, a spherical particle with a diameter of 1.0 μm and a specific gravity of 3.0 settles a few millimeters per year in the medium with a viscosity

Table 15.4 Electrophoretic Mobilities and Acidities of the Titanium Dioxides Used and Saturated Adsorption Amounts of LES-6 and LAN-4 for the Titanium Dioxides

Titanium dioxide[a]	Mobility (μ sec^{-1}/V cm^{-1})	Acidity ($\times 10^{10}$ mol/cm^2)	Saturated adsorption amount ($\times 10^{10}$ mol/cm^2)	
			LES-6[b]	LAN-4[c]
R-630	1.46	2.26	9.85	4.85
R-780	-1.82	2.08	0.0	5.00
R-820	-0.47	0.98	4.05	3.20
R-1	1.68	—	7.50	5.40
A-100	-1.69	2.44	0.0	5.50
A-220	1.33	1.46	6.85	3.45

[a]R and A indicate rutile and anatase, respectively, and figures are commercial grades.
[b]Hexaethoxylauryl ester.
[c]Tetraethoxylaurylamine.
Source: Ref. 29.

Table 15.5 Electrophoretic Mobility of Titanium Dioxides in Aqueous Solution of LES-6 and LAN-4

	Mobility (μ sec^{-1} V^{-1} cm^{-1})	
Titanium dioxide[a]	LES-6[b]	LAN-4[c]
R-630	−1.28	1.28
R-780	−3.00	0.0
R-820	−1.40	1.87
R-1	−0.12	1.87
A-100	−2.14	1.66

[a]R and A indicate rutile and anatase, respectively, and figures are commercial grades.
[b]Hexaethoxylauryl ester.
[c]Tetraethoxylaurylamine.

of 10,000 Hz. The viscosity of a nail lacquer most easily and readily applied is said to be between 200 and 300 Hz. Under such low viscosity, the same particle will settle as much as 10 cm in a year.

An ingeneous way to avoid settling of particles without making the viscosity of a system higher is to give thixotropic property to the

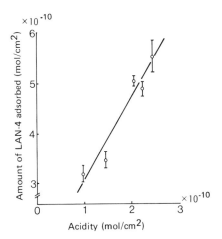

Figure 15.1 Relationship between the saturated adsorption amount of LAN-4 and the acidity of the titanium dioxide surfaces.

medium. The downward force working on a particle in a medium under the action of gravity is $\pi D^3 (d - d_o)g/6$, where D is the diameter of the particle, d and d_o are the specific gravities of the particle and medium, respectively, and g is the acceleration of gravity. Since the force works on the cross-sectional area of the particle, $\pi D^2/4$, the particle will be immobilized if the medium has a yield value greater than $2D(d - d_o)g/3$ [30]. For a particle of 1.0 μm diameter and 3.0 specific gravity, 0.1 dyn/cm^2 is sufficient. This technique provides good cosmetics with a proper viscosity and an excellent dispersion stability, and two methods have been developed for this purpose.

Synthetic Hectorite

Bentonites are well known as swelling clays. The particle is platelike and has a layered structure, as shown in Fig. 15.2. The unit layer thickness is approximately 1 nm, and each layer has cation-exchangeable sites.

The chemical formula of swelling clay is generally expressed as

$$\{Si_8[Mg_{6-x}X_x]O_{20}[(OH)_{4-y}Y_y]\}^{x(-)} \frac{x}{n} M^{n(+)}$$

where X is a mono- or bivalent cation, Y is a monovalent anion, x is a value between 0 and 6, y is a value between 0 and 4, and $M^{n(+)}$ is an n-valent cation.

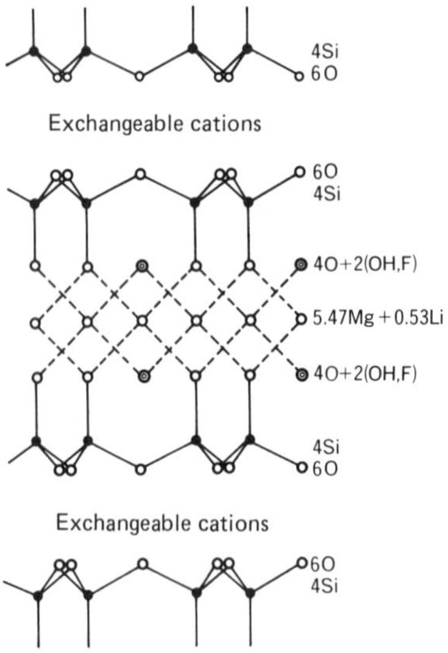

Figure 15.2 Structure of the synthetic hectorite.

When $x = 0$ (i.e., X is also Mg) the particle is electrostatically neutral and has no ion exchangeability. However, if a part of the magnesium ions are replaced by monovalent cations, the particle will have a negative charge and be neutralized by the equivalent number of counterions, $M^{n(+)}$.

In water, since the counterions diffuse into it, the particles acquire a negative charge. In contrast, the particle edge is positively charged due to the exposed partially unbound magnesium ions [31,32]. Such particles flocculate and build flocs of face-to-edge or card-house structure with positive edges bounded to negative faces. Thus the whole system acquires a small yield value.

Natural hectorite, belonging to the bentonite family, is an excellent gel-forming clay but is rare and is contaminated with impurities such as dolomite and quartz. In the early 1960s, Neuman succeeded in synthesizing an analog of hectorite [33]. The formula of the synthetic hectorite is

$$\{Si_8[Mg_{5.47}Li_{0.53}]O_{20}[(OH)_2F_2]\}0.53Na$$

This clay is superior to the natural one in gelling ability as well as the external appearance. Hence it is used in various cosmetics of aqueous media in which pigments are suspended.

Organophilic Bentonite

A gelling agent that builds a gel in an organic medium is also used in many pigmented cosmetics of oily vehicles [34]. Organophilic bentonite has been developed for such use and is widely used in the cosmetic field.

The organophilic bentonite is made from bentonite by treating with an aqueous solution of an organic ammonium salt or a cationic surface-active agent [35]. It makes a gel in organic media which have a degree of polarity. Organic media consisting primarily of hydrocarbons are deficient in the polarity for the organophilic bentonite to make a gel. Therefore, low-molecular-weight alcohols such as ethanol are added to the medium.

There are no references discussing the gelling mechanism in terms of electrical phenomena at the particle surface. However, it is believed that any electrostatic forces may contribute to the gellation.

The organophilic bentonite is applied to oily-base cosmetics such as nail lacquer [34,36], dry spray aerosol perspirant [37], and so on for preventing dispersed pigments from settling.

REFERENCES

1. H. L. Greenwald, *J. Soc. Cosmet. Chem.*, 6, 164 (1955).
2. S. Fukushima, M. Takahashi, and M. Yamaguchi, *J. Colloid Interface Sci.*, 57, 201 (1976).

3. S. Fukushima, M. Yamaguchi, and F. Harusawa, J. Colloid Interface Sci., 59, 159 (1977).
4. S. Fukushima and M. Yamaguchi, J. Jpn. Oil Chem. Soc., 29, 106 (1980).
5. S. Fukushima and M. Yamaguchi, Yakugaku Zasshi, 101, 1010 (1981).
6. J. A. Kitchener and P. R. Musselwhite, in Emulsion Science (P. Sherman, ed.), Academic Press, London, 1968, p. 77.
7. W. I. Higuchi and J. Misra, J. Pharm. Sci., 51, 459 (1962).
8. S. S. Davis and A. Smith, in Theory and Practice of Emulsion Technology (A. L. Smith, ed.), Academic Press, London, 1976, p. 325.
9. T. M. Riddick, Am. Perfum. Cosmet., 85, 31 (1970).
10. M. V. Smoluchowski, Bull. Acad. Sci. Cracovie, 1903, 182 (1903).
11. P. Debye and E. Hückel, Z. Phys., 25, 49 (1924).
12. E. Hückel, Z. Phys, 25, 204 (1924).
13. D. C. Henry, Proc. R. Soc. A, 133, 106 (1931).
14. P. Becher and S. Tahara, Proc. 6th Int. Congr. Surf. Activ., II-2, 519 (1972).
15. M. K. Sharma and S. N. Srivastava, Rocz. Chem. Ann., 49, 2047 (1975).
16. P. Becher, S. E. Trifiletti, and Y. Machida, in Theory and Practice of Emulsion Technology (A. L. Smith, ed.), Academic Press, London, 1976, p. 271.
17. T. F. Tadros, in Theory and Practice of Emulsion Technology (A. L. Smith, ed.), Academic Press, London, 1976, p. 281.
18. P. H. Elworthy, A. T. Florence, and J. A. Rogers, J. Colloid Interface Sci., 35, 34 (1971).
19. P. H. Elworthy and A. T. Florence, J. Pharm. Pharmacol., 21s, 70S (1969).
20. P. Sherman, in Emulsion Science (P. Sherman, ed.), Academic Press, London, 1968, p. 177.
21. R. P. Varma and P. Bahadur, Cellul. Chem. Technol., 9, 381 (1975).
22. D. Rambhau, D. S. Phadke, and A. K. Dolre, J. Soc. Cosmet. Chem., 28, 183 (1977).
23. W. H. Dawes and M. J. Groves, Int. J. Pharm., 1, 141 (1978).
24. P. Becher, in Emulsion: Theory and Practice, 2nd ed., Reinhold, New York, 1966, p. 95; Encyclopedia of Emulsion Technology, Vol. 1 (P. Becher, ed.), Marcel Dekker, New York, 1983.
25. P. Sherman, J. Phys. Chem., 67, 2531 (1963).
26. E. J. W. Verwey and J. Th. G. Overbeek, Theory of the Stability of Lyophobic Colloids, Elsevier, Amsterdam, 1948.
27. H. C. Hamaker, Physica, 4, 1058 (1937).

28. S. Fukushima and S. Kumagai, *J. Colloid Interface Sci.*, 42, 539 (1973).
29. S. Kumagai and S. Fukushima, *J. Colloid Interface Sci.*, 56, 227 (1976).
30. J. S. Wolff and R. J. Meyer, *Soap Chem. Spec.*, 1961(4), 89 (1961).
31. H. Van Olphen, *An Introduction to Clay Colloid Chemistry*, Wiley-Interscience, New York, 1963, p. 89.
32. B. J. R. Mayes, *Int. J. Cosmetic Sci.*, 1, 329 (1979).
33. B. S. Neumann, Br. Pat. 1,054,111 (1967).
34. F. W. Busch, Jr., U.S. Pat. 3,864,294 (1975).
35. J. W. Jordan, B. J. Hook, and C. M. Finlayson, *J. Phys. Chem.*, 54, 1196 (1950).
36. M. L. Schlossman, *J. Soc. Cosmet. Chem.*, 31, 29 (1980).
37. C. Kahn and C. J. R. Eichhorn, *Cosmet. Perfum.*, 89(12), 31 (1974).

16
Antirusting

Isao Tari Okayama University, Tsushima-naka, Okayama, Japan

16.1. Equi-acid-base Point 388
16.2. Ion-Exchange Adsorption 388
16.3. Isoelectric Point 390
16.4. Composite Plating 392
16.5. Electrophoretic Deposition 394
    References 395

Rusting and antirusting in aqueous solutions can be approached on the basis of interfacial electrophenomena. Corrosion on a metal proceeds with the formation of the local cell. The potential difference between the local cathode and anode is the driving force of corrosion. The cathode and anode are short-circuited through the bulk metal. The potential difference is compensated mainly by the induced concentration polarization, which reduces the corrosion rate. In an unbuffered solution such as seawater, the pH of the solution adjacent to the anode decreases greatly, while that near the cathode increases as the local cell reactions progress. The variation results in a high concentration polarization and thus retards the corrosion rate. The variation depends not only on the corrosion rate, but also on the pH buffer capacity of solutions and corrosion products existing around the local cathode and anode. Antirusting is achieved by inhibiting the local cathodic or the local anodic reaction. There are many methods of inhibiting both reactions, one of which is to form a protective film on the cathode and anode. The formation is related to interfacial electrophenomena.

A metal in a corrosive environment will generally be covered with rust, mainly oxide, oxyhydroxide, and hydroxide. Thus the characteristics of oxide-electrolyte interfaces such as zero point of charge (ZPC), isoelectric point (IEP), equi-acid-base point (EABP), ion-exchange adsorption, and so on, can also be studied in a corrosion system. Here such characteristics are related to electrode potential and antirusting.

16.1 EQUI-ACID-BASE POINT

The pH titration method can be used to determine the pH at which the amount of protons adsorbed equals that of hydroxyl ions adsorbed on the surface of an oxide. The pH should be denoted as EABP rather than ZPC, because some ions other than H^+ and OH^- ions might behave as potential-determining ions and adsorb specifically on the oxide surface. Hence the surface charge at the ZPC determined using the titration method is not necessarily zero [1].

The pH shift method is more applicable than the pH titration method to determine the EABP of solid products in corrosion, especially crevice and pit corrosion, due to the small amount of the solutions [2].

Since a metal covered with an oxide film as a corrosion product is regarded as a type of oxide electrode, its potential is greatly dependent on the pH of the solutions adjacent to the metal surface. As the pH of solutions in contact with the metal shifts toward the EABP of the oxide of the metal surface, the potential drifts toward a certain value at the EABP.

The corrosion versus pH diagram for iron and soft steel was examined in tap water acidified with hydrochloric acid and alkalized with sodium hydroxide [3]. In the neutral pH range, 4-9, corrosion was almost independent of the pH. This is explained as follows: in near-neutral and unbuffered solutions with small volumes, the pH rapidly varies to a certain value with time and then remains at rest as a result of the buffer action of the corrosion product. The steady pH value may be regarded as the EABP of the product.

16.2 ION-EXCHANGE ADSORPTION

Ion-exchange adsorption generally occurs on an oxide surface in a solution. It is a typical heteropolar adsorption which proceeds on an oxide having a considerably large zeta (ζ) potential. Since metal ions generated by corrosion are multivalent, they easily adsorb on the hydrated oxide surface with release of protons. Such ion-exchange adsorption occasionally occurs even in an acidic solution.

Table 16.1 Potential Versus pH Slopes

Electrolyte	Slope ($V \cdot pH^{-1}$)
HCl	−0.058
1.0 M LiCl	−0.069
1.0 M KCl and 1.0 M KNO_3	−0.065
0.5 M $BaCl_2$	−0.062
0.5 M $NiCl_2$	−0.066
0.5 M $CuCl_2$	−0.096
0.5 M $CoCl_2$	−0.104
0.33 M $LaCl_3$	−0.065

As the solution becomes more acidic with the ion-exchange reaction, the potential of an anode becomes more noble. Such acidity makes further corrosion possible.

Manganese dioxide is well established as a cation-exchange material. The electrode potential of an electrolytic manganese dioxide was measured as a function of pH in solutions containing several kinds of cations [4]. The pH effect obtained is summarized in Table 16.1. The potential versus pH slopes for all solutions except HCl are greater than the theoretical slope of -0.059 $V \cdot pH^{-1}$ at 25°C required by the following equation:

$$MnO_2 + H^+ + e^- = MnOOH \qquad (16.1)$$

Such high slopes are explained by the ion-exchange adsorption because replacement of hydrogen in surface hydroxyls by other cations in an ion-exchange process must cause a change in the chemical potential of surface hydroxyls and also of the surface elements, such as Mn^{4+}, Mn^{3+}, and O^{2-} ions. Such effects of ion-exchange adsorption will also be found in the pH response of a metal potential covered with oxide in a corrosive environment.

Corrosion products such as ferric oxide and ferric oxyhydroxide are cation exchangers. When α-ferric oxide is in contact with a solution containing ferric ions, the pH of the solution drifts to a lower value, which varies with the concentration of the ions. The pH of 50 ml of 10^{-2} M $FeCl_3$ was varied from 2.7 to 2.0 by dispersing 4 g of hematite powder prepared by thermal decomposition of ferrous oxalate at 600°C in air. Such variation might be caused by the ion-exchange adsorption of ferric ions [5].

Lukomski and Bohnenkamp [6] found for chromium steels that the pH value of 3 M KCl in a crevice model decreased to almost zero with increase in the concentration of chromium ions. They could not clearly explain the reason of this dependency. Such pH lowering was also observed when an electrolytic manganese dioxide was put in a solution containing zinc or manganous ions [7,8]. This is explained by the ion-exchange adsorption of the metal ions on the dioxide with release of protons. This shows that the mechanism of pH lowering observed in a crevice or a pit might also be elucidated from the standpoint of an ion-exchange reaction between metal ions and oxides as corrosion products.

16.3 ISOELECTRIC POINT

The measurement of an electrokinetic phenomenon may be used to obtain data for the electrical double layer of a solid-liquid interface. Since the phenomenon provides important information on the electrical double layer and also on its change due to adsorption of ionic or nonionic species at the interface, the phenomenon should also be measured in a corrosion system in order to investigate the adsorption of corrosive ions such as chloride ion and of surfactants as a corrosion inhibitor. Among the techniques of measuring the phenomenon, the streaming potential method may be most suitable for wire or plate samples. There is a serious problem if the Helmholtz-Smoluchowski equation is applied to estimate the ζ potential from the observed streaming potential in a metal-electrolyte system because the equation requires a very poor electronic conductor as a solid sample. However, a smooth curve fitting on the plots of the ζ potential estimated from the streaming potential against the pH provides an IEP where the polarity of ζ potential turns from negative to positive or positive to negative, and hence the IEP should be completely independent of the conductance of a solid. This is regarded as one of characteristics, even at a metal-electrolyte interface.

Let us consider an electrode system in which an IEP can be determined by the pH scale. The potential of the electrode can generally be measured against an auxiliary electrode such as a saturated calomel electrode (SCE). On the other hand, this may also be expressed, on the basis of surface chemical studies on potential difference at oxide-electrolyte interface, by the following equation [9]:

$$E = E_{pH} - E_{IEP} = k(pH - IEP) \tag{16.2}$$

where E represents the potential at a given pH relative to the bulk solution, E_{IEP} and E_{pH} the potential of an electrode measured at IEP and a given pH, respectively, against a SCE, and k is a constant which is sometimes equal to -0.059 V·pH^{-1} at 25°C. E in Eq. (16.2) can be commonly equated with the Volta potential difference Ψ_0 of the electrode surface relative to the bulk solution. The quantity E is negative or positive, depending on whether the experimental pH of the solutions is greater than or less than the pH at the IEP. The polarity of E controls the adsorption of ionic species in the solutions.

Manganese dioxide was found to be one of electrodes, with a potential represented by Eq. (16.2), because both the potential-pH relation and IEP in the pH scale can be determined on the same manganese dioxide sample, using a modified Gortner's streaming cell [10].

Chloride ion is known to play an important role in metal corrosion. Both adsorption of the ion on a metal and its oxide, and migration of the ion through the oxide film on a metal are greatly inhibited when their surface charge is negative [11]. In order to obtain some information on the surface charge and to check the pH dependency of the potential of the metal in solutions, Hirai and Tari [12] measured the streaming potential and electrode potential of aluminum and austenitic stainless steel (SUS 27) in both 0.5 M KCl and 0.5 M KNO_3 as a function of pH. Both metals had an apparent surface area of 2×5 cm^2. The streaming potential measurement was carried out by utilizing a small gap between a pair of parallel metal plates. The IEP values of aluminum were about 7.0 and 8.5 in KCl and KNO_3 solutions, respectively. The change in IEP from 8.5 to 7.0 might be caused by the specific adsorption of chloride ions, providing a negative charge on the aluminum surface in solutions with a pH below 8.5. The surface of the SUS 27 plates was charged negatively in the testing pH range as

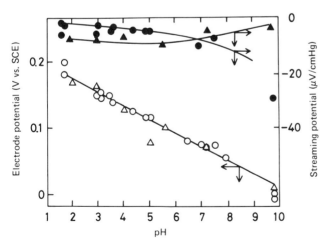

Figure 16.1 pH dependence of the streaming potential and the electrode potential of an austenitic stainless steel in 0.5 M KNO_3 and 0.5 M KCl. ○ and ●, 0.5 M KNO_3; △ and ▲, 0.5 M KCl.

shown in Fig. 16.1 and thus the IEP should occur at a pH value less than 1.5. The negatively charged surface repels chloride ions electrostatically and hence might protect the passive film on the steel from corrosion caused by the ions. Plots of electrode potential against pH fell along a straight line, as expected from Eq. (16.2). The line is useful for estimating changes in the pH of the solutions in the vicinity of the metal from the potential, which slowly varies with time when there is no agitation of the solutions.

Heterocoagulation is closely related to the ζ potential and IEP. Colloidal particles are known to deposit on metallic and nonmetallic materials present in colloidal solutions. Such deposition is used to form a protective film on a metal. Colloidal ferric oxide particles deposit on copper and brass in a colloidal solution prepared by adding ferrous sulfate to boiling water. The deposit is very effective for protecting condenser tubes made of brass from corrosion and errosion. It has been demonstrated that anodic pits previously formed were rendered inactive by the formation of a film of ferric oxide. The deposition mechanism is not well understood, but the deposition is surely caused by electrostatic attraction between the surface of the condenser tubes and the colloidal particles. The tubes possess a positive surface charge owing to the presence of a film of Cu_2O, whose ζ potential is positive up to pH 10, while the particles in the colloidal solution are negatively charged with a pH value of about 7.5 [13]. Hirai and Tanaka [14] proposed that a thin layer on the surface of copper alloy is composed of an oxide such as $CuFe_2O_4$. Hurd and Hackerman [15] determined the ζ potential of platinum, gold, and silver in the form of capillaries by means of the streaming current method. Levy and Fritsch [16] also determined the ζ potential of a stainless steel by the same method.

16.4 COMPOSITE PLATING

Ceramics and metal powders are known to codeposit with a metal during electrodeposition in an electroplating bath containing the powders. The codeposition brings about an improvement in the wear properties, strength, hardness, and corrosion behavior. A codeposit of aluminum particles with zinc was more resistant to corrosion than pure zinc in the form of electroplated coatings on iron [17].

A codeposit of plastics particles with metal provides excellent adhesion of paint or plastics coatings to the metal surface. Such adhesion results in prominant antirusting. Composite films such as styrene-butadine copolymer/zinc, epoxy resin/copper, and phenol resin/zinc were obtained during electrodeposition in electroplating baths containing polymer powders and suitable surfactants [18]. The latter provided the positively charged suspended polymer particles.

The adhesion of polyethylene or polyamide layer to the phenol resin/ zinc composite coatings was examined by applying the 180°C peel test, and the fractured surfaces were observed by means of a scanning electron microscope (SEM) [19]. A maximum peel strength was obtained for the polymer films attached to the composite coating covered with the phenol resin particles to the extent of 17-25%. The SEM observations of the fractured surfaces showed that the cohesive failure took place in the polyethylene layer adjacent to the interface in the polyethylene and composite substrate system, but the adhesive failure took place at the interface in the polyamide and composite substrate system. Such improvement in peel strength is explained by the presence of phenol resin particles on the composite film surfaces, and also by an increase in real contact area between the polymer films and the substrates due to increased roughness of the substrate surface.

The codeposit of cerium oxide particles with chromium was obtained during electrodeposition in a chromic acid bath containing cerium oxide particles or soluble cerium sulfate [20]. When tetraethylene pentamine was added to the plating bath as a promoter for the codeposition, the codeposits contained a large amount of interconnected porosity. The microporous coating improved the corrosion protection provided by the coating and facilitated hydrogen embrittlement relief.

The codeposition of metallic and nonmetallic powder with metal is strongly controlled by the surface charge density and the ζ potential of both the cathode and the particles suspended in a plating bath. The surface of the cathode charges negatively because of its cathodic polarization. Particles with positive charges thus adhere to the cathode electrostatically and are then embedded in the growing metal deposit. There are many methods of determining the surface charges of particles [21-25], one of them being to measure electrokinetic phenomena such as streaming potential, electrophoresis, electroosmosis, and sedimentation potential. The measurement of the phenomena, however, is very difficult because electroplating baths are so concentrated that the double layer at the solid-electrolyte interface is greatly compressed and the specific electric conductivity is very high. The streaming potential method may be most suitable for such a concentrated solution. Tari and Hirai [25] determined the ζ potential of titania in a Watt-type nickel bath with pH 3-6 by means of the streaming potential measurement, in order to clarify that the ζ potential plays an important role in the codeposition of titania with nickel during electroplating. The bath contained 250 g of $NiSO_4 \cdot 6H_2O$, 46 g of $NiCl_2 \cdot 6H_2O$, 30 g of H_3BO_3, and 50 g of titania powder per liter. Figure 16.2 shows the variation of the ζ potential and the amount of deposited titania with pH. The IEP was found to be 5.2. The amount of electrodeposited particles increased as the pH of the bath became lower. Even in the pH region where the ζ potential was negative, a considerable number of titania particles were codeposited. The particles codeposited in

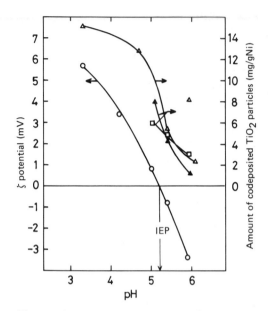

Figure 16.2 pH dependency of the ζ potential and the amount of codeposited titania particles. △, 50 mA·cm^{-2}; □, 25 mA·cm^{-2}; ▲, 15 mA·cm^{-2}.

the pH region below 5.2 were well dispersed in the nickel matrix, while those in the pH region above 5.2 were aggregated to form a cluster in the nickel matrix, as observed through the SEM. Since in the latter pH region titania powder agglomerated in the solution, the agglomerated particles occasionally collided with the cathode under mechanical agitation so that they overcame the electrostatic force of repulsion, and then were incorporated into the nickel matrix to form the clusterlike deposit.

16.5 ELECTROPHORETIC DEPOSITION

Fusion of electrophoretic deposits of aluminum and plastics powders provides high-quality antirusting coatings for substrate metals.

Electrodeposition of paint is an attractive technique for applying organic coating to metal. The charged pigment and resin particles are electrophoretically transported to and deposited on the oppositely charged electrode. The deposition process offers unique advantages in uniformity of coating appearance and protection over normally unpaintable areas. Suspended plastics particles also deposit on metal electrophoretically.

Electrophoresis can be used to produce deposits on the steels of metal powder such as aluminum, zinc, nickel, and iron [26,27]. Aluminum is often picked up as a powdered sample for electrophoretic deposition because of its availability in various powder forms, its low density, and its low fusion temperature. Pearlstein, Wick, and Gallaccio [26] found that the greatest deposition rate and most coherent deposits were produced when butylamine was used as the suspending medium for aluminum flake. Brown and Salt [27] thought that the theory of colloidal stability developed by Verwey and Overbeek [28] should be applicable to electrophoretic deposition, and compared the observed electrophoretic behavior of metallic and nonmetallic powders with that predicted by the theory.

REFERENCES

1. T. Hirai and I. Tari, in *Surface Electrochemistry* (T. Takamura and A. Kozawa, eds.), Japan Scientific Societies Press, Tokyo, 1978, p. 124.
2. I. Tari, K. Fujii, and T. Hirai, *Denki Kagaku*, 49, 517 (1981).
3. G. W. Whitman, R. P. Russel, and V. J. Altieri, *Ind. Eng. Chem.*, 16, 665 (1924).
4. P. Benson, W. B. Price, and F. L. Tye, *Electrochem. Technol.*, 5, 517 (1967).
5. T. Hirai, unpublished results.
6. N. Lukomski and K. Bohnenkamp, *Werkst. Korros.*, 30, 482 (1979).
7. J. J. Morgan and W. Stumm, *J. Colloid Sci.*, 19, 347 (1964).
8. A. Kozawa, *J. Electrochem. Soc.*, 106, 552 (1959).
9. S. M. Ahmed, in *Oxides and Oxide Films*, Vol. 1 (J. W. Diggle, ed.), Marcel Dekker, New York, 1972, p. 418.
10. I. Tari and T. Hirai, in *Extended Abstracts of 2nd International Symposium on Manganese Dioxide*, Electrochemical Society of Japan, Tokyo, 1980, p. 69.
11. M. Sakashita and N. Sato, *Denki Kagaku*, 45, 238 (1977).
12. T. Hirai and I. Tari, *Proc. 5th Int. Congr. Met. Corros.*, National Association of Corrosion Engineers, Houston, 1974, p. 106.
13. R. Gasparini, C. Della Rocca, and E. Ioannilli, *Corros. Sci.*, 10, 157 (1970).
14. Y. Hirai and M. Tanaka, *Ishikawajima-Harima Tech. Rep.*, 12, 222 (1972).
15. R. M. Hurd and N. Hackerman, *J. Electrochem. Soc.*, 102, 594 (1955); 103, 316 (1956).
16. B. Levy and A. R. Fritsch, *J. Electrochem. Soc.*, 106, 730 (1959).

17. H. Kimura, K. Yoshihara, and S. Harada, 57th Symp. Met. Finish. Jpn., 1980, Ext. Abstr. B-27, p. 108; B-28 p. 110.
18. K. Naitoh, K. Deguchi, M. Kubo, and S. Kurosaki, J. Met. Finish. Soc. Jpn., 28, 39 (1977).
19. K. Naitoh, K. Deguchi, M. Kubo, and S. Kurosaki, J. Met. Finish. Soc. Jpn., 28, 29 (1977).
20. G. R. Kamat, Plating Surf. Finish., 66, 56 (1979).
21. F. K. Sautter, Metall., 18, 596 (1964).
22. T. W. Tomaszewski, L. C. Tomaszewski, and H. Brown, Plating, 56, 1234 (1969).
23. D. K. Ramanauskene, E. S. Mikailene, and Y. Y. Matulis, Electrodepos. Met. (RSSR), p. 34 (1970).
24. D. W. Snaith and P. O. Groves, Trans. Inst. Met. Finish., 50, 95 (1972).
25. I. Tari and T. Hirai, Denki Kagaku, 46, 544 (1978).
26. F. Pearlstein, R. Wick, and A. Gallaccio, J. Electrochem. Soc., 110, 843 (1963).
27. D. R. Brown and F. W. Salt, J. Appl. Chem., 1965, 40.
28. E. J. W. Verwey and J. Th. G. Overbeek, Theory of the Stability of Lyophobic Colloids, Elsevier, Amsterdam, 1948.

17
Electrokinetic Phenomena in Biological Systems

Tamotsu Kondo Science University of Tokyo, Shinjuku-ku, Tokyo, Japan

17.1. Introduction 397
17.2. Electrophoretic Behavior of Blood Cells 398
17.3. Electrokinetic Properties of Some Biological Particulates 399
    17.3.1. Estimation of surface pK value of bacteria 399
    17.3.2. Electrophoretic distribution of white cells 401
    17.3.3. Isoelectric point of hair 401
17.4. Interactions of Blood Cells with Small Ions and Polymers 403
    17.4.1. Cation binding to white cells 403
    17.4.2. Calcium ion binding to blood cell surfaces 404
    17.4.3. Adsorption of polymers to red cell surfaces 405
17.5. Aggregation of Biological Cells by Simple Electrolytes 407
    17.5.1. Aggregation of white cells 408
    17.5.2. Aggregation of bacteria 410
17.6. Concluding Remarks 411
    References 412

17.1 INTRODUCTION

It would be useful to try to explain various interfacial phenomena in biological systems in terms of electrical properties of the cell surface in view of the fact that the human body consists of a large number of cells whose surfaces are electrically charged. The metastasis of cancer cells or the phagocytic actions of white cells, for instance, should have a close relation to the surface electrical properties of the cells involved, at least in the initial stages of the process.

This chapter deals with electrokinetic phenomena observed in various biological systems with special emphasis on the electrophoretic behavior of blood cells. The electrokinetic behavior of blood cells and some other biological particulates is described first, followed by discussions on the interactions of blood cells and bacteria with small ions and/or polymers, including applications of the theory of colloid stability to the aggregation of blood cells. This kind of information is very useful for elucidating the surface structure of biological particles and the mechanisms of aggregation in biological systems.

17.2 ELECTROPHORETIC BEHAVIOR OF BLOOD CELLS

Cell electrophoresis is a very convenient method for obtaining information about the surface structures of mammalian blood cells without causing significant alteration or destruction of the cellular organization. Electrophoretic mobility of a blood cell is due to the presence of fixed ionogenic groups and adsorbed ions on the surface. The fact that all cells investigated so far are negatively charged is ascribed to the presence of terminal carboxyl groups of N-acetyl and/or glyconeuraminic acids, as demonstrated by neuraminidase treatment [1,2].

Among various types of blood cells, red cells have been examined most extensively [3]. Far less effort has been expended on electrokinetic study of other types of blood cells, due to the difficulty with which pure samples of these cells can be prepared from blood. Consequently, there have been few reports of electrophoretic mobilities of various types of blood cells under identical conditions.

Figure 17.1 gives electrophoretic mobility as a function of the bulk pH for platelets, red cells, lymphocytes, and granulocytes which were obtained from fresh human peripheral blood and suspended in Michaelis buffer [4]. Platelets have a low positive electrophoretic mobility at low pH, showing an isoelectric point at pH 3.25. Lymphocytes exhibit no true isoelectric point, even though some cells have a positive mobility at pH 2 because the mean mobility is zero at this pH.

A mobility of -1.1 μm sec^{-1}/V cm^{-1} is reported for human red cells at pH 7.4. This is very close to the value of -1.07 μm sec^{-1}/V cm^{-1} reported for sheep red cells [5]. Granulocytes also give zero electrophoretic mobility at pH 2. The scatter in the mobility values for lymphocytes and granulocytes strongly suggests the heterogeneity of these two populations.

It has been shown that for different cells the relation between the electrophoretic mobility and the amount of neuraminic acid released is by no means constant [5]. In fact, granulocytes having more neuraminic acid per square micrometer of surface, have less charge than the other types of cells. These data and the curves in Fig. 17.1 suggest the presence of ionogenic groups other than the carboxyl group, such as the amino group, which produces a positive charge at low pH.

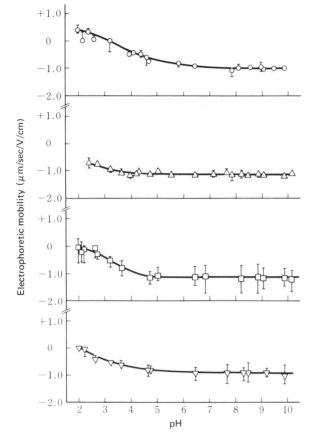

Figure 17.1 Electrophoretic mobility of various blood cells as a function of the bulk pH. ○, Platelets; △, red cells; □, lymphocytes; ▽, granulocytes. (From Ref. 4.)

17.3 ELECTROKINETIC PROPERTIES OF SOME BIOLOGICAL PARTICULATES

17.3.1 Estimation of Surface pK Value of Bacteria

An interesting electrokinetic study was reported by Schott and Young [6], who measured the electrophoretic mobility of *Streptococcus faecalis* as a function of the pH of the medium and estimated the surface pK value of the bacteria from the electrophoresis data. Figure 17.2 shows the plot of electrophoretic mobility versus pH for *S. faecalis*.

For pH lower than 7, the curve is quite similar in shape to that for the titration of an acid with an alkali. The initial increase in the elec-

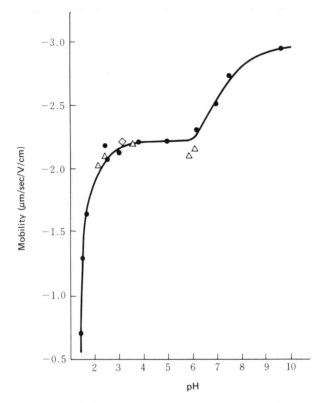

Figure 17.2 Electrophoretic mobility of *S. faecalis* as a function of the bulk pH. (From Ref. 6.)

trophoretic mobility corresponds to dissociation of the acidic groups on the cell surface. The surface pK value can be estimated in the following way. First, the zeta (ζ) potential is obtained from the electrophoretic mobility by use of the Smoluchowski equation. It is then used to calculate the surface pH according to Eq. (17.1):

$$pH_s = pH_b + \frac{e\zeta}{2.303kT} \qquad (17.1)$$

where ζ is the ζ potential of the cell; pH_s and pH_b the values of the pH at the cell surface and in the bulk solution, respectively; e the electronic charge; k the Boltzmann constant; and T the absolute temperature. Since the cell surface bears a negative charge, the hydrogen concentration is higher on the cell surface than in the bulk, and the surface pH is lower than the bulk pH by a factor of $e^{-(e\zeta/kT)}$. Substitution of pH_b and ζ at 50% dissociation of the surface acidic groups into Eq. (17.1) gives a pK value of 2.3. This indicates that the surface negative charge of S. faecalis is due to dissociation of carboxyl groups.

17.3.2 Electrophoretic Distribution of White Cells

A major disadvantage of microelectrophoresis is the substantial period of time needed to measure the electrophoretic mobilities of a statistically significant number of cells. A second disadvantage is the necessity of relying on the judgment of the experimenter for visual adjustments, cell selections, and measurements. Electrophoretic light scattering is a relatively new technique [7] in which the velocity of cells migrating in an electric field is determined spectroscopically by measuring the Doppler shift of the laser light scattered from the cells in a suspension. This takes a very short time, usually just a few minutes. Furthermore, only a few drops of a dilute suspension are required and the instrument allows for a simultaneous measurement of the velocities of all the cells being analyzed.

Using this technique for electrophoretic analysis, Smith, Ware, and Weiner [8] performed experiments on mononuclear white cells from human peripheral blood of normal subjects and leukemic patients. They found that mode mobilities (mobilities at which maximum light-scattering intensities are observed) of cells from leukemic patients are 7-28% below the average mode mobility of normal cells. Typical electrophoretic distributions of normal lymphocytes and leukemic lymphocytes suspended in a buffer of pH 7.1 are shown in Fig. 17.3. The intensity of scattered light is plotted against the electrophoretic mobility at which the maximum intensity is observed.

The leukemic distribution has its peak at a lower mobility than the normal distribution. The lower electrophoretic mobility of leukemic cells suggests a lower surface charge density, which may arise from a decrease in the net negative charge or an increase in the cell surface area, or both. Future studies on the magnitude and significance of the surface charges of normal and leukemic cells will be facilitated by the technique of electrophoretic light scattering.

17.3.3 Isoelectric Point of Hair

Streaming potential is sometimes useful for studying the electrical properties of biological systems. An example is given by recent work of Parreira [9], who measured the ζ potential of human hair at various bulk pH by means of a streaming potential technique. Figure 17.4 shows the experimentally determined ζ potential of brown hair in 10^{-4} M KCl as a function of pH, together with the ζ potential calculated using the Gilbert-Rideal theory for protein titration [10].

The experimentally determined and theoretically calculated ζ potentials are in good agreement with each other, demonstrating the validity of the theory under the conditions used in this work. It is

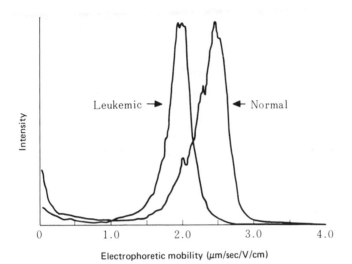

Figure 17.3 Plots of spectral intensity versus magnitude of electrophoretic mobility for two samples, one containing cells from a normal subject and the other containing cells from a subject with acute lymphocytic leukemia. (From Ref. 8.)

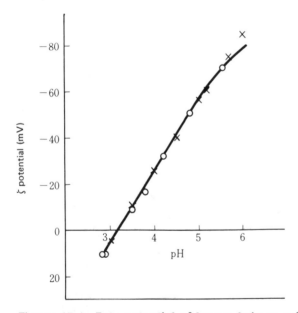

Figure 17.4 Zeta potential of human hair as a function of the bulk pH. ○, experimentally observed; X, theoretically calculated. (From Ref. 9.)

suggested that the rather low isoelectric point of 3.17 for human hair is possible due to the presence of oxyacids of sulfur which are formed by surface oxidation of the hair keratin.

17.4 INTERACTIONS OF BLOOD CELLS WITH SMALL IONS AND POLYMERS

17.4.1 Cation Binding to White Cells

Cell electrophoresis provides valuable information on the binding of cations to the cell surface. In Fig. 17.5, the ζ potentials of sheep cells in 10^{-2} M NaCl solutions, which contain 5% w/v sorbitol and various cations, are plotted against the cation concentration [11].

The sign of the ζ potential reverses at high cation concentrations, indicating excess cation binding in the Stern layer of the electrical double layer. The cation concentration at which the ζ potential becomes zero is dependent not only on the valence but also on the chemical nature of the cation. This implies that a specific interaction, in addition to the electrostatic interactions, between the anionic sites on the

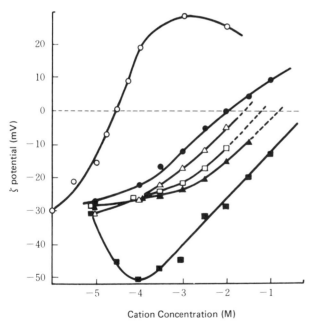

Figure 17.5 Zeta potentials of white cells in the presence of different cations in various amounts. ○, Th^{4+}; ●, La^{3+}; △, Cu^{2+}; □, Ba^{2+}; ▲, Ca^{2+}; ■, Na^+. (From Ref. 11.)

cell surface and cation is involved in cation binding. The free energy of cation adsorption obtained by Wilkins, Ottewill, and Bangham [11], based on Stern's adsorption model [12], seems to support this view.

17.4.2 Calcium Ion Binding to Blood Cell Surfaces

A systematic study of cation binding to blood cell surfaces was made by Seaman, Vassar, and Kendall [13]. They investigated the uptake of calcium ions by red cells and by polymorphonuclear white cells from human peripheral blood both before and after neuraminidase treatment by cell electrophoresis. Using Stern's adsorption model [12], they derived the following equation:

$$\frac{1}{\Delta \sigma_{Ca}} = \frac{1}{2eN_a} + \frac{1}{C} \frac{1}{2eN_a K} \tag{17.2}$$

where $\Delta \sigma_{Ca}$ is the decrease in the electrokinetic charge density of the cell in a solution at constant osmolarity containing C moles per liter of calcium ions, e the electronic charge, N_a the number of sites available for adsorption, and $K = \exp(\Delta G/kT)/55.6$, where ΔG is the electrochemical free energy of calcium ion adsorption, k the Boltzmann constant, and T the absolute temperature. The plot of $1/\Delta \sigma_{Ca}$ versus $1/C$ will give a straight line if calcium ion binding proceeds according to this mass action mechanism. N_a and ΔG may be evaluated from the intercept and the slope, respectively.

In Fig. 17.6, the calcium ion binding plots for both normal and neuraminidase treated blood cells are given. Values for N_a and ΔG are listed in Table 17.1, which also includes the number of anionic sites per cell and the concentration of calcium ion at which the ζ potential changes sign. The number of anionic sites per square centimeter of cell surface is calculated from electrophoretic mobilities by use of the Gouy equation for monovalent electrolytes and converted to the number of sites per cell by assuming the appropriate surface area.

Approximately 65% of the anionic sites of normal red cells and about 60% of the sites of normal white cells are occupied by calcium ions at maximum binding, while only about 45% and 40%, respectively, of the sites are occupied by calcium ions after neuraminidase treatment. Although fewer calcium ions are bound to the neuraminidase-treated cells, they are bound more strongly than to normal cells, as is clear from the ΔG values in Table 17.1. In order to account for the observed changes in calcium ion binding after neuraminidase treatment, the presence of at least three types of anionic sites on the cell surface should be assumed, two of which bind calcium ions. The binding sites comprise both neuraminate groups and a set of unidentified sites which bind calcium ions more strongly than neuraminate groups. A third set of anionic sites with no affinity for calcium ions is also present. The ΔG values for normal blood cells given in the table represent mean values for both types of sites, whereas the values of ΔG for the

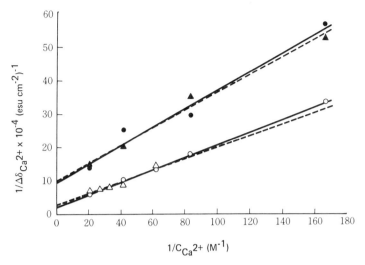

Figure 17.6 Calcium ion binding plots for normal and neuraminidase-treated blood cells. ○, Normal red cells; ●, neuraminidase-treated red cells; △, normal white cells; ▲, neuraminidase-treated white cells. (From Ref. 13.)

neuraminidase treated cells are only for those with a higher affinity for calcium ions than the neuraminate groups. The absence of charge reversal concentration of calcium ions for neuraminidase-treated cells is an indication of the high proportion of sites which do not bind calcium ions.

17.4.3 Adsorption of Polymers to Red Cell Surfaces

The electrokinetic properties of red cells are noticeably affected by the adsorption of positively charged polymers such as polylysine. Katchalsky et al. [14] studied the influence of a variety of polybases on red cell surface charge and aggregation by electrophoresis and found that red cell aggregation takes place at a critical ζ potential for each cell-polybase system. The value of the critical ζ potential is negative and large in magnitude. Thus the polycation-induced aggregation seemed to arise from the active binding of the polycations to neighboring cells with resulting bridging among the cells, instead of being caused by a drastic decrease in the electrokinetic charge of the cells. It was found that, in general, the polybases with higher molecular weights aggregate red cells at lower concentrations and at ζ potentials very close to the original negative values for the cells.

Rapid aggregation of red cells was also observed in dilute solutions of polybases by Katchalsky et al. [14], who found that the aggregates

Table 17.1 Calcium Ion Binding Data for Human Polymorphonuclear Leucocytes and Red Cells

System	Anionic sites per cell	N_a	ΔG (kcal mol^{-1})	Charge reversal conc. of Ca^{2+} (M)
Normal red cells	12.5×10^6	8.09×10^6	−3.82	0.300
Neuraminidase-treated red cells	3.90×10^6	1.79×10^6	−4.47	—
Normal white cells	17.8×10^6	10.9×10^6	−4.01	0.284
Neuraminidase-treated white cells	7.14×10^6	2.99×10^6	−4.53	—

Source: Ref. 13.

are not redispersed by vigorous shaking or by thorough washing with saline, but they can be dispersed by the addition of a polyacid. The red cells obtained after redispersion of the aggregates were found to have the same appearance in electron micrographs as the untreated cells. The examinations by both optical and electron microscopy of red cell aggregates revealed marked changes in the cell membrane and in a "seam" of aggregation consisting of a polymer film in which the polycations are anchored at each end onto neighboring cell membranes, with the maximum width of the bonding polymer film nearly equal to that of the fully stretched polycations.

Ponder [16] found that the addition of dextran, a neutral polymer, to red cells or platelets in saline solution produces an increase in the electrophoretic mobilities of the cells. A model for the increase in the ζ potential was proposed by Brooks [17], who calculated the effect of adsorbed polymer on the chemical potential of counterions in the electrical double layer adjacent to the particle surface. According to this model, the double layer is expected to expand as a result of generalized excluded volume effects in the presence of adsorbed neutral polymer to raise the surface potential under the condition of constant surface charge. He derived an expression for the quantitative evaluation of the relative ζ potential in the presence of neutral polymer as a function of the difference in polymer concentration between the surface and bulk phases, the depth of the adsorbed layer, the location of the shear plane, and the bulk ionic strength. The expression predicts that the ζ potential will increase in the presence of adsorbed neutral polymer if the shear plane is not shifted too far from the particle surface.

17.5 AGGREGATION OF BIOLOGICAL CELLS BY SIMPLE ELECTROLYTES

It is well known that the addition of a simple electrolyte to a hydrophobic colloid causes aggregation of the colloid particles. Quantitative explanation to the aggregation phenomena is given by the colloid stability theory [15] in terms of long-range van der Waals attractive forces and electrostatic repulsive forces. Similarly, biological cells aggregate upon addition of electrolytes since they are electrically charged. In these cases, however, the colloid stability theory will only be applicable for cells separated by distances which are large compared with the effective thickness of the electrical double layer surrounding them because present-day knowledge of ionic arrangements, water structure, and so on, on and around the cell surface is still too scanty to provide a meaningful evaluation of short-range interactions.

17.5.1 Aggregation of White Cells

Slow aggregation of white cells is observed when small amounts of a cation are added to a cell suspension. The rate of cell aggregation can be evaluated by counting the total number of particles irrespective of their size in the samples taken at suitable time intervals after mixing a cell suspension and an electrolyte solution. In general, not all of the collisions between the cells are effective, so only a fraction of them will lead to aggregation of the cells. In other words, the rate of cell aggregation is lowered by a factor of 1/W due to the presence of a potential barrier between the cells compared with that of the case where all the collisions are effective. W is defined by the following equation and is called the stability ratio:

$$W = 2 \int_2^\infty \exp\left(\frac{V}{kT}\right) \frac{ds}{s^2} \qquad (17.3)$$

where V is the intercellular potential energy, s the center-to-center intercellular distance divided by the cell radius, k the Boltzmann constant, and T the absolute temperature.

The Smoluchowski theory gives the relationship between the total number of cell aggregates and time as

$$\frac{1}{N_t} = \frac{1}{N_0} + \frac{k_0 t}{W} = \frac{1}{N_0} + k_1 t \qquad (17.4)$$

where N_0 and N_t are the total numbers of cell aggregates at $t = 0$ and $t = t$, respectively, and k_0 and k_1 are rate constants. A plot of $1/N_t$ versus time will yield a straight line whose slope is k_1. In Fig. 17.7, the data for the aggregation of white cells in 10^{-2} M $CuCl_2$ solution are shown [18].

The value of k_1 depends not only on the species but also on the concentration of the cation. In general, it increases and reaches a maximum as the cation concentration increases. However, it decreases after passing through the maximum when polyvalent cations such as Th^{4+} are used because these ions cause redispersion of the cells. W may be assumed to be unity at the maximum value of k_1, which equals k_0. The values of W at different cation concentrations can be calculated by using Eq. (17.4). In Table 17.2, the values of k_0 for several cations are listed, together with their critical aggregation concentrations and the ζ potentials of the cells at W = 1 [18].

According to the colloid stability theory [15], the interaction energy V between two equal spherical particles of radius a and surface potential ψ_0 is approximately given for large κa by

$$V = \frac{\varepsilon a \psi_0^2}{2} \ln[1 + \exp(-2\kappa h_0)] - \frac{Aa}{24 h_0} \qquad (17.5)$$

Table 17.2 Aggregation Data for Sheep White Cells

Cation	$k_0 \times 10^9$ cm^3/sec	Critical aggregation conc. (M)	ζ potential at W = 1 (mV)
Th^{4+}	3.1	3.8×10^{-5}	+6.5
Cu^{2+}	3.0	3.9×10^{-3}	−20.5
Ba^{2+}	0.4	1.0×10^{-2}	−8.0
Ca^{2+}	1.2	1.0×10^{-1}	−3.5

Source: Ref. 18.

where ε is the dielectric constant of the medium, κ the reciprocal double-layer thickness, $2h_0$ the minimum separation between two spheres, and A the Hamaker constant of attraction in the medium. For each of the cations listed in Table 17.2, the critical aggregation concentration can be calculated from simultaneous equations of dV/dh_0 and $V = 0$ at $W = 1$ if ψ_0 and A are known. Alternatively, if ψ_0 is known or assumed to be equal to ζ and the experimentally determined critical aggregation concentration is used, the Hamaker constant will be obtained

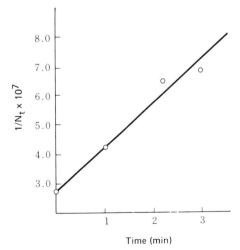

Figure 17.7 Plot of $1/N_t$ against time for aggregation of white cells in CuCl$_2$ solutions. (From Ref. 18.)

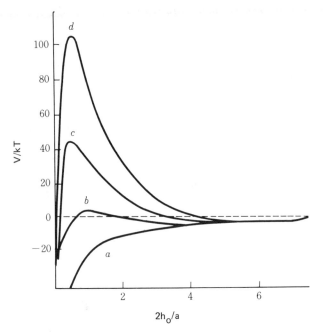

Figure 17.8 Potential energy curves for sheep white cells. a, $\psi_0 = 1$ mV; b, $\psi_0 = 5$ mV; c, $\psi_0 = 7.5$ mV; d, $\psi_0 = 10$ mV. (From Ref. 18.)

by substituting them into the simultaneous equations. The values of A obtained in this manner are 4.3, 9.9, and 32.4 × 10^{-15} erg for Th^{4+}, Cu^{2+}, and Ca^{2+}, respectively.

The potential curves shown in Fig. 17.8 are obtained by plotting the value of V/kT calculated using Eq. (17.5) against $2h_0/a$. Here values of 10^{-14} erg and 3 × 10^{-4} cm are assumed for A and a, respectively.

It is predicted that fast aggregation of white cells would begin when the potential barrier becomes lower than 15kT or the ζ potential lower than 6 mV. In fact, the values of the ζ potential at W = 1 in Table 17.2 seem to confirm this prediction. The secondary minimum on the potential curve is not deep enough to cause cell aggregation.

17.5.2 Aggregation of Bacteria

Aggregation of many types of bacteria is initiated by the addition of cations since bacteria are negatively charged in water. This aggregation phenomenon can also be explained in terms of Eq. (17.5). Figure 17.9 shows potential energy curves between two *Escherichia coli* cells in the presence of various concentrations of Mg^{2+} [19]. The ordinate represents the interaction energy in ergs, V, and the abscissa the minimum separation between the cells in centimeters, $2h_0$. The

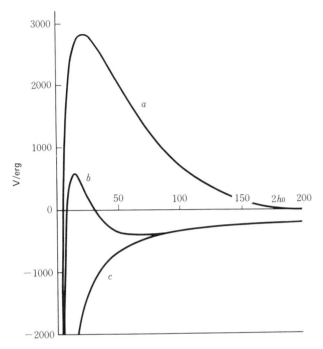

Figure 17.9 Potential energy curves for *E. coli* in the presence of Mg^{2+}. Mg^{2+} concentration: a, 10^{-3} M; b, 10^{-2} M; c, 10^{-1} M. (From Ref. 19.)

degree of aggregation and the ζ potential of the cells are determined by means of turbidimetry and microelectrophoresis, respectively. A value of 5×10^{-13} erg is assumed for the Hamaker constant.

As is evident from curve a in the figure, no aggregation is observed at a Mg^{2+} concentration of 10^{-3} M. The repulsive forces are higher than the attractive forces. When Mg^{2+} concentration reaches a value of 5×10^{-1} M, the cells approaching each other aggregate readily due to the dominant attractive forces between them (curve c). The case represented by curve b corresponds to an intermediate state between the states represented by curves a and c.

17.6 CONCLUDING REMARKS

Parsegian and Gingell [20] applied some of the recent advances in the theory of electrostatic and electrodynamic forces to biological cell interactions. They concluded that cells undergo a nonspecific attraction at long distances arising mainly from the lipid bilayers in the membrane. The glycoprotein of the peripheral zone is able to exert a specific attraction at distances shorter than 5×10^{-7} cm and glycoprotein-glycoprotein interactions dominate the electrodynamic energy in this range. There is a secondary potential minimum with a depth of about 5×10^{-4} erg cm^{-2} at intercellular distances of 5 to 8×10^{-7} cm, and deep primary energy minima of about 0.1 erg cm^{-2} may occur in the limit of close cell contact. In view of this, care must be taken when applying colloid stability theory to biological cell interactions, especially at short intercellular distances, even though electrokinetic study of biological cells proves itself a useful tool in obtaining information on the surface structure and aggregation properties of the cells.

REFERENCES

1. E. H. Eylar, M. A. Madoff, O. V. Brody, and J. L. Oncley, J. Biol. Chem., 237, 1992 (1962).
2. G. V. F. Seaman and G. Uhlenbruck, Arch. Biochem. Biophys., 100, 493 (1963).
3. G. V. F. Seaman, in The Red Blood Cells (D. MacN. Surgenor, ed.), 2nd ed., Vol. 2, Academic Press, New York, 1975.
4. A. Zerial and D. J. Wilkins, Experientia, 28, 1435 (1972).
5. A. D. Bangham, B. A. Pethica, and G. V. F. Seaman, Biochem. J., 69, 12 (1958).
6. H. Schott and C. Y. Young, J. Pharm. Sci., 61, 182 (1972).
7. B. R. Ware, Adv. Colloid Interface Sci., 4, 1 (1974).
8. B. A. Smith, B. R. Ware, and R. S. Weiner, Proc. Natl. Acad. Sci. USA, 73, 2388 (1976).
9. H. C. Parreira, J. Colloid Interface Sci., 75, 212 (1980).
10. G. A. Gilbert and E. K. Rideal, Proc. R. Soc. Lond. A, 182, 355 (1944).
11. D. J. Wilkins, R. H. Ottewill, and A. D. Bangham, J. Theor. Biol., 2, 165 (1962).
12. O. Stern, Z. Elektrochem., 30, 508 (1924).
13. G. V. F. Seaman, P. V. Vassar, and M. J. Kendall, Experientia, 25, 1259 (1969).
14. A. Katchalsky, D. Dannon, A. Nevo, and A. DeVries, Biochim. Biophys. Acta, 33, 120 (1959).
15. E. J. W. Verwey and J. Th. G. Overbeek, Theory of the Stability of Lyophobic Colloids, Elsevier, Amsterdam, 1948.
16. E. Ponder, Rev. Hematol., 12, 11 (1957).
17. D. E. Brooks, J. Colloid Interface Sci., 43, 687 (1973).
18. D. J. Wilkins, R. H. Ottewill, and A. D. Bangham, J. Theor. Biol., 2, 176 (1962).
19. J. Dirkx, J. Beumer, and P. Beumer-Jochmas, Biochim. Biophys. Acta, 27, 442 (1958).
20. V. A. Parsegian and G. Gingell, J. Adhes., 4, 283 (1972).

18
Reproduction in Copying and Electrophoretic Display

Shuichi Karasawa   Ricoh Co., Ltd., Tokyo, Japan

18.1. Introduction   413
18.2. Electrophotography   414
    18.2.1. Principles of electrophotography   414
    18.2.2. Mechanism of liquid development   415
    18.2.3. Composition of liquid developer   419
    18.2.4. Characteristics of liquid developer   420
    18.2.5. Conclusion   423
18.3. Application of Electrophoresis to Displays   424
    18.3.1. Electrophoretic displays   424
    18.3.2. Principles of electrophoretic displays   427
    18.3.3. Materials   428
    18.3.4. Characteristics of electrophoretic display devices   431
    18.3.5. Photoelectrophoretic displays   433
    18.3.6. Conclusions   434
    References   434

18.1 INTRODUCTION

A current method for the development of electrophotographic images in office photocopying machines makes use of electrophoretic particle deposition in a colloidal suspension of carbon black or colored pigment particles. For nonemmisive electrophoretic display units, which will be important in the office automation machines of the near future, a reversible display technique also utilizes an electrophoretic process in which the charged pigment particles are transported by means of an applied electric field. This chapter describes the present state of the technology in electrophotographic liquid development and electrophoretic displays in relation to interfacial electrical phenomena, and indicates trends for future development.

18.2 ELECTROPHOTOGRAPHY

18.2.1 Principles of Electrophotography

The history of electrophotography—the essence of the modern photocopying industry—begins with the inventions of Carlson [1] in 1938. Carlson proposed methods for utilizing a photoconductive insulating surface to produce electrostatic latent images. The processes by which the electrostatic images are formed on the surface of a photosensitive layer, made of, for example, selenium evaporated on, or an organic photoconductor coated on, the conductive substrate, were given the name "xerography." A modification of the xerographic process, in which a photoconductive coating consisting of a ZnO binder on paper was used, was given the name "electrofax." For making a print by xerography, five basic steps are involved:

1. Sensitizing the photoconductive layer by corona discharging
2. Exposing the photoconductive layer to form an electrostatic latent image
3. Developing the latent image with fine particles
4. Transferring the developed image to the paper
5. Fixing the image by fusing

Multiple copies are made by cleaning the photoconductive plate and repeating the five steps. These steps are illustrated in Fig. 18.1.

Metcalfe [2,3] proposed a liquid-development process which develops the electrical latent image by electrophoretic particle deposition from colloidal suspension. His early liquid developers were prepared by dispersing finely ground pigments such as carbon black or magnesium oxide with charge control agents such as linseed oil, oleic acid, alkyd resin, or polystyrene resin, in an insulating liquid such as kerosene or carbon tetrachloride. For a good review of the history, principles, and applications of electrophotography, the reader is referred to the publications by Shaffert [4] and Dessaur and Clark [5].

Liquid development has several advantages compared with dry development using fine particles charged triboelectrically:

1. High resolution and halftones are obtained by using a fine-grained suspension.
2. The edge effect, by which the stronger electric field of the edge portion in a latent image causes a partially higher image density, is softened by controlling the conductivity of the developer.

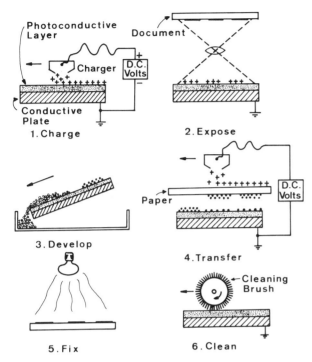

Figure 18.1 Basic steps in xerography.

3. A compact unit is designable and machine troubles are decreased because of the simple construction of the developing unit.

A large number of firms manufactured copying machines using ZnO-coated paper and liquid developer. In recent years, copying machines using ZnO-coated paper have been replaced by the plain paper copier, in which images are developed with liquid toner or dry powder, and in which the photoconductive layer, usually amorphous selenium, is reusable.

Practical liquid developing methods are illustrated as examples: in Fig. 18.2a, the dipping method, in which ZnO-coated paper is dipped in liquid developer; and in Fig. 18.2b, a method in which liquid developer is supplied to the surface of the photoconductor.

18.2.2 Mechanism of Liquid Development

We shall now consider the process of liquid development following the formation of the electrostatic latent image on a photoconductive layer, a negatively charged ZnO paper, or a positively charged amorphous selenium plate.

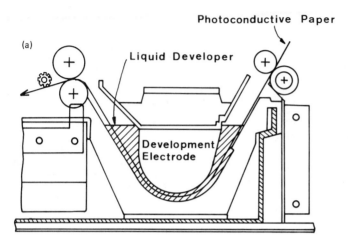

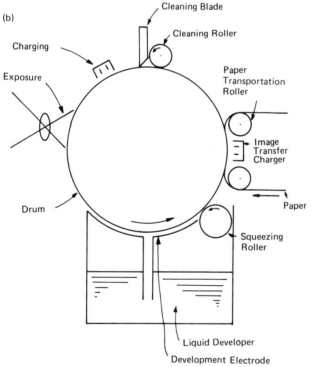

Figure 18.2 (a) Development tank. (From Ref. 6.) (b) Development tank. (From Ref. 7.)

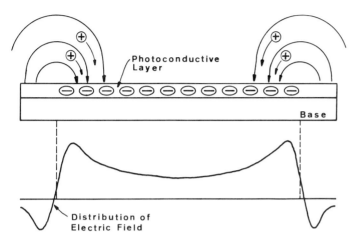

Figure 18.3 Electric field on the surface of the photoconductor.

When a charge pattern is formed on the surface of either of the above-mentioned photoconductors, electric fields are formed near the edges of the charge pattern as shown in Fig. 18.3, although the surface charge is uniform in density within the charge pattern. By means of the electric fields, positive particles are attracted to the charge pattern as shown in Fig. 18.3. With a microscopic method using a corona-discharged transparent Mylar film [8], we can directly observe the developing process, including migration of pigment particles, the possible zone in which pigment particles are capable of migration, and the edge effect.

The mechanism of liquid development has been investigated by Kurita and co-workers [9,10], Anfilov et al. [11,12], Inoue et al. [13], Schaffert [4], and many others.

The electrophoretic rate v of a fine particle under the electric field E formed by charge Qs on the photoconductor can be calculated from Huckel's equations:

$$v = \frac{\varepsilon \zeta E}{6 \pi \eta} \tag{18.1}$$

and

$$\theta = \frac{\varepsilon \zeta}{6 \pi \eta} \tag{18.2}$$

where ε is a dielectric constant of the medium, ζ is the zeta (ζ) potential, η is viscosity of the medium, and θ is the electrophoretic mobility.

Kurita and co-workers [10] treat the phenomena of liquid development from the standpoint of surface charge neutralization and takes into account the effect of the conductivity of the liquid. The rate of discharge of surface charge becomes

$$-\frac{dq_s}{dt} = (avq_p + \sigma)E \qquad (18.3)$$

$$= (a\kappa + \sigma)E \qquad (18.4)$$

where q_s is the surface charge in C/cm^2, a is the concentration of particles in g/cm^3, q_p is the particle charge in C/g, σ is the conductivity of the liquid, and $\kappa = q_p \theta$.

Kurita and co-workers [10] also assume a direct proportionality between the optical density D of the developed layer and the amount of particle deposition:

$$\frac{dD}{dt} = D_0 av \qquad (18.5)$$

$$= D_0 a \frac{\varepsilon \zeta E}{6\pi\eta} \qquad (18.6)$$

where D_0 is a proportionality constant.

Solutions to Eqs. (18.4) and (18.6) lead to the following relationship between developed density and developing time:

$$D = \frac{D_0 C_s V_0}{q_p(1 + \sigma/a\kappa)} \left\{ 1 - \exp\left[-\frac{\alpha}{C_s}(a\kappa + \sigma)t\right] \right\} \qquad (18.7)$$

where α is a proportionality constant relating the field to surface popotential $E = \alpha V$, $q_s = C_s V$, C_s is the capacitance of the photoconductor, $V = V_0 \exp(-t/\tau)$, and

$$\tau = \frac{C_s}{\alpha(a\kappa + \sigma)} = \frac{q_s}{E_0(a\theta q_p + \sigma)}$$

where E_0 is the initial electric field.

Schaffert [4] treats the effect of a development electrode schematically shown in Fig. 18.4. Considering the electrostatic conditions, the electric field is

$$E_0 = \frac{Q_s L_s + V_a \varepsilon_s}{\varepsilon_s L_\ell + \varepsilon_\ell L_s} \varepsilon_0 \qquad (18.8)$$

The relationship between density and developing time is

$$D = D_0 mn\theta\tau E_0 \left[1 - \exp\frac{-t}{\tau}\right] \qquad (18.9)$$

where m is the average mass per particle, n is the number of particles deposited per unit area, and

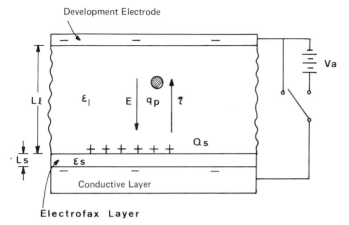

Figure 18.4 Model of liquid development. L_s, thickness of the xerographic layer; L_ℓ, distance from the xerographic layer to the development electrode; ε_ℓ, dielectric constant of the liquid; ε_s, dielectric constant of the xerographic layer; Q_s, surface charge of the xerographic layer; E, electric field in the liquid; q_p, particle charge; η, viscosity of the suspending liquid; V_a, applied voltage.

$$\tau = \frac{\varepsilon_s L_\ell / L_s + E_\ell}{nq_p \theta + \sigma} \varepsilon_0$$

For the image density, D_1, of the initial development state and the maximum density, D_m, we get Eqs. (18.10) and (18.11), approximately:

$$D_1 = D_0 a \theta E_0 t_1 \qquad (18.10)$$

$$D_m = \frac{D_0 C_s V_0}{q_p} \qquad (18.11)$$

18.2.3 Composition of Liquid Developer

We must take into consideration several restrictive conditions for the material design of liquid developer: (a) the toner must have stable polarity and stable charges on the particles and possess sufficient stability against settling over a long period; (b) the liquid must have a low dielectric constant and be of low ion concentration in order to prevent electrical leakage of the electrostatic image; (c) color and concentration of the pigment must be selected so as to give high image quality; (d) the resin must be selected from the standpoint of quick drying and fixing; and (e) safety must be preserved.

Pigment

Usually carbon black is used for commercial liquid developers in office copiers. Development in color copiers utilizes colored pigments of yellow, magenta, and cyan. The important parameters for pigment choice are color tone; specific surface area and the functional groups on the surface of a pigment which influence adsorption of the charge control agents; and specific gravity, distribution of particle size, and affinity to the solvent, which influence the stability of the dispersion.

Solvent

The solvent has to be a highly insulating hydrocarbon solvent with a resistivity that should be in the order of more than 10^{12} Ω cm and with a boiling point that should be more than 180°C. Isopar G, H (Esso) and Shell Sol (Shell), among others, have been used for commercial developers.

Charge Control Agent

The charge control agent, which is believed to be ionized in the dielectric media, must be adsorbed preferentially onto the pigment surface, with one of the ions of the material attracted to the pigment. The number of free ions in the bulk solution must be minimized, since excess ions give a high conductivity and cause electrical leakage of the electrostatic latent image. The mechanism of charge controlling in liquid developers is not well understood. Therefore, no clear rules regarding a choice of charge control agent can be advanced, and materials have been selected experimentally. Many charge control agents have been reported and patented independently since their invention by Metcalf [2,3]. Natural materials such as linseed oil, soybean oil, oleic acid, and alkyd resin mixed with soybean oil were used as charge control agents in earlier experiments. In order to improve on the unstable characteristics of the natural materials, many synthesized materials, such as polymers [14], block or graft copolymers [15], and graft carbon black [16], have been investigated. Charge control agents also include combinations of a polymer and a monomer [17], and a polymer and a cationic or an anionic surfactant [18], for example.

18.2.4 Characteristics of Liquid Developer

In order to understand the mechanism of liquid development, we must consider the physicochemical properties of the liquid toner. Typical properties of liquid toners consisting of carbon black-methacrylate copolymer-n-hexane and carbon black-Aerosol OT-n-hexane are shown in Table 18.1. These properties depend on the ingredients of the toner and on the conditions of preparation of the toner. Therefore, many experimental results by Envall [20], Vijayenadran [21,22], Kondo and Yamada [23], and other investigators must be referred to.

Electrical Charge on Toner Particle

In order to control polarity of the toner particles selectively, it is necessary to understand the mechanism responsible for the charging of suspended particles in nonpolar media. In general, charges on particles in nonpolar media are produced by the following phenomena:

1. Preferential adsorption of ions generated from polymers, surfactants, and other ionic additives on suspended particles [24-26]
2. Proton transfer with the particle surface [27]
3. Contact electrification

These mechanisms responsible for the charge on the particle in nonpolar media have been described in Chap. 5.

Mechanisms responsible for the charging of toner systems containing polymer charge control agents effective to promote stability of electric charges and stability of suspended particles will be dealt with here as examples. In general, carbon black particles washed or extracted with polar and nonpolar solvents have both a negative and a positive charge in nonpolar solvents such as n-hexane and Isopar H. When a copolymer of lauryl methacrylate (LMA) and dimethylaminomethyl methacrylate (DAMA) is adsorbed on the carbon black surface, the polarity of the carbon black particles is controlled so as to be negative, as shown in Table 18.2 [28]. Experimental results indicate that the adsorption of LMA-DAMA onto carbon black (Mogul A) from n-hexane closely follows the Langmuir equation, and interaction between a dimethylamino function group and polar function groups such as —COOH and —OH on the surface of carbon black affects the adsorption of LMA-DAMA, the negative charge measured from the infrared spectrum and the electrophoretic mobility. The effects expected by changing the various polar groups are indicated in Table 18.3. The proton transfer mechanism may be important in systems containing polymers capable of undergoing a proton transfer, such as a dimethylamino group, with the surface of carbon black.

Positive charges on suspended carbon black can be controlled by a small amount of metal soap, such as metal naphthenate [31] and Ni(2-ethylhexoic acid)$_2$ [29]. The effects of metal soap are shown in Fig. 18.5. The following mechanism may be tentatively proposed, with reference to Fig. 18.5, to explain the effect of metal soap. The negative adsorption sites such as >C$^{\delta +}$ = O$^{\delta -}$ on the carbon black surface or in copolymers adsorbed on the carbon black are neutralized by the positive metal ions. Therefore, the positive sites near the interface of the suspended particles may be increased. In these systems consisting of two kinds of charge control agents, it is necessary to follow the competitive adsorption phenomena that cause changes with time of the ζ potential.

Table 18.1 Observed Values of Liquid Developer

	Toner number						
	Methacrylate series						
Characteristic	M0.5	M1	M2	M3	M5	M7	M10
Resin/carbon black	0.5	1	2	3	5	7	10
Average toner charge, Q_p	1.8	2.8	3.0	3.0	5.1	6.6	10.7
Dielectric constant, ε	1.89	1.90		1.90	1.90		1.90
Viscosity, η	0.308	0.308	0.310	0.311	0.312	0.314	0.315
Zeta potential, ζ	77	105	108	111	116	118	124
Electrophoretic mobility, r	2.82	3.85	3.96	4.06	4.25	4.32	4.54
Conductivity, δ	1.26	2.53	3.33	4.00	4.55	6.0	7.0
Particle size, $2r_0$	0.19	0.14	0.12	0.10	0.12		0.11

Source: Ref. 19.

It is important to investigate the effect of water in toner suspension on the ζ potential and the stability of the suspended particle. The ions controlling the charging of suspended particles may be formed by the partition of water between the interface of the carbon black and the bulk solution [25,30].

Stability of Toner Suspensions

The concentrated toner suspension is prepared by mixing carbon black, a polymer, and the other additives in a ball milling machine and diluting the mixture with a solvent. An example of such a composition is shown in Table 18.4. Carbon black particles dispersed in a solvent are aggregates with a mean diameter of $0.1\text{-}2\mu m$. It is necessary to stabilize the dispersions over a long period of time under conditions of temperature and humidity of 10°C 20% relative humidity (RH) to 30°C 90% RH and of repeated electrophoretic development.

It is known that there is a close relationship between the electric charges of suspended particles and the stability of the suspension. According to the Derjaguin-Landau-Verwey-Overbeek (DLVO) theory [32], the stability of a lyophobic colloid is determined by the attractive van der Waals force (VA) that exist between the particles, and the electrostatic repulsive potential (VR) of the diffuse electrical double layer around the particles. To stabilize the dispersion, a repulsive potential must be introduced between the particles such that VR > VA.

The adsorption of a polymer from solution onto the surface of a pigment gives rise to a repulsive barrier due to the steric hindrance generated by the interpenetration of adsorbed polymers [33-36].

		Aerosol OT series				
A0.1	A0.5	A1	A2	A3	A5	Unit g/g for methacrylate series; M/g for Aerosol OT series
0.1	0.5	1	2	3	5	
3	5	9	12	13	18	$\times 10^{-4}$ C/g
1.90					1.91	
0.308	0.308	0.309	0.310	0.311	0.312	$\times 10^{-2}$ P
42	44	49	42	41	36	$\times 10^{-3}$ V
1.5	1.6	1.8	1.5	1.5	1.3	$\times 10^{-5}$ cm^2/V·sec
0.91	4.55	9.10	17.6	31.0	47.6	M $\times 10^{-14}$/A $\times 10^{-12}$ $\Omega^{-1}\cdot$cm^{-1}
0.70	0.85	0.90	0.95	1.20	1.50	μm

The important characteristics that control the stability of carbon black in liquid toner are polymer adsorption and steric hindrance. To prevent desorption of a polymer from the surface of carbon black into solution, caused by a change of temperature, imbalance of contents, or other factors, graft carbon black that is polymerized with the functional groups on the surface of the carbon black has been developed [16]. It is recognized that graft carbon black is effective for stabilization. In the liquid toner system, particles seem to be predominantly stabilized by entropic repulsion. But a relationship between the stability and behavior of polymers adsorbed or polymerized at interface is as yet unclear.

18.2.5 Conclusion

Various interfacial phenomena between a liquid and suspended particles control the characteristics of electrophotographic liquid development and liquid toner. However, in practice, since numerous constituents are used in the suspension, this prevents clear understanding of the mechanism of liquid development, charging of the particles, and stability of the dispersion. Therefore, we must design simple material constituents for the dispersion.

At present, it would seem that improved copy quality and more maintenance-free liquid development units are reasonable objectives in the field of copying machines. To achieve these objectives, we must understand the fundamental behavior of a complicated liquid toner and must, at the same time, introduce new ideas into the toner system, for example, controlling wettability of liquid toner on the electrostatic latent image and controlling size distribution of suspended particles.

Table 18.2 Polarity of Carbon Black Particles in Nonaqueous Dispersion System[a]

Carbon black	Carbon black-n-hexane		Carbon black-copolymer-n-hexane	
	Positive particle	Negative particle	Positive particle	Negative particle
Mogul A	0.90	0.97	0.06	1.53
Carbolac 2	0.77	1.00	0.15	0.95
Elf 0	1.04	1.08	0.14	1.64
Monarch 81	1.18	1.18	0.03	1.73
XC-550	1.12	1.08	0.10	1.60
Mitubishi 44	0.92	0.90	0.05	1.50
Vulcan XC-72R	0.80	1.15	0.26	1.59
Elftex 5	1.16	1.17	0.04	1.67
Stering MT	0.90	0.90	0.01	1.02

[a]The values were obtained by measuring the density of reflectance of carbon black deposited on an electrode to which an electric field of 534 V/cm was applied for 30 sec.

The concentration of carbon black was 0.01 g/100 ml and the concentration of the copolymer was 0.012 g/100 ml. Carbon black was washed by Soxhlet extraction method, using water, acetone, benzene, and n-hexane successively in that order.
Source: Ref. 28.

18.3 APPLICATION OF ELECTROPHORESIS TO DISPLAYS

18.3.1 Electrophoretic Displays

Display devices are interfaces between human and machine, such as computers and word processors, and, by their nature, produce "soft copies," as compared with hard copies produced by electrophotography.

Various display devices grouped according to whether they emit light or modulate it have been investigated vigorously. The performance of each type of display is summarized in Table 18.5. The reader can refer to the books regarding various display apparatus edited by Pankove [37] and Metz and von Willisen [38].

It was first pointed out by Evans et al. [39] in 1971 that the electrophoretic concept could also form the basis for a reversible, nonemissive display. This is known as the electrophoretic image display

Table 18.3 Effects of Polar Groups on ζ Potential

Structure of copolymer	Copolymer number	X: polar group	Zeta potential (mV)	
			Furnace black	Channel black
CH_3 | $CH_2\!=\!\!C\!\!-\!\!COOC_{12}H_{25}$ (LMA)	S-1	—Cl	−13	−12
+	S-2	—NO_2	−14	−17
$CH\!=\!CH_2$	S-3	—COOH	−61	−10
	S-4	—$N(CH_3)_2$	−96	−51
⟨phenyl⟩	S-5	—NH_2	−77	−39
| X	S-6	—H	−30	−11
	S-7	—CH_3	−12	−10
LMA + $CH_2\!=\!CHCOX$ || O	A-1	—H	−14	−65
LMA + $CH_2\!=\!\!C\!\!-\!\!COX$ | || CH_3 O	M-1	—H	−10	−32
	M-4	—$C_2H_4N(CH_3)_2$	−87	−110

Source: Ref. 30.

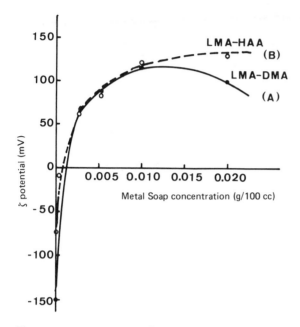

Figure 18.5 Amount of metal soap added and ζ potential. LMA-HAA, copolymer of lauryl methacrylate and acrylic acid; LMA-DMA, copolymer of lauryl methacrylate and dimethylaminoethyl methacrylate; Me-soap; Ni-(2-ethylhexoic acid)$_2$.

(EPID) or the electrophoretic display (EPD), as summarized by Dalisa [40].

There is another electrophoretic display which utilizes the phenomena of photoelectrophoresis. Photoelectrophoresis is based on the photo-induced migration of pigment or dye particles suspended in an insulating liquid. The principle of this imaging process was pointed out by Tulagin [41] and was applied to display devices by Ohta [42], Nozaki et al. [43], and Segawa et al. [44].

Table 18.4 Example of Liquid Toner Composition

Carbon black (Mogul A)	0.01 g
Copolymer (LMA-DAMA)	0.01 g
Metal soap (Mn-naphthenate)	3×10^{-3} g
Isopar H	100 ml

18.3.2 Principles of Electrophoretic Displays

An electrophoretic display cell consists of two electrodes, one of which must be transparent, between which is contained a stable dispersion of pigment particles in a dyed-fluid vehicle. The application of an electric field between the two electrodes, which are separated by a spacer of about 50 μm, causes the particles to migrate to the transparent front electrode and causes a change in optical reflectivity. Reversal of the field then causes the particles to migrate to the rear electrode. If the concentration of the dyed solution is controlled sufficiently to absorb the light incident on the transparent electrode, the particles on the rear electrode will be hidden. Therefore, a high-contrast image consisting of the color of the dye and the color of the pigment is formed. A schematic diagram of an electrophoretic display cell is shown in Fig. 18.6.

To address a display device electrically, each display element can be individually connected to the electronic drive circuitry. As an example, a varistor capacitor array driven electrophoretic device by Chiang and Fairbairn [45] is shown in Fig. 18.7. This device is built on a 17-mil wafer cut from a 2-in.-diameter commercial varistor device. The 31 lines/in. 1-in.^{2} active area contains a 32 × 32 matrix or 1024 pixels. The insulators are dry film photoresists and aluminum is used in all electrodes. The capacitors are formed by fabricating a second

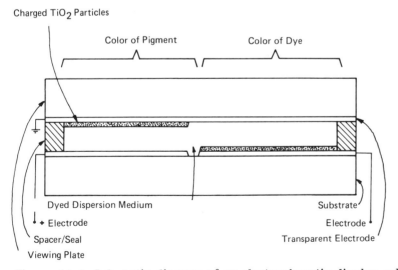

Figure 18.6 Schematic diagram of an electrophoretic display cell.

Table 18.5 Summarized Performance of Display Devices

Type of display properties	LED	EL dc	EL ac	Plasma dc	Plasma ac
Spectral output	Red to blue	Yellow	Yellow	Red	Red
Luminance (fL)	50,000	100	1500	20	700
Gray scale	>64		16		64
Contrast	>50:1		50:1	36:1	30:1
Electrical					
Operating voltage	2-20V	−100 V	−100 V	250 V	250 V
Power, current consumption	10 mW/mm^2	mA/cm^2	mA/cm^2	mA/cm^2	mA/cm^2
Maximum-power efficiency	10%	0.1	0.1		
Luminous efficiency	2000	0.5	0.5	0.07	0.3
Response					
Time (on/off)	μsec	0.5 msec	0.1 msec	10 μsec	
Storage, memory	No	Yes	Yes	No	Yes
Degradation lifetime	10^6 hr	10^3 hr	>2 × 10^4 hr		
Geometrical					
Element size (min./max.)	50 μm/10 mm		0.5 mm	0.5 mm	0.4 mm
Maximum panel size practical	30 cm^2	100 cm^2		200 lines	

[a]B/W, black and white (all colors possible).
Source: Ref. 37.

set of electrodes under each row of pixel electrodes and separating them by a layer of dielectrics using Al_2O_3, Ta_2O_5, or photoresist. The 2-mil-thick EP layer consists of a dispersion of positively charged TiO_2 white particles in a blue dielectric fluid. These were written one row at a time by applying 20-μs, ±70-V pulses to the selected row and column electrodes. A 1000-line display would require a 20-ms frame time or be driven at 50 frames/sec.

18.3.3 Materials

At present, in display technology, it would seem that there are problems to be solved from the material and physicochemical standpoints. The colloidal dispersion used must be well stabilized over the lifetime of the display. There are other problems, such as the irreversible deposition of contamination in the colloidal dispersion on the electrodes, and dye bleaching by electrochemical and photochemical effects. These points must be considered in the selection of materials. Croucher et al. [46] summarized the criteria for selection of materials as related to the performance of the electrophoretic display. Certain criteria for selection of each constituent will be described.

LCD	ECD	EPID	Light valve	Electro-plating	CL	PLZT
B/W[a]	Blue/W		B/W[a]	B/W[a]	B/W[a] 200	B/W[a] 150
10:1	8 3:1	40:1	10^3:1	Yes 5:1	64 50:1	Yes 10:1
2-10 V_2 $\mu A/cm^2$	0.5 V mC/cm^2	> 40 V $\mu A/cm^2$		0.5 V 100 mA/cm^2 1%	5 V-20$_2$ kV $\mu A/cm^2$ 30%	100 V 0.3 mW
msec/0.1 sec >2 × 10^4 hr	0.1 sec Yes >10^7 cycles	10 msec Yes >10^7 cycles	0.1 sec Yes >10^{10} cycles	0.1 sec Yes ~10^3 cycles	μ sec	msec Yes >10^9 cycles
0.25 mm	0.2 mm	0.2 mm m^2			30 mm 0.1 m^2	50 μm

Pigments

The type of pigment particle used will depend on the color of the image; the scattering coefficient, which depends strongly on particle size; and the surface characteristics that influence the stability of the suspended particles. For a white image display, mainly titanium dioxide (TiO_2), which is easily controlled as to optimum particle size (i.e., the scattering coefficient), has been investigated. The other inorganic pigment that has been used for display devices is iron oxide [47]. Organic pigments include Benzidin Yellow [48], Hansa Yellow [48], and newly synthesized pigments [49]. After these pigments are selected, methods of purifying them by eliminating the irreversible impurities therefrom must be established.

Solvent

The solvent used must be an insulating fluid with a resistivity that should be more than 10^{10} Ω cm. In practice, an aliphatic hydrocarbon solvent such as n-hexane or n-heptane, an aromatic solvent such as benzene or toluene, or a halogenated hydrocarbon solvent such as tetrachloroethylene can be selected. When selecting a solvent, a

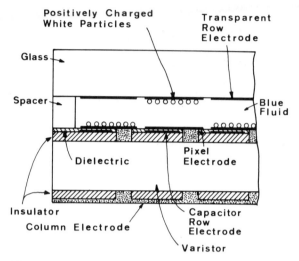

Figure 18.7 VCEP device structure.

number of factors should be considered, including the solubility of additives; the wettability of the pigment surface; the stability of the suspended particles; electrophoretic mobility, which depends on the viscosity and dielectric constant of the solution; and photochemical stability.

Charge Control Agent and Stabilizing Agent

These materials were described in the section on electrophotography. For an electrophoretic image display device, a charge control agent and a stabilizing agent are quite important because they strongly influence the life and the access time of the display device. Therefore, simple, purified constituents are desirable for the charge control agent and stabilizing agent.

Dye

A dye is used to obtain optical contrast with the deposited particles. Selection of the dye is based on the following requirements: stable chemical structure against photochemical and electrochemical reactions, solubility in the suspended liquid, chemical stability, chemical compatibility with suspended constituents, and ease of purification.

The dyes that may be used inlcude Oil Blue N, Oil Blue A, Oil Red XO, Sudan Black B, Sudan Red 4B, Macrolex Blue FR, and Hytherm Blue [46].

18.3.4 Characteristics of Electrophoretic Display Devices

Electrical Characteristics

The potential applied across a typical electrophoretic cell is about 1 V/μm. The current density is about 0.2-1 μA/cm^2, corresponding to a suspension resistivity of 10^{11}-10^{12} Ω cm. Under this electric field, the particles in suspension migrate to the electrode. They travel at a velocity given by

$$v = \theta E = \frac{\varepsilon \zeta E}{6\pi\eta} \tag{18.1}$$

where v is the velocity, θ the mobility, E the electric field, ε the dielectric constant, η the viscosity, and ζ the ζ potential. When the particles travel to the electrode from the pigment layer, a relatively long period of time is required for it to rise. This is due to the packing of the pigment layer. Hence the liquid with the highest ε/η ratio and the charge control agent capable of obtaining the highest ζ potential should be chosen for a fast rise time.

The pigment layer remains on the electrode after removal of the applied voltage. These particles move away from the electrode under reversal of the electric field. The process of the reversible dispersion affects the fall time of the display operation, the stability of the suspension, and the lifetime of the display.

Optical Characteristics

A typical optical response of TiO_2 dispersion stabilized and charged with Aerosol OT is shown in Fig. 18.8. The changes in the scattered light detected using a photomultiplier positioned normal to the cell surface is related to the changes in the spatial distribution of particles and to the response time of the device. The magnified optical response curves show a delay time (2-3 msec) after the polarity of a driving field was reversed. This delayed optical response arises from the nature of the particle packing and removal processes from the electrode and from the layers of imaged particles. Optical response time of TiO_2/toluene systems stabilized and positively charged with various charge control agents are shown in Table 18.6. The results show that the magnitude of the response times are very dependent on the charge control agent used and that the fall times t_{off} are always greater than the rise times t_{on}. This result reflects the different processes involved in forming and erasing the image. To achieve a fast response in the device, the packing and the removal mechanism of the pigment layer near the electrode must be understood.

Stability of Suspension

The charging and stabilizing mechanism, including interactions of the charging and stabilizing agents with the pigment surface, has been

Table 18.6 Summary of Response Times[a]

Charge control agent/surfactant	t_{on} (msec)	t_{off} (msec)
AOT	30	120
Lecithin	100	300
Triton-X100	130	600
Cobalt napthenate	140	500
Polystyrene/AOT	30	140
Polyisobutylene/AOT	50	300
Poly(vinyl methyl ether)/AOT	70	300

[a]These were measured using 0.25 weight fraction of particles, using a cell thickness of 50 µm with a driving frequency of 100 V.
Source: Ref. 46.

described in relation to liquid toner in electrophotography. In display systems, compared with electrophotographic systems, the stability of the suspensions is critical. This is because the deposition of particles on the surface of the electrode and formation of the packing particle layer, and the redispersion thereof into the bulk liquid, are reversed in repetition by the conversed electric field. If the stability is not

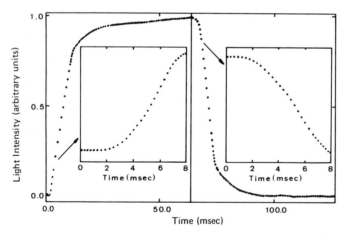

Figure 18.8 Optical response of TiO_2 dispersion in a 64-µm cell with an applied voltage of 100 V. The inserts show details of the short time response of the dispersion. (From Fef. 50.)

enough, the particles tend to coagulate. Furthermore, substantially no agitation takes place in the suspended particles of the narrow cells employed in the display systems. A new approach for stability of suspensions from the electrochemical and colloidal chemical standpoints will be required to provide devices with significantly longer lifetimes.

18.3.5 Photoelectrophoretic Displays

Tulagin [41] employed photoelectrophoretic phenomena in a new direct color-imaging system. He found that many organic pigments, when dispersed in a highly insulating liquid and exposed simultaneously to the action of an electric field and radiant energy, show typical photoelectrophoretic behavior. In Fig. 18.9, photoelectrophoretic imaging is shown schematically. The particles exposed to radiant energy have become positive, exchanged charges, and migrated to the electrode of opposite polarity. The polarity will be reversed by the charge exchange at the electrode, which is controlled by employing light-sensitive particles whose charge-exchange properties are altered by the absorption of radiant energy.

By projecting positive or negative images of the picture or print, it is evident that either positive or negative images, consisting of pigment color and dye, can be obtained. Similar display processes, utilizing migration of photoconductive particles, have been described by Nozaki et al. [43] and Segawa et al. [44].

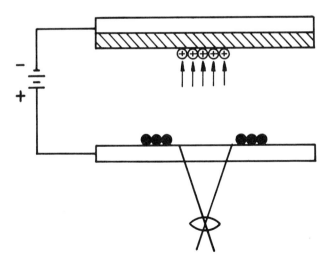

Figure 18.9 Photoelectrophoresis upon exposure to radiant energy in an electric field; conducting particles exchange charge and migrate to the opposite electrode.

Segawa et al. [44] found a different phenomenon: that positively charged TiO_2 particles nearest the front electrode and exposed to radiant energy remain on the front electrode, and the unexposed TiO_2 particles migrate to the rear electrode by simultaneous reversal of the electric field. It is surmised that the polarity of the TiO_2 particles nearest the front electrode will be neutralized by charge injection from the electrode and will be reversed by exchange of charges formed by exposure. By utilizing this process, positive images in the color of the particles are obtained from exposure through the original positive pattern.

18.3.6 Conclusions

The electrophoretic image display and the photoelectrophoretic display are being investigated again from the physicochemical and engineering standpoints because both methods are expected to be applicable to the flat reflective-type display devices, which have the advantages of high contrast, high resolution, and low power comsumption.

To develop practical devices, however, interfacial phenomena, including stability of colloidal suspensions, which is of course thermodynamic instability, and the mechanisms of charge exchange between an electrode and the suspended particles, must be understood. Material design and purification methods for eliminating impurities that cause irreversible deposition on the surface of the electrodes must be established. When this fundamental technology is established, electrophoretic displays will become suitable for practical use.

REFERENCES

1. C. F. Carlson, U.S. Pat. 2,221,776 (1938; 1940).
2. K. A. Metcalfe, *J. Sci. Instrum.*, 32, 74 (1955).
3. K. A. Metcalfe, *J. Sci. Instrum.*, 33, 194 (1956).
4. R. M. Schaffert, *Electrophotography*, Focal Press, London, 1965.
5. J. H. Dessaur and H. E. Clark, *Xerography and Related Processes*, Focal Press, London, 1965.
6. Service Manual of Ricopy BS 399, Ricoh Co. Ltd.,
7. T. Hayashi, Jpn. Pat. 49-91461 (1974).
8. I. Tashiro and T. Kawanishi, *Imaging*, 9, 10 (1973).
9. T. Kurita, *Electrophotography*, 3(2), 16 (1961).
10. Y. Ohyama, T. Kurita, and Y. Takahashi, *Electrophotography*, 3(3), 26 (1961).
11. I. V. Anfilov et al., *Zh. Nauch. Prikl. Foto. Kinematogr.*, 7, 220 (1961).
12. I. V. Anfilov et al., *Zh. Nauch. Prikl. Foto. Kinematogr.*, 5, 367 (1960).

13. E. Inoue et al., *Electrophotography*, 3(3), 20 (1961).
14. Jpn. Pat. 40-14155 (1965); Jpn. Pat. 45-33478 (1970); Jpn. Pat. 46-6156 (1971); Jpn. Pat. 4618117 (1971); Jpn. Pat. 47-4438 (1972); Jpn. Pat. 47-8120 (1972); and Jpn. Pat. 411847 (1972).
15. Jpn. Pat. 45-12712 (1970); Jpn. Pat. 46-3715 (1971); Jpn. Pat. 46-8279 (1971); Jpn. Pat. 46-18117 (1971); Jpn, Pat. 47-2760 (1972).
16. Jpn. Pat. 43-27597 (1968); Jpn. Pat. 44-19196 (1969); Jpn. Pat. 46-6155 (1971); Jpn. Pat. 46-6157 (1971); Jpn. Pat. 46-8278 (1971); Jpn. Pat. 47-14958 (1972).
17. Jpn. Pat. 47-4439 (1972); Jpn. Pat. 48-5062 (1973).
18. Jpn. Pat. 47-8756 (1972).
19. T. Kawanishi, *Electrophotography*, 12(3), 93 (1973).
20. A. D. Envall, *Rev. Pure Appl. Chem.*, 20, 118 (1970).
21. B. R. Vijayendran, *Colloidal Dispersions and Micellar Behavior* (K. L. Mittal, ed.), ACS Symposium Series No. 9, American Chemical Society, Washington, D.C., 1975.
22. B. R. Vijayendran, *IAS '75 Annual*, 6-D, 145 (1975).
23. A. Kondo and J. Yamada, *J. Chem. Soc. Jpn.*, 1972(4), 716 (1972).
24. J. L. Van der Minne and R. H. J. Hermanie, *J. Colloid Sci.*, 8, 38 (1953).
25. A. Kitahara, S. Karasawa, and H. Yamada, *J. Colloid Interface Sci.*, 25, 490 (1967).
26. H. C. Parreira, *J. Electroanal. Chem.*, 48, 125 (1970).
27. J. Lyklema, *Adv. Colloid Interface Sci.*, 2, 65 (1968).
28. S. Karasawa ana I. Tashiro, *RICOH Tech. Rep.*, 680,474, unpublished (1968).
29. A. Kitahara, S. Karasawa, et al., *RICOH Tech. Rep.* 680,014, unpublished (1968).
30. H. Ushiyama and T. Kawanishi, *RICOH Tech. Rev.*, 7, 36 (1974).
31. H. Machida and Z. Okuno, Jpn. Pat. 46-6157 (1971).
32. E. J. W. Verwey and J. Th. G. Overbeek, *Theory of Stability of Lyophobic Colloids*, Elsevier, Amsterdam, 1940.
33. M. Van der Waaden, *J. Colloid Sci.*, 6, 443 (1951).
34. E. J. Clayfield, *Discuss. Faraday Soc.*, 42, 285 (1966).
35. E. J. Clayfield, *J. Colloid Interface Sci.*, 22, 269 (1968).
36. A. Kitahara and M. Hasunuma, *J. Colloid Interface Sci.*, 41, 383 (1972).
37. J. I. Pankove, in *Topics in Applied Physics*, Vol. 40: *Display Devices* (J. I. Pankove, ed.), Springer-Verlag, Berlin, 1980.
38. A. R. Metz and F. K. von Willisen, *Nonemissive Electrooptic Display*, Plenum Press, New York, 1975.
39. P. R. Evans, H. D. Lees, M. S. Maltz, and J. L. Dailey, U.S. Pat. 3,612,758 (1971).

40. A. L. Dalisa, in *Topics in Applied Physics*, Vol. 40: *Display Devices*, (J. I. Pankove, ed.), Springer-Verlag, Berlin, 1980.
41. V. Tulagin, *J. Opt. Soc. Am.*, *59*(3), 328 (1969).
42. I. Ohta, Jpn. Pat. 49-35459 (1974).
43. H. Nozaki, Y. Toyoshima, H. Komidawa, and T. Iida, *Symp. Electrophotogr.*, Japan, 1947.
44. H. Segawa, I. Hamaguchi, W. Tokuda, and Z. Okuno, *Symp. Electrophotogr.*, Japan, 1975.
45. A. Chiang and D. G. Fairbairn, *SID Int. Symp., Dig. Tech. Pap.*, Vol. XI, 1980.
46. M. D. Croucher, J. Harbour, M. Hopper, and M. L. Hair, *Photogr. Sci. Eng.*, *25*(2), 80 (1981).
47. Jpn. Pat. 55-163515 (1980).
48. I. Ota, M. Tsukamoto, and T. Ohtsuka, *Proc. SID*, *18*, 243 (1977).
49. Jpn. Pat. 45-33236 (1970).
50. V. Novotny and M. Hopper, *J. Electrochem. Soc.*, *126*, 2211 (1979).

Author Index

Numbers in parens are reference numbers and indicate that an author's work is referred to although the name may not be cited in text. Italic numbers give the page on which the complete reference is listed.

Abel, R., 158(71), *179*
Abendroth, R. P., 42(40), *46*
Absolom, D. R., 78(60), *96*
Abson, D., 330(68), *336*
Adams, G. E., 227(25), 232(25), *263*
Agar, G. E., 290(9), 292(24), *296*
Ahmed, S. M., 42(35), *45*, 170(115), *181*, 390(9), *395*
Albers, W., 133(22), *142*
Alder, B. J., 89(118), 90(119), *98*
Alince, B., 327(47), 333(136, 137), *335, 338*
Allan, G. G., 327(46), *335*
Allen, L. H., 328(55), 333(132, 133), *336, 338*
Altieri, 388(37), *395*
Alty, A., 294(33), *297*

Amano, M., 123(9), 124(9), 135(9), *142*
Anderhoff, J. W., 83(101), *97*
Andersen, T. N., 33(20), *45*, 149(14), 153(14), *177*
Anderson, F. W., 141(42), *143*
Anderson, R. G., 324(14,22) 326(14,22), 327(14), *334*
Anderson, R. G., 324(22), 326(22), *334*
Anfilov, I. V., 417(11), *434*
Angelova, S. M., 80(83), *97*, 295(53), *298*
Anson, F. C., 158(63,64,65,66, 68,71), *179*
Aplan, F., 288(25), 292(25), *297*
Arakawa, M., 348(12), *354*
Arcelus, M., 324(7), *334*
Arledter, H. F., 332(99), *337*

437

Author Index

Armstrong, R. D., 154(36,39, 42), 155(58), 156(58), 158(69) *177, 178, 179*
Arno, J. N., 331(95), 332(95), *337*
Arvela, P., 332(100), *337*
Atkinson, R. J., 13(2), *14*, 42(37), *46*
Auhorn, W., 326(28), 328(28), 333(28), *335*
Avery, L. P., 332(97), *337*

Bach, N., 294(35), *297*
Baddi, N. T., 303(16), *321*
Bahadur, P., 373(21), 374(21), 376(21), 377(21), *384*
Bailey, R. M., 330(68), *336*
Balodis, V., 324(18), 326(18,19), 330(19), *334*
Bangham, A. D. 398(5), 403(11), 404(11), 408(18), 409(18), 410(18), *412*
Bard, A. J., 158(67), *179*
Barouch, E., 70(41), 79(41), *95, 97*
Barradas, R. G., 149(15), *177*
Barskii, A. A., 80,(83), *97*, 295(53), *298*
Bates, N. A., 327(44), *335*
Bayley, C. H., 318(59), *322*
Becher, P., 371(14,16), 372(14), 376(14,16,24), *384*
Beck, U., 328(57,58), 333(57, 58), *336*
Beckmann, W., 308(28), 315(28), *321*
Bell, G. M., 55(4), 71(46), 74(46,54), *94, 95*, 130(16), 138(16), *142*
Bell, M. F., 158(69), *179*
Benninga, H., 333(117), *338*
Benson, P., 389(4), *395*
Benzel, W., 200(21), 201(21), *223*, 317(44), 319(44), *322*
Bergatt, M., 333(140), *338*
Bernalin, B., 131(19), *142*

Bérubé, Y. G., 42(31), *45*, 170(116), 173(116), 175(136), *181, 182*
Beumer, J., 410(19), 411(19), *412*
Beumer-Jochmas, P., 410(19), 411(19), *412*
Bicu, I., 326(26), 327(49), 331(26), 332(16), *335*
Bijsterbosh, B. H., 19(8), *44*
Bikerman, J. J., 214(44,45,46), 215(44), *224*
Birkner, F. B., 363(27), *366*
Black, A. P., 363(27), *366*
Black, W., 365(36), *367*
Blake, T. D., 294(30), *297*
Blanchin, B., 331(88), *337*
Blank, M., 158(74), 161(74), *179*, 340(2), *354*
Bleier, A., 79(75,76), *97*, 294(29), *297*
Block, L., 42(34), *45*, 175(132), *181*
Boardman, W. W., 300(2), *320*
Bockrls, J. O'M., 33(20), 41(27), 42(27), *45*, 148(1,12), 151(12), 154(1,12,29,43), 156(12), *176, 177, 178*
Böhm, J., 326(32), *335*
Bohme, G., 271(9), *283*
Bohnenkamp, K., 390(6), *395*
Bolt, G. H., 42(28), *45*, 175(139), *182*
Booth, F., 108(1), *117*
Borchers, B., 331(90), *337*
Branion, R., 324(7), *334*
Bredin, R., 295(38), *297*
Breeuwsma, A., *14*(3), 42(42), *46*, 172(125), *181*, 291(15), *296*
Brenner, H., 65(35), *95*
Breunig, A., 333(138), *338*
Brezina, M., 165(104), 166(104), *180*
Briant, J., 131(19), *142*
Brintzinger, H., 360(14), 361(14), *366*

Britt, K., 327(48), 330(48), 335
Britt, K. W., 324(6), 328(6,61, 62,63), 330(67,81), 332(81), 333(63), 334, 336
Brody, O. V., 398(1), 412
Brooks, D. E., 407(17), 412
Brown, A. S., 204(28), 223
Brown, D. R., 395(27), 396
Brown, F. E., 160(94,95), 161(95), 164(95), 180
Brown, H., 393(22), 396
Brown, J. C., 84(110), 98
Bryson, H. R., 330(79), 336
Buchanan, A. S., 212(43), 214(50), 224
Buess-Herman, CL., 149(16,17), 158(16), 177
Burden, J., 207(34), 223
Buscall, R., 83(97), 97
Busch, F. W., Jr., 383(34), 385

Cain, F. W., 228(30), 263
Campanella, L., 153(26), 177
Campbell, A. B., 175(130), 181
Carlson, C. F., 414(1), 434
Carlsson, G., 326(31), 335
Carter, M. N. A., 215(54), 217(54), 224
Chan, D., 74(55,56), 76(55, 56), 96
Chan, D. Y. C., 70(42), 95
Chander, S., 74(49), 95
Chapman, P. L., 25(16), 45
Chiang, A., 427(45), 436
Choi, H. S., 295(45), 297
Christie, J. H., 158(65), 179
Churaev, N. V., 80(87), 97, 228(28), 263
Cichos, C., 83(92), 97, 294(37), 297
Clark, H. E., 414(5), 434
Clayfield, E. J., 422(34,35), 435
Cling, J. H., 78(64), 96

Clint, G. E., 78(64), 96
Clunie, J. S., 83(96), 97
Cluxton, D. H., 192(13), 223
Coad, K., 204(26), 223
Coffin, E. M., 153(24), 177
Collins, G. L., 294(36), 297
Colombo, A. F., 286(1), 295(47), 296
Cooke, S. R. B., 286(1), 295(45), 296
Cooper, W. D., 131(20), 135(31), 139(31), 142
Corkill, J. M., 78(64), 96
Corrin, M. L., 318(50), 322
Counter, R., 330(77), 331(77,86), 332(86,108), 336, 337
Cremer, M., 363(29), 367
Croucher, M. D., 428(46), 430(46), 432(46), 436
Crouse, B. W., 332(102), 337
Crowl, V. T., 365(34), 367
Cummings, J. I., 153(24), 177
Cummins, H. Z., 194(19), 223
Curtis, H. J., 215(49), 224

Dailey, J. L., 424(39), 435
Dalisa, A. L., 426(40), 436
Damaskin, B. B., 36(23), 45
Dannon, D., 405(14), 412
Das, B. S., 326(34), 335
Daucik, K., 333(127), 338
Davies, G. W., 330(78), 331(78), 336
Davies, J. T., 161(96), 162(98), 180
Davis, H. T., 120(3), 141
Davis, J. A., 175(137), 176(137), 182
Davis, S. S., 370(8), 384
Davison, R. W., 324(20,21), 326(20,21), 327(20), 328(20,21), 330(20), 332(101,111), 333(112), 334, 337
Dawes, W. H., 375(23), 376(23), 384
DeBacker, R., 216(56), 224

De Bruyn, H., 175(131), 181
De Bruyn, P. L., 19(9,10,11), 42(29,31,33,34), 44, 45, 66(39), 95, 141(40), 143, 170(116), 173(116), 175(132, 133,134,136), 181, 182, 204(27), 205(31), 207(27), 223, 226(4), 230(4), 263, 290(12,13), 291(14,18), 292(18,19,24), 296
Debye, P., 371(11), 384
DeGuchi, K., 392(18), 393(19), 396
Delahay, P., 17(2), 44, 148(4), 154(4,32), 158(62), 176, 177, 179
De Levie, R., 158(62), 179
Della Rocca, C., 392(13), 395
Deme, J., 333(113), 338
Derjaguin, B. V., 48(1), 54(3), 65(37), 66(38), 70(3), 78(38), 80(87), 81(88), 94, 95, 97, 201(25), 208(37), 214(53), 215(53), 223, 224, 226(2,3), 227(18), 228(28), 263, 273(13), 283, 292, 294(27,28), 297
De Smet, M., 209(39), 211(39), 214(48), 223, 224
Dessaur, J. H., 414(5), 434
De Sturler, A. A., 333(117), 338
Deutch, J. M., 65(34), 95
Devanathan, M. A. V., 17(4), 34(21), 36(21,24), 38(24), 40(25), 41(27), 42(27), 44, 45, 148(88), 154(29), 176, 232(53), 264
Devereux, O. F., 66(39), 95, 226(4), 230(4), 263
De Vooys, D. A., 24(14), 45
De Vries, A., 405(14), 412
Dibbs, H., 295(38), 297
Dietz, R., 84(110), 98
Dillon, A. G., 328(63), 333(63), 336

Dirkx, J., 410(19), 411(19), 412
Ditter, W., 365(35), 367
Dobbins, R. J., 333(142), 338
Dobronauteanu, V., 327(49), 335
Dolre, A. K., 375(22), 376(22), 384
Donnay, G., 174(127), 181
Douglas, H. W., 207(34), 223
Drost, Hansen, W., 216(61), 224
Duff, A., 333(140), 338
Dukhin, S. S., 65(37), 95, 208(37), 214(53), 215(53), 223, 224, 292(26,27), 297
Dunn, V. K., 141(37), 143
Dupeyrat, M., 158(76,77,78), 160(78), 161(76,77,78), 179, 340(6), 354
Durham, K., 276(16), 281(16,25), 283
Dutkiewiz, E., 17(5), 44
Dzyaloshinski, I. E., 57(8), 94, 227(13), 263

Eda, K., 41(26), 45
Edwards, G. R., 318(64), 322
Eichhorn, J. R., 383(37), 385
Eklund, D., 326(30), 327(30), 329(30), 332(4), 334, 335
Elworthy, P. H., 371(18,19), 374(18,19), 375(19), 376(19), 378(18), 384
Engel, D. J. C., 70(44), 95
Envall, A. D., 420(20), 435
Epelboin, I., 154(47), 155(47), 178
Etter, D. O., 324(13), 334
Evans, H. C., 318(55), 322
Evans, L. A., 328(63), 333(63), 336
Evans, P. R., 424(39), 435
Eversole, W. G., 300(2), 320
Exerowa, D., 81(89,90), 97,

[Exerowa, D.]
227(19,20), *263*
Exner, M. P., 328(59), *336*
Eylar, E. H., 398(1), *412*
Eyring, H., 318(53), *322*

Fairbairn, D. G., 427(45), *436*
Faulkner, L. R., 158(67), *179*
Fava, A., 318(53), *322*
Feat, G. R., 133(21), 140(21), *142*
Feig, S., 158(74,75), 160(75), 161(74,75), *179*, 340(2), *354*
Felderhof, B. U., 65(34), *95*
Finlan, J. M., 70(41), 79(41), *95*
Finlayson, C. M., 383(35), *385*
Flanagan, J. B., 158(68), *179*
Fleer, G. J., 236(64), 237(64), 238(64), 241(72), *265*
Fleishmann, M., 155(50), *178*
Florence, A. T., 371(18,19), 374(18,19), 375(19), 376(19), 378(18), *384*
Florus, G., 364(31), *367*
Flygare, W. H., 192(8,9,10), 194(9), *222*
Forrester, A. T., 194(18), *223*
Fowkes, F. M., 57(7), *94*, 141(42,43), *143*
Fox, R. J., 162(97), *180*
Frankel, S., 83(94), *97*
Frankle, W. E., 330(82), 331(82, 93,95), 332(95), 333(127, 143), *336, 337, 338*
Franklin, M. J. B., 219(66), *224*, 364(32), *367*
Frens, G., 70(44,45), *95*, 226(6), *263*
Freyberger, W. L., 19(9), *44*, 290(12), *296*
Friberg, S., 332(110), *337*
Fricke, H., 215(49), *224*
Fridriksberg, D. A., 207(33), *223*

Friend, J. P., 295(50), *298*
Fries, B. A., 318(51), *322*
Friesen, W., 333(123,124), *338*
Fritsch, A. R., 392(16), *395*
Frommer, D. W., 295(47), *297*
Frumkin, A. N., 36(23), *45*, 135(25), *177*
Fuchs, N., 233(61), *265*
Fuerstenau, D. W., 42(32,38), *45, 46*, 66(40), 79(40), *95*, 171(119), 172(119), *181*, 204(29), *223*, 226(5), *263*, 273(12), 275(12), *283*, 286(2, 3,4), 287(3,4,5,6), 288(6,7,25), 289(6,7), 290(2), 291(16), 292(20,25), 295(43,44), *296, 297*
Fujii, A., 158(89), 164(89), 168(89), *180*
Fujii, K., 388(2), *395*
Fujii, T., 134(25), *142*
Fujioka, Y., 330(83), 331(83), 332(83), 333(83), *337*
Fukushima, S., 370(2,3,4,5), 378(2,3,4,5,28,29), 379(28,29), 380(29), *383, 384, 385*
Fuller, M. J., 176(141), *182*
Fuoss, F. M., 120(4), *141*
Furusawa, K., 61(25), 188(4), *222*, 241(68,69), 242(68), 243(69), 244(70), 245(69,71), 246(71), 247(71), 249(76,78), 250(81), 251(81), 254(81), 255(71), 256(71,76,88,90), 258(90), 259(90), 262(90), *266*

Gallaccio, A., 395(26), *396*
Gaonkar, A. G., 232(52), *264*
Gasparini, R., 392(13), *395*
Gaudin, A. M., 287(5), 295(43, 44), *296, 297*
Gavach, C., 158(79), 161(79), 165(101), *179, 180*

Gerjuoy, E., 194(18), *223*
Gervason, G., 331(88), *337*
Gierst, L., 149(16,17), 158(16), *177*
Gilbert, G. A., 401(10), *412*
Gill, R. A., 333(135), *338*
Gillies, G. A., 318(64), *322*
Gilman, A., 294(35), *297*
Gingell, G., 411(20), *412*
Giuliani, A. M., 158(62), *179*
Glenz, O., 308(28), 315(28), *321*
Goddard, E. D., 294(29), *297*
Goetz, P. J., 192(16,17), *223*
Goldsbrough, K., 364(32), *367*
Goldsmith, H. L., 61(27), *95*
Goodman, J. F., 83(96), *97*
Goossens, J. W. S., 327(42), 328(42,57), 333(57), *335*
Gordon, B. E., 318(64), *322*
Gotoh, R., 60(18), *94*, 158(80, 81,82), 160(80), 162(81), 163(81), *179*, 229(39), 230(43), 231(47,48,49), *264*, 340(3,4), *354*
Gotshal, Y., 318(56), *322*
Götte, E., 270(6), *283*
Gouy, L., 25(15), *45*
Grahame, D. C., 17(3), 20(12), 21(3), 23(13), 24(13), 31(12), 32(12), 40(12), 41(12), *44, 45,* 148(10), 153(23,24), 154(10), 155(55), *176, 177, 178*
Grantham, D. H., 149(13), 154(13), *177*
Greene, B. W., 333(120), *338*
Greenwald, H. L., 370(1), *383*
Gregory, J., 57(10), 74(53), *94, 95,* 226(8), *263,* 275(15), *283*
Griggs, W. H., 324(13), 332(102), *334, 337*
Groves, P. O., 393(24), *396*
Guastalla, J., 158(72,73), 161(72,73), *179,* 231(46), *264,* 340(1), *354*

Guggenheim, E. A., 17(1), *44*
Guidelli, R., 154(30), *177*
Gunder, W., 326(28), 328(28), 333(28), *335*

Hachisu, S., 61(25), 84(106, 107,108,109,111,115), 86(106), 91(107,108,109), *95, 98,* 188(4), *222,* 241(69), 243(69), 244(70), 245(69), 256(88), *265, 266*
Hackerman, N., 212(41,42), *224,* 392(15), *395*
Hair, M. L., 192(15), *223,* 428(46), 430(46), 432(46), *436*
Hakozaki, T., 79(69), *96,* 318(49), *322*
Halabisky, D. D., 332(96), *337*
Ham, A. J., 200(23), 201(23), *223*
Hamaguchi, I., 426(44), 433(44), 434(44), *436*
Hamaker, H. C., 55(5), 56(5), *94,* 226(10), *263,* 271(10), *283,* 376(26), *384*
Hamann, K., 357(3), 364(31), *366, 367*
Hamilton, J. D., 192(7), *222*
Hansen, F. K., 79(77), *97*
Hansen, R. S., 149(13), 154(13, 27), *177*
Harada, S., 392(17), *396*
Harbour, J., 428(46), 430(46), 432(46), *436*
Harding, R. D., 79(82), *97,* 362(24), *366*
Harkins, W. D., 160(94,95), 161(95), 164(95), *180,* 318(50), *322*
Harraway, D. H., 295(45), *297*
Harrison, J. A., 158(70), *179*
Harsveldt, A., 333(117), *338*
Hartmann, H.-J., 331(90,91), *337*

Hartwig, G. M., 318(64), 322
Harusawa, F., 370(3), 378(3), 384
Hasegawa, F., 232(51), 264
Hasunuma, M., 422(36), 435
Hatsutori, K., 175(140), 182
Haug, R., 360(14), 361(14), 366
Hayashi, K., 158(92), 180
Hayashi, M., 318(60,61), 322
Hayashi, T., 416(7), 434
Hayes, J. W., 157(59), 179
Healy, T. W., 42(32,43), 43(43), 45, 46, 66(40), 74(55,56), 76(55,56), 79(40), 95, 96, 171(119,122), 172(119), 173(122,123), 176(123), 181, 226(5,7), 263, 273(12), 275(12), 283, 357(4), 358(5), 362(25,26), 366
Heinegard, C., 326(24), 333(24), 335
Heins, R. W., 201(24), 223
Hengst, J. H., 175(129), 181
Hengst, J. H. Th., 169(112), 173(112), 181
Henry, D. C., 110(2), 117, 371(13), 384
Hentschel, H., 333(121), 338
Hermanie, P. H., 126(18), 131(18), 135(18), 142
Hermanie, R. H. J., 421(24), 435
Hernádi, S., 327(39), 333(113), 335, 338
Herring, A. P., 171(119), 172(119), 181
Hesselink, F. Th., 141(34), 143, 228(31), 248(73), 263, 265, 365(36), 367
Heyman, E., 212(43), 224
Hickson, D. A., 154(27), 177
Higashitsuji, K., 158(84,89,90), 164(89,90), 165(103), 167(103), 168(89,107,108,109, 110,111), 180, 181, 340(8,9,

[Higashitsuji, K.]
 10,11), 342(9,10), 354
Higuchi, W. I., 370(7), 384
Hills, G. J., 155(57), 178
Hiltner, P. A., 84(104), 98
Hilton, P. A., 84(105), 86(105), 98
Hirai, T., 388(1,2), 389(5), 391(10,12), 393(25), 395, 396
Hirai, Y., 392(14), 395
Hirata, M., 301(11,12), 302(12), 320
Hodgson, W. 200(23), 201(23), 223
Hoffman, F., 324(5), 326(5), 327(5), 333(5), 334
Hogg, R., 66(40), 79(40), 95, 226(5), 263, 273(12), 275(12), 283
Hohl, H., 42(44), 46
Hollick, C. T., 296(55), 298
Homola, A., 76(58), 96
Homolka, D., 165(105), 180
Honig, E. P., 65(33), 95, 169(112), 173(112), 175(129), 181
Hook, B. J., 383(35), 385
Hoover, H. G., 90(119), 98
Hopper, M., 428(46), 430(46), 432(46,50), 436
Hopstock, D. M., 290(9), 296
Horn, D., 324(9), 326(9,35), 327(9,35), 328(9), 334, 335, 365(35), 367
Horn, J. M. Jr., 205(32), 206(32), 223
Horn, R. G., 233(58), 265
Hornof, V., 324(11), 334
Hough, D. B., 57(15), 94
Howard, G. J., 326(33), 331(33), 335
Huang, C. T., 216(61), 224
Huber, O., 330(70), 331(70), 333(114,118,119), 336, 338
Hückel, E., 371(11,12), 384
Hudson, F. L., 326(33), 331(33), 335

Huggenberger, L., 330(80), 336
Hull, M., 78(63), 96
Hunter, R. J., 333(122), 338
Hurd, M., 212(41,42), 224
Hurd, R. M., 392(15), 395

Iida, T., 426(43), 433(43), 436
Ikeda, H., 133(24), 142
Imagawa, H., 359(9,10), 366
Imamura, T., 78(66,67), 96, 274(14), 276(19), 278(20,21,22), 279(23), 280(22,23), 281(19), 282(19,22,23), 283
Imanaga, H., 175(138,140), 182
Ince, C. R., 295(41), 297
Ingram, B. T., 83(96), 97
Inoue, E., 417(13), 435
Inoue, Y., 330(83), 331(83), 332(83), 333(83), 337
Ioannilli, E., 392(13), 395
Israelachvili, I. N. 57(11), 61(23), 94, 227(16,17,25,27), 228(32), 232(25), 233(58), 263, 264, 265
Iwadare, Y., 319(65, 66), 322
Iwai, S., 301(12), 302(12), 320
Iwanow, S. N., 330(73), 331(73), 336
Iwasaki, I., 19(10), 44, 286(1), 288(8), 290(13), 291(18), 292(18,23), 295(45), 296, 297
Iyer, S. R. S., 300(3), 320

Jacobson, C. F., 163(100), 180
James, A. M., 215(54), 216(58),
[James, A. M.] 217(54), 224
James, R. O., 76(58), 96, 173(123), 175(137), 176(123,137), 181, 182
Jameson, G. J., 294(36), 297
Jayaram, R., 300(3), 320
Jaycock, M. J., 324(8), 330(71, 74,75,76,77), 331(71,75,77,86), 332(108), 334, 336, 337
Jinnai, H., 141(42), 143
Jones, A. Y., 331(94), 337
Jones, J. E., 74(51), 96
Jones, T. G., 276(17), 278(17), 283
Joppien, G. R., 357(3), 366
Jordan, J. W., 383(35), 385
Joseph-Petit, A. M., 260(92), 266

Kahn, C., 383(37), 385
Kakiuchi, T., 149(20), 177
Kakutani, T., 165(106), 180
Kale, P. D., 308(32), 321
Kamada, K., 168(107), 180
Kamat, G. R., 393(20), 396
Kanamaru, K., 301(8,9,11), 303(8), 308(30,31), 320, 321
Kandori, K., 141(41), 143
Kar, G., 74(49), 95
Karasawa, S., 123(8), 126(8), 127(8), 135(8), 138(8), 139(8), 142, 362(23), 366, 421(25,28,29), 422(25), 424(28), 435
Karnis, A., 331(94), 332(94), 337
Katano, S., 134(25), 142
Katchalsky, A., 405(14), 412
Kavanagh, B. V., 79(71), 96, 362(25), 366
Kawai, H., 127(44), 135(44), 143, 364(33), 367
Kawakami, K., 313(39), 315(39), 321
Kawanishi, T., 417(8), 420(19),

[Kawanishi, T.]
 422(30), 425(30), *434*
Kawasaki, S., 123(9), 124(9),
 135(9), *142*
Keddam, M., 154(47,48),
 155(47), *178*
Kelsh, D. J., 149(13), 154(13),
 177
Kenaga, D. L., 327(45), *335*
Kendall, M. J., 404(13),
 405(13), 406(13), *412*
Kikkawa, A., 192(14), *223*
Kimmerle, F. M., 149(15), *177*
Kimura, H., 392(17), *396*
Kindler, W. A., 327(45), *335*
King, A. M., 216(59), *224*
King, C. A., 328(65), *336*
Kirkwood, I. G., 89(117), *98*
Kitagawa, N., 303(14), 305(14),
 310(14), 311(14), *320*
Kitahara, A., 79(73), *96*, 123(8,
 9), 124(9), 126(8), 127(8,45),
 133(24), 134(25), 135(8,9,45),
 138(8), 139(8,33), 140(33),
 141(33,41), *142, 143*, 222(67),
 224, 270(4), *283*, 362(23),
 366, 421(25,29), 422(25,36),
 435
Kitamura, A., 301(10), *320*
Kitchener, J. A., 78(62,63),
 80(84,85), 83(95), *96, 97*,
 256(89), *266*, 281(26), *283*,
 294(30), 295(48,49,50), *297,
 298*, 318(57,63), *322*, 370(6),
 376(6), *384*
Kito, M., 165(102), *180*
Kittaka, S., 174(128), 210(40),
 212(40), 217(62), 218(62),
 224
Klein, J., 228(34), *264*
Klier, M., 84(102), 86(102),
 98
Kling, W., 200(21), 201(21),
 223, 270(6), 271(9), *283*,
 317(44), 319(44), *322*
Klungness, J. H., 328(59),
 336

Kobayashi, S., 221(69), *224*
Kobayashi, Y., 84(106,107,109),
 86(106), 91(107,109), *98*,
 133(24), *142*
Koelmans, H., 127(17), 131(17),
 142
Koishi, M., 359(12), *366*
Kojima, K., 194(20), *223*
Kolakowski, J. E., 79(79), *97*
Kolthoff, I. M., 204(26), *223*,
 249(79), *265*
Komagata, S., 190(5), *222*,
 294(34), *297*
Komatsuzawa, T., 127(45),
 135(45), *143*
Komidawa, H., 426(43), 433(43),
 436
Komori, S., 318(62), *322*
Komura, A., 175(140), *182*
Komura, T., 175(138), *182*
Kondo, A., 219(65), 220(65),
 224, 420(23), *435*
Kondo, T., 348(12), *354*
Kon-no, K., 123(9), 124(9),
 127(45), 135(9,45), 141(41),
 142, 143
Korpi, G. K., 205(31), *223*
Koryta, J., 165(104), 166(104),
 180
Kose, A., 84(106,107,108),
 86(106), 91(107,108), *98*
Koshiyama, Y., 133(24), *142*
Kotera, A., 249(76), 256(76,90),
 258(90), 259(90), 262(90), *265,
 266*
Kovak, Z., 148(12), 151(12),
 154(12), 156(12), *177*
Kowalewska, J., 332(110), *337*
Kozawa, A., 390(8), *395*
Kraus, C. A., 120(4), *141*
Krieger, I. M., 84(103,104,105),
 86(105), *98*
Krupp, H., 57(9), *94*, 271(9),
 283
Kruyt, H. R., 234(62), 238(62),
 256(85), *265*
Kubo, M., 392(18), 393(19), *396*

Kudo, K., 256(90), 258(90), 259(90), 262(90), *266*
Kulick, R. J., 332(106), *337*
Kulkarnie, R. D., 294(29), *297*
Kumagai, S., 378(28,29), 379(28,29), 380(29), *385*
Kummert, R., 176(142), *182*
Kuo, J., 79(81), *97*
Kuo, S., 219(63), *224*
Kurita, T., 417(9,10), 418(10), *434*
Kurosaki, S., 392(18), 393(19), *396*

Landau, L., 48(1), *94*
Landau, L. D., 226(2), *263*
Lane, J. E., 232(55), *264*
Lange, H., 271(9,11), 276(18), 281(24), *283*
Lapin, V. V., 328(53), *336*
Larsen, R. P., 153(23), *177*
Laseur, W. J. J., 333(128), *338*
Laskowski, J., 294(31), *297*
Lason, L., 331(87), *337*
Latimer, J. J., 333(135), *338*
Lauer, G., 158(71), *179*
Lawrence, J., 149(18), *177*
Leblanc, D., 324(11), *334*
Leckie, J. O., 175(137), 176(137), *182*
Lees, H. D., 424(39), *435*
Lenderman, C. D., 330(68), *336*
Leob, A. L., 219(64), *224*
Lessard, R. R., 232(50), *264*
Levin, S., 42(39), *46*
Levine, S., 55(4), 74(51), *94, 96*, 130(16), 133(21), 138(16), 140(21), *142*, 154(33), *177*
Levy, B., 392(16), *395*
Lewis, K. E., 361(21), *366*

Li, H. C., 204(27), 207(27), *223*, 292(19), *296*
Lifshits, E. M., 57(8), *94*, 227(12, 13), *263*
Lind, E. L., 318(50), *322*
Lindström, T., 326(23,24,29,30, 31), 327(23,29,30), 329(30), 331(23), 332(111), 333(23,24), *335, 337*
Linke, W. F., 327(41), *335*
Lipic, B., 331(92), *337*
Lisse, M. W., 191(6), *222*
Ljadowan, W., 330(73), 331(73), *336*
Loeb, A. L., 114(4), *117*, 255(84), *266*
Lokhande, H. T., 308(32,33), 309(33), 313(37), 315(40), 316(40,41), *321*
Lomas, H., 326(34), *335*
London, F., 226(11), 252(11), *263*
Longsworth, L. G., 163(100), *180*
Lorenz, W., 154(37,38), *178*
Los, J. M., 149(22), *177*
Luck, W., 84(102), 86(102), *98*
Lui, Y. K., 138(29), *142*
Lukomski, N., 390(6), *395*
Luner, P., 326(32), 327(42), 328(42), *335*
Lyklema, J., 13(1), 14(3), 19(8), 42(36,41,42), *44, 45, 46*, 82(91), *97*, 120(1), 124(1), 131(1), 132(1), 138(1), 141(1), 171(118, 121), 172(125), 175(118), *181*, 208(36), *223*, 227(22), 228(32), 236(64), 237(64), 238(64), 241(72), 248(74), 249(78), *263, 264, 265*, 291(15), *296*, 421(27), *435*

Machida, H., 421(31), *435*
Machida, Y., 371(16), 374(16), *384*

Madan, G. L., 303(16,17), *321*
Madoff, M. A., 398(1), *412*
Maeda, Y., 84(115), *98*
Mahanty, J., 57(13), *94*
Mahl, H., 270(6), *283*
Maksimov, D., 42(35), *45*
Maltz, M. S., 424(39), *435*
Manev, E., 229(36), *264*
Marecek, V., 165(105), *180*
Marshall, J. K., 78(62), *96*, 281(26), *283*
Martin, R. F., 158(63), *179*
Martin-Lof, S., 326(24), 333(24), *335*
Marton, J., 327(37,38), 333(131), *335, 338*
Marton, T., 327(37), *335*
Mason, S. G., 61(27), *95*
Matijevic, E., 70(41), 79(41, 75,76), *95, 97*, 332(109), *337*
Matsuda, K., 157(61), *179*
Matsuda, Y., 359(9,10,11), *366*
Matsukawa, H., 295(40), *297*
Matsumoto, M., 60(18,19,20), *94*, 158(80,81,82,83), 160(80), 162(81), 163(81,83), *179*, 229(40), 230(43), 231(45,47,48,49), 232(52), *264*, 340(3,4), *354*
Matsumura, S., 139(33), 140(33), 141(33), *143*
Matsunaga, T., 141(43), *143*
Matsuoka, I., 290(11), *296*
Mattes, O. R., 154(48), *178*
Matulis, Y. Y., 393(23), *396*
Mayer, A., 332(99), *337*
Mayes, B. J. R., 383(32), *385*
McBain, J. W., 216(59,60), *224*
McCartney, L. N., 55(4), *94*
MacDonald, D. D., 155(53), *178*
McDowell, M., 83(99), *98*
McGown, D. N. L., 126(28), 127(28), 131(26), 132(26),

[McGown, D. N. L.]
135(28), 138(26,28), 139(32), *142*, 361(20), 362(20), *366*
MacInnes, D. A., 120(5), *141*
McKague, J. J., 324(13), *334*
McKenzie, A. W., 324(1)., 326(19), 330(19,78), 331(78), *334, 336*
McLean, A. W., 170(117), *181*
McTaggart, H. A., 294(32), 295(32), *297*
Meader, A. L., 318(51), *322*
Meadus, F. M., 138(30), *142*
Melzer, J., 324(9), 326(9,27,35), 327(9,27,35), 328(9), 331(27), 333(27), *334, 335*
Metcalfe, A. A., 158(69), *179*
Metcalfe, K. A., 414(2,3), 420(2,3), *434*
Metz, A. R., 424(38), *435*
Meyer, F. J., 327(45), *335*
Meyer, R. J., 382(30), *385*
Miaw, H. L., 295(43,44), *297*
Mica, T. S., 74(49), *95*
Micale, F. J., 138(29), *142*
Michel, J., 158(76), 161(76), *179*, 340(6), *354*
Mikailene, E. S., 393(23), *396*
Milgrom, A., 326(19), 330(19), *334*
Miller, I. K., 249(79), *265*
Minday, R. M., 120(3), *141*
Minturn, R. E., 154(27), *177*
Mishra, S. K., 292(21), *296*
Misra, J., 370(7), *384*
Mitchell, D. J., 232(57), *264*
Mizuhiro, Y., 310(34), 311(34), 313(24), *321*
Möckel, F., 154(38), *178*
Modi, H. J., 286(2,3,4), 287(3, 4), 290(2), *296*
Mody, N. R., 308(33), 309(33), *321*
Mohilner, C. M., 149(18), *177*
Mohilner, D. M., 19(6), *44*, 148(6), 149(18,19,20), *176, 177*

Mohilner, P. R., 149(19), *177*
Mohn, E., 222(68), *224*
Moore, E. E., 327(50), 328(60), *335, 336*
Moore, R. J., 141(42), *143*
Morgan, J. J., 363(27), *366*, 390(7), *395*
Morimoto, T., 174(128), *181*, 210(40), 212(40), 217(62), 218(62), *224*
Morris, H. H., 333(116), *338*
Mostafa, M. A., 141(42), *143*
Motarjemi, M., 294(36), *297*
Mugler, G., 324(15), *334*
Mugler, P., 324(15), *334*
Muhonen, J. M., 328(64), *336*
Mukai, T., 305(21), 310(21), 311(21), *321*
Müller, F., 324(5), 326(5), 327(5), 328(57,58), 333(57, 58), *334, 336*
Müller, K., 41(27), 42(27), *45*, 148(12), 151(12), 154(12,29), 156(12), *177*
Muller, V. M., 74(50), *96*
Murata, T., 359(9,10,11), 360(13), *366*
Musselwhite, P. R., 370(6), 376(6), *384*
Mysels, K. J., 82(91), 83(94), *97*, 227(22), *263*

Nagai, R., 290(10), *296*
Nagareda, K., 310(34), 311(34), 313(34), *321*
Naitoh, K., 392(18), 393(19), *396*
Nakache, E., 158(78), 160(78), 161(78), *179*, 340(7), *354*
Nakadomari, H., 149(19), *177*

Nakamura, M., 83(93), *97*
Nakamura, Y., 353(13), *354*
Nakanishi, Y., 127(44), 135(44), *143*, 364(33), *367*
Nakano, J., 324(16), *334*
Nakatsuka, K., 290(10,11), *296*
Nakazawa, Y., 318(60), *322*
Napper, D. H., 141(36), *143*
Neale, G. H., 324(11), *334*
Neale, S. M., 311(36), *321*
Neihof, R., 188(3), *222*
Nelson, J. A., 330(68), *336*
Nelson, S. M., 120(6), *141*
Neumann, A. W., 78(60), *96*
Neumann, B. S., 383(33), *385*
Nevo, A., 405(14), *412*
Nicke, R., 331(90,91), *337*
Nicol, S. K., 333(122), *338*
Nieuwenhius, E. A., 84(112), *98*
Ninham, B. W., 57(13), 75(57), *94, 96*, 227(14,15), 232(57), *263*
Nishikawa, K., 330(83), 331(83), 332(83), 333(83), *337*
Nishizawa, K., 158(84), 165(103), 166(104), 168(107,108), *180*, 340(8,10), 342(10), *354*
Norde, W., 248(74), 255(83), *265*
Noro, S., 359(12), *366*
Novotny, V., 192(15), *223*, 432(50), *436*
Nozaki, H., 426(43), 433(43), *436*
Nugent, H. M., 328(55), *336*

O'Conner, D. J., 214(50), *224*
Ohshima, H., 74(52), *96*
Ohta, I., 426(42), *436*
Ohtsuka, T., 429(48), *436*
Ohyama, Y., 417(10), 418(10), *434*
Okuno, Z., 421(31), 426(44), 433(44), 434(44), *435, 436*

Oldham, J. B., 208(38), 209(38), 210(38), *223*
Olivier, J. P., 333(116), *338*
Onabe, F., 324(16), 327(43), *334, 335*
Oncley, J. L., 398(1), *412*
O'Neill, F. M., 84(103), *98*
Onoda, G. Y., 42(33), *45*
Onoda, G. Y. Jr., 175(134), *181*, 205(32), 206(32), *223*
Onsager, L., 115(5), *117*
Osmond, D. W. J., 130(15), *142*
Oster, G., 234(63), *265*
Osterle, F., 219(63), *224*
Osterle, J. F., 208(38), 209(38), 210(38), *223*
Osteryoung, R. A., 158(65), *179*
Ota, I., 429(48), *436*
Ota, S., 308(30,31), *321*
Ottewill, R. H., 64(31), 83(97), *95, 97*, 141(35), *143*, 185(2), *222*, 228(29,30), 260(91), *263, 266*, 304(20), *321*, 403(11), 404(11), 408(18), 409(18), 410(18), *412*
Overbeek, J. Th. G., 13(1), 14(4), 19(7), 26(17), 27(17), *44, 45*, 48(2), 54(2), 56(6), 58(2), 60, 61(24,28), 63, 64(29,30), 65, 70, 83(101), 84(114), *95, 98*, 111(3), 113(3), 114(4), *117*, 127(17), 131(17), 133(22), 135(17), 140(17), 141(34,40), *142, 143*, 148(3), 169(114), 171(118), 173(124), 175(118), *176, 181*, 184(1), 207(35), 208(36), 216(57), 219(64), *222, 223, 224*, 226(1), 228(31), 233(59), 237(65), 248(73), 250(80), 254(80), 255(84,85), *262, 263, 265, 266*, 270(3), 271(3), *283*, 376(26), *384*,

[Overbeek, J. Th. G.]
395(28), *396*, 407(15), 408(15), *412*, 422(32), *435*
Oyabu, Y., 127(44), 135(44), *143*, 364(33), *367*
Ozaki, K., 84(106), 86(106), *98*
Ozaki, M., 194(20), *223*

Pailthorpe, B. A., 232(57), *264*
Pallmann, H., 169(113), *181*
Pankove, J. I., 424(37), 428(37), *435*
Papier, Y. S., 84(105), 86(105), *98*
Parfitt, G. D., 120(2), 126(28), 127(28), 131(26), 132(26), 133(23), 135(28), 138(26,28), 139(32), 141(2), *142*, 358(6,7), 359(6), 361(20,21), 362(20), 364(32), *366, 367*
Parkins, W. E., 194(18), *223*
Parks, G. A., 19(11), 42(29,30), *44, 45*, 174(126), 175(133), *181*, 291(14), *296*, 357(2), 358(2), *366*
Parreira, H. C., 126(14), 128(14), 135(14), *142*, 401(9), 402(9), *412*, 421(26), *435*
Parsegian, V. A., 57(12), 75(57), *94, 96*, 227(14,15), *263*, 411(20), *412*
Parsons, R., 17(5), 33(19), *44, 45*, 148(7), 154(28), 155(49), *176, 177, 178*
Pash, J. A., 204(29), *223*
Pashley, R. M., 61(23), *94*, 232(54), *264*
Payne, R., 148(8,9), 149(9), 153(8), 155(56,57), 158(56), *176, 178*
Peacock, J., 120(2), 141(2), 364(32), *367*
Peaker, C. R., 216(59,260), *224*
Pearlstein, F., 395(26), *396*

Pearson, J. L., 324(8), 330(71, 74,75,76,77), 331(71,75,77, 86), 332(86,108), *336, 337*
Peat, R., 154(28), *177*
Pelton, R. H., 328(55), *336*
Peniman, J. G., Jr., 192(16), *223*
Penniman, J., 324(14,22), 326(14,22), 327(14), 328(51, 54), 330(82), 331(82), 333(51, 82,125,127,143), *334, 335, 336, 338*
Penniman, J. G., 330(82), 331(82), 333(82,125,127,143), *336, 338*
Peri, J. B., 175(135), *182*
Perkins, R. S., 149(14), 153(14), *177*
Perram, J. W., 74(55), 76(55), *96*
Perrin, J., 83(98), *98*
Peter, R. H., 311(36), *321*
Peterson, G. C., 71(46), 74(46, 54), *95*
Pethica, B. A., 398(5), *412*
Phadke, D. S., 375(22), 376(22), *384*
Phansalkar, A. K., 318(52), *322*
Pieper, J. H. A., 24(14), *45*
Pilgrim, J. O., 324(13), *334*
Pink, R. C., 120(6), *141*
Pitaevskii, L. P., 57(8), *94*, 227(13), *263*
Ponder, E., 407(16), *412*
Poppel, E., 326(26), 327(49), 331(26), 332(26,204), *335, 337*
Posner, A. M., 13(2), *14*, 42(37), *46*
Poth, M. A., 153(23,24), *177*
Powe, W. C., 269(1), *283*
Price, W. B., 389(4), *395*
Prieve, D. C., 76(59), *96*
Puddington, I. E., 138(30), *142*
Pugh, R. J., 80(84), *97*, 141(43), *143*, 295(48), *297*
Pusey, P. N., 84(110), *98*

Quirk, J. P., 13(2), *14*, 42(37), *46*
Quarin, G., 149(17), *177*

Rabideau, J., 333(140), *338*
Race, W. P., 154(36,39,42), 155(58), 156(58), *177, 178*
Ramanauskene, D. K., 393(23), *396*
Rambhau, D., 375(22), 376(22), *384*
Ramsbotham, J., 358(6), 359(6), *366*
Rastogi, M. C., 304(20), *321*
Read, A. D., 295(51), 296(55), *298*
Rechmann, H., 356(1), 358(1), 360(1), *366*
Reddy, A. K. N., 148(1), 154(1, 43), *176, 178*
Reder, A. S., 333(120), *338*
Reerink, H., 63(29), 64(29), *95*, 233(59), *265*
Reif, W. M., 327(46), *335*
Reilley, C. N., 157(59), *179*
Retajczyk, T. F., 157(60), *179*
Reyerson, L. H., 204(26), *223*
Riccick, T. M., 371(9), 372(9), 376(9), *384*
Rideal, E. K., 161(96), *180*, 401(10), *412*
Ring, T. A., 70(41), 79(41), *95*
Robenfeld, L., 318(56), *322*
Roberts, A. D., 227(23,24), 229(38), *263, 264*
Roberts, K., 332(110), *337*
Robertson, A. A., 333(137), *338*
Robinson, J., 155(50), *178*
Robinson, M. D., 204(29), *223*
Roebersen, G. J., 65(33), *95*
Rogers, J. A., 371(18), 374(18),

[Rogers, J. A.]
378(18), *384*
Roginsky, A., 318(50), *322*
Rohloff, E., 324(5), 326(5), 327(5), 328(57), 333(5,57), *334, 336*
Romo, L. A., 125(12), 127(12), 138(27), *142*, 360(16), 362(22), 364(16), *366*
Rooy, N. D., 141(40), *143*
Ror, D. K., 157(60), *179*
Rose, W., 201(24), *223*
Ruch, R., 141(39), *143*
Ruckenstein, E., 76(59), *96*
Russel, R., 388(3), *395*
Rutgers, A. J., 209(39), 211(39), 214(47,48), *223, 224*
Rutland, D. F., 331(94), *337*

Sachs, G., 360(14), 361(14), *366*
Saito, T., 300(4,5), 305(4,5), 310(4,5), 311(4), *320*
Sakai, C., 318(62), *322*
Sakamori, Y., 158(84,89), 164(89), 168(89), *180*
Sakashita, M., 391(11), *395*
Salam, T., 174(127), *181*
Saleeb, F. Z., 318(57), *322*
Salt, F. W., 395(27), *396*
Salvi, A. S., 313(37), 315(40), 316(40,41), *321*
Samec, Z., 165(105), *180*
Samygin, V. D., 80(83), *97*, 295(53), *298*
Sandstede, G., 271(9), *283*
Sandstrome, E. R., 333(141), *338*
Sany, Z., 359(8), *366*
Sasaki, H., 79(78), *97*, 295(39, 40), *297*
Sasaki, S., 232(51), *264*
Sato, N., 391(11), *395*
Sato, T., 125(13), 141(13,39),

[Sato, T.]
142, 143, 295(46), *297*
Sauret, G., 331(88), *337*
Sautter, F. K., 393(21), *396*
Sawada, Y., 155(54), *178*
Schaffert, R. M., 414(4), 417(4), *434*
Schausberger, A., 324(2), *334*
Scheludko, A., 80(86), 81(89,90), 97, 227(19,20,21), 229(21,35, 36), *263, 264*
Schempp, W., 333(115,123,124), *338*
Schenkel, J. H., 256(89), *266*
Schiesser, R. H., 331(89), *337*
Schlossman, M. L., 383(36), *385*
Schmidt, L. D., 120(3), *141*
Schmidt, P. G., 363(30), *367*
Schott, H., 270(7), *283*, 399(6), 400(6), *412*
Schufle, J. A., 216(61), *224*
Schulze, H. J., 83(92), *97*
Schurz, J., 324(2), 333(115,123, 124), *334, 338*
Schweizer, G., 333(139), *338*
Seaman, G. V. F., 398(2,3,5), 404(13), 405(13), 406(13), *412*
Seeton, F. A., 295(52), *298*
Segawa, H., 426(44), 433(44), 434(44), *436*
Senda, M., 165(106), *180*
Sennett, P., 333(116), *338*
Seta, P., 165(101), *180*
Sexsmith, F. H., 318(56), *322*
Shallhorn, P. M., 331(94), 332(94), *337*
Sharma, M. K., 371(15), 372(15), 373(15), 376(15), 377(15), *384*
Shaw, J. N., 64(31), *95*, 185(2), *222*, 260(91), *266*
Shebs, W. T., 318(64), *322*
Sheridan, J. L., 331(93,95), 332(95), *337*
Sherman, P., 371(20), 376(25), *384*

Shibamoto, A., 301(10), 320
Shilov, V. N., 65(37), 95
Shimoiizaka, J., 60(21), 78(21), 94(21), 94, 229(41), 264, 290(10,11), 295(46), 296, 297
Shinada, N., 361(18), 366
Shinoda, K., 83(94), 97
Shinohara, H., 300(4), 305(4), 310(4), 311(4), 320
Shirahama, H., 79(68), 96
Shishin, V. A., 80(87), 97, 228(28), 263
Shore, W. S., 163(99), 180
Shrivastava, S. K., 303(16,17), 321
Shukakidse, N. D., 294(28), 297
Shuttleworth, T. H., 276(17), 278(17), 283
Sigg, L., 176(142), 182
Simons, P. B., 330(68), 336
Sirianni, A. F., 138(30), 142
Sirois, L. L., 295(38), 297
Sivaramakrishnan, N. H., 318(54), 322
Sluyters, J. H., 154(46), 178
Sluyters-Rehbach, M., 154(46), 178
Small, C. E., 158(70), 179
Smit, G., 332(98), 333(129), 337, 338
Smith, A., 370(8), 384
Smith, B. A., 401(8), 402(8), 412
Smith, D. E., 154(44,45), 157, (44,45), 178
Smith, L., 42(39), 46
Smith, M. E., 191(6), 222
Smith, M. L., 124(11), 142, 360(15), 366
Smith, R. W., 296(17), 296
Smitham, J. B., 228(30), 263
Smolders, C. A., 149(21), 177
Smoluchowski, M., 233(60), 265, 371(10), 384
Snaith, D. W., 393(24), 396

Snook, I. K., 232(56), 264
Soderberg, B. A., 17(3), 21(3), 44
Somasundaran, P., 292(20,22), 296(56), 296, 298
Sonntag, H., 230(4), 264
Söremark, C., 326(23,24,29,30, 31), 327(23,29,30), 329(30), 331(23), 332(111), 333(23,24), 335, 337
Sørensen, P., 361(19), 366
Sparks, B. D., 138(30), 142
Sparnaay, M. J., 148(5), 154(5), 176
Spielman, L. A., 65(32), 95
Spurlin, H. M., 332(107), 337
Spurling, T. H., 232(55), 264
Srivastava, S. N., 371(15), 372(15), 373(15), 376(15), 377(15), 384
Stanley, J. S., 317(43), 319(43), 321
Steiner, R. F., 83(100), 98
Stern, O., 30(18), 45, 253(82), 265, 404(12), 412
Stevens, T. J., 192(7), 222
Stewart, D. R., 205(30), 223
Stigter, D., 200(22), 223
Stone-Masui, J., 249(77), 265
Stoutjesdijk, I. P. G., 332, 333 (98,129), 337, 338
Stoutjesdijk, P. G., 332(98), 337
Strassner, E. A., 216(55), 224
Stratton, R. A., 324(3), 327(3), 332(100), 334, 337
Strazdins, E., 324(10,12), 326(10, 12,25), 327(10,12,40), 332(103, 105), 333(25,40,105,126), 334, 335, 337, 338
Street, N., 205(30), 214(50,51, 52), 215(51), 223, 224
Stryker, L. J., 332(109), 337
Stumm, W., 176(142), 182, 390(7), 395
Subba Rao, V. V., 162(97), 180

Sugimoto, K., 155(54), *178*
Sümegi, M., 333(113), *338*
Sumimoto, M., 327(36), *335*
Sun, S. C., 295(42), *297*
Suzawa, T., 79(68,69), *96*,
  300(1,4,5,6,7), 303(14,15,18),
  304(18,19), 305(4,5,14,15,21,
  22,23,24,25,26), 308(24),
  310(4,5,6,14,19,21,34,35),
  311(4,14,19,21,25,34),
  313,34,38,39), 315(15,22,39),
  316(42), 317(45,46,47),
  318(45,48,49), 319(65), *320,
  321, 322*
Suzuki, K., 328(52), *336*
Swanson, J. W., 324(3),
  327(3), 332(100), *334, 337*
Swanson, S. A., 163(99),
  180
Szwarcsztajn, E., 331(87),
  337

Tabor, D., 57(11), *94*, 227(23,
  26,27), *263*
Tachibana, T., 270(2), *283*,
  318(58), *322*
Tachiki, K., 327(36), *335*
Tadros, Th. F., 42(36), *45*,
  371(17), 374(17), *384*
Tagawa, M., 79(70), *96*
Tagaya, H., 158(93), 163(93),
  165(102,103), 167(103),
  168(93), *180*
Tahara, S., 371(14), 372(14),
  376(14), *384*
Takahashi, H., 300(6), 310(6),
  320
Takahashi, K., 154(34,35,40,
  41), 155(40,41), 157(61),
  158(68), 175(138), *177, 179*
Takahashi, M., 370(2), 378(2),
  383
Takahashi, S., 330(83),
  331(83), 332(83), 333(83),
  337

Takahashi, T., 318(48), *322*
Takahashi, Y., 417(10), 418(10),
  434
Takamura, A., 359(12), *366*
Takamura, K., 61(27), *95*,
  155(52), *178*
Takamura, T., 155(52), *178*
Takano, K., 84(106,111), 86(106),
  98
Takeda, Y., 249(76), 256(76),
  265
Takenaka, T., 232(52), *264*
Takenouti, H., 154(48), *178*
Tamai, H., 79(68,69), *96*, 158(80,81,
  82,83,85,86,88,89,90,91,92),
  160(80), 161(85,86), 162(81,
  86), 163(81,83) 164(89,90),
  168(88,89), *179*, 231(47,48,49),
  264, 340(4), *354*
Tamamushi, R., 148(2), 154(34,
  35), 157(61), *176, 177*
Tamaribuchi, K., 124(11), *142*,
  360(15), *366*
Tamura, T., 139(33), 140(33),
  141(33), *143*
Tanaka, H., 327(36), 328(52),
  335, 336
Tanaka, M., 392(14), *395*
Tandon, R. K., 228(33), *264*
Tanford, C., 163(99), *180*
Tang, W., 359(8), *366*
Tappe, M., 331(90), *337*
Tari, I., 360(17), 363(28), *366,
  367*, 388(1,2), 390(10), 391(12),
  393(25), *395, 396*
Tashiro, I., 417(8), 421(28),
  424(28), *434, 435*
Tauksk, J. M., 83(101), *98*
Tausk, R. J. M., 250(80),
  254(80), *265*
Tay, C. H., 328(56), *336*
Tewari, P. H. 170(117),
  175(130), *181*
Tezuka, Y., 245(71), 246(71),
  247(71), 255(71), 256(71),
  265
Thies, C., 141(38), *143*

Thirsk, H. R., 154(39,42), 155(58), 156(58), *178*
Thomas, B. D., 332(109), *337*
Tichy, J., 331(94), 332(94), *337*
Tilak, B. V. K. S. R. A., 34(21), 36(21), *45*, 148(11), *176*, 232(53), *264*
Titievskaya, A. S., 227(18), *263*
Titijevskaya, A. S., 81(88), *97*
Toda, M., 89(116), 90(116), *98*
Tokiwa, F., 78(66,67), *96*, 270(5), 274(14), 276(19), 278(20,21,22), 280(22), 281(19), 282(19,22), *283*
Tokuda, W., 426(44), 433(44), 434(44), *436*
Tomaszewski, L. C., 393(22), *396*
Tomaszewski, T. W., 393(22), *396*
Tomiyama, T., 361(18), *366*
Topham, A., 365(36), *367*
Touma, K., 305(22), 315(22), *321*
Toyoshima, Y., 426(43), 433(43), *436*
Trasatti, S., 154(31), *177*
Tretter, H., 324(5), 326(5), 327(5), 328(57), 333(5,57), *334, 336*
Trifiletti, S. E., 371(16), 374(16), *384*
Troelstra, S. A., 234(62), 238(62), *265*
Tsubomura, M., 318(58), *322*
Tsuji, F., 35(22), 36(22), *45*
Tsukamoto, M., 429(48), *436*
Tulagin, V., 426(41), 433(41), *436*
Tye, F. L., 389(4), *395*

Uchida, K., 83(93), *97*
Ueda, S., 35(22), 36(22), *45*
Uhlenbruck, G., 398(2), *412*
Unbehend, J. E., 324(6), 328(6, 62,65), 329(66), 330(66,67), *334, 336*
Urban, F., 216(55), *224*
Usher, F. L., 83(99), *98*
Ushiyama, H., 79(73), *96*, 422(30), 425(30), *435*
Usui, S., 60(21), 69(43), 71(43), 74(47), 75(47), 78(21,43), *94, 95, 96*, 226(9), 229(41), 230(42), 232(51), *263, 264*, 288(8), 292(23), 295(39,40), *296, 297*
Uzgiris, E. E., 189(11), 192(11, 12,13), *222, 223*
Vallette, G., 155(51), *178*
Vallette, P., 331(88), *337*
Van Del Hul, H. J., 83(101), *98*, 249(75), 250(75,80), 254(80), 260(75), *265*
Vandenberg, E. J., 332(107), *337*
Vanderhoff, J. W., 249(75), 250(75,80), 254(80), 260(75), *265*
van der Minne, J. L., 126(18), 131(18), 135(18), *142*, 421(24), *435*
Van der Waaden, M., 422(33), *435*
van de Ven, T. G. M., 61(26), *95*
Van Helden, A. K., 84(113), *98*
Vanlaethem-Meuree, N., 149(16), 158(16), *177*
van Megen, W., 232(56), *264*
Van Olphen, H., 383(31), *385*
Van Oss, C. J., 78(60), *96*
van Vliet, T., 228(32), *264*
Vanysek, P., 165(104), 166(104), *180*
Varma, R. P., 373(21), 374(21), 376(21), 377(21), *384*
Vassar, P. V., 404(13), 405(13),

[Vassar, P. V.]
  406(13), 412
Verwey, E. J. W., 26(17),
  27(17), 45, 48(2), 54(2), 58(2),
  94, 124(10), 142, 184(1), 222,
  226(1), 255(85), 262, 266,
  270(3), 271(3), 283, 376(26),
  384, 395(28), 396, 407(15),
  408(15), 412, 422(32), 435
Vijayendran, B. R., 121(7),
  134(7), 142, 420(21,22), 435
Vinther, A., 361(19), 366
Visser, J., 57(14), 94, 78(61),
  96, 271(8), 283
Vold, R. D., 141(37), 143,
  318(52,54), 322
Völgyi, P., 327(39), 335
von Buzagh, A., 256(87), 266
von Stackelberg, M., 200(21),
  201(21), 223, 317(44),
  319(44), 322
von Willisen, F. K., 424(38),
  435
Vos, H., 149(22), 177
Vrij, A., 84(112,113), 98,
  141(34), 143, 228(31),
  229(37), 248(73), 263, 264,
  265

Wadachi, M., 89(116), 90(116),
  98
Wainright, T. E., 89(118), 98
Walker, T., 78(64), 96,
  141(35), 143, 228(29), 263
Walkush, J. C., 324(17),
  328(17), 334
Walslau, H., 84(102), 86(102),
  98
Walter, G., 271(9), 283
Waltner, G., 330(72), 333(72,
  134), 336, 338
Wang, X., 359(8), 366
Ware, B. R., 192(8,9,10), 194(9),
  222, 401(7,8), 402(8), 412
Warren, L. J., 295(54), 298

Waser, R., 155(50), 178
Watanabe, A., 35(22), 36(22), 45,
  60(18), 79(70), 94, 96, 158(80,
  81,82,83,84,85,86,87,88,89,90,
  91,92,93), 160(80), 161(85,86,
  87), 162(81,86,87), 163(81,83,
  93), 164(89,90), 165(102,103),
  167(103), 168(88,89,93,107,108,
  109,110,111), 179, 180, 181,
  229(39), 230(43), 231(47,48,49),
  238(66,67), 239(66), 245(71),
  246(71), 247(71), 255(71),
  256(71), 264, 265, 304(20),
  321, 340(3,4,5,8,9,10,11),
  342(9,10), 353(13), 354
Watanabe, N., 360(17), 363(28),
  366, 367
Watillon, A., 216(56), 224,
  249(77), 260(92), 265, 266
Weatherburn, A. S., 318(59),
  322
Weber, J., 165(105), 180
Weigl, J., 330(70,72,80,84),
  331(70), 333(72,114,134,139),
  336, 338
Weiner, R. S., 401(8), 402(8),
  412
Werdouschegg, F. MK., 330(69),
  336
West, J., 326(33), 331(33), 335
Westall, J., 42(44), 46
Weyh, E., 330(72), 333(72,134),
  336, 338
White, H. J., 318(56), 322
White, H. L., 216(55), 224
White, L. R., 42(43), 43(43),
  46, 57(15), 70(42), 74(55,56),
  76(55), 77(56), 94, 95,
  228(33), 264
Whitehead, A., 295(51), 298
Whitman, W., 388(3), 395
Wick, R., 395(26), 396
Wiersema, P. H., 65(33), 95,
  114(4), 117, 255(84), 266
Wiese, G. R., 79(71,72), 96,
  171(122), 173(122), 181,
  358(5), 362(25,26), 366

Wiesema, P. H., 219(64), *224*
Wijga, P. W., 207(35), *223*
Wilke, F., 200(21), 201(21), *223*, 317(44), 319(44), *322*
Wilkins, D. J., 394(4), 399(4), 403(11), 404(11), 408(18), 409(18), 410(18), *412*
Williams, D. G., 324(17), 328(17,64,65), 331(85), *334, 336, 337*
Willis, E., 126(28), 127(28), 135(28), 135(28), *142*, 361(20), 362(20), *366*
Windhager, R. H., 333(130), *338*
Winterton, R. H. S., 227(26), *263*
Wolff, J. S., 382(30), *385*
Wright, P., 131(20), 135(31), 139(31), *142*
Wroblowa, H., 148(12), 151(12), 154(12), 156(12), *177*

Yaba, A., 318(58), *322*
Yamada, B. J., 291(16), *296*
Yamada, H., 123(8), 126(8), 127(8), 133(24), 135(8), 138(8), 139(8), *142*, 362(23), *366*, 421(25), 422(25), *435*
Yamada, J., 219(65), 220(65), *224*, 420(23), *435*
Yamaguchi, M., 370(2,3,4,5), 378(2,3,4,5), *383, 384*
Yamaki, K., 308(29), 317(29), *321*
Yamamoto, K., 79(68), *96*

Yamaoka, T., 310(35), *321*
Yamasaki, T., 60(21), 78(21, 65), *94, 96*, 229(41), 230(42), *264*
Yamashita, Y., 303(15), 305(15), 315(15), *320*
Yarar, B., 80(85), *97*, 295(49), *297*
Yarinitzky, C., 158(63,64), *179*
Yates, D. E., 79(71), *96*, 173(123), 126(123), *181*, 357(4), 362(25), *366*
Yen, Y., 194(19), *223*
Yoon, R. H., 174(127), *181*
Yopps, J., 42(38), *46*
Yoshihara, K., 392(17), *396*
Yoshimura, T., 192(14), *223*
Yoshizaki, O., 303(13), 308(27), *320, 321*
Yoshizawa, S., 360(17), 363(28), *366, 367*
Young, C. Y., 399(6), 400(6), *412*
Young, D. A., 90(119), *98*
Young, F. J., 208(38), 209(38), 210(38), *223*
Yuzawa, M., 317(47), *322*

Zerial, A., 398(4), 399(4), *412*
Zettlemoyer, A. C., 138(29), *142*, 162(97), *180*
Zhang, S., 359(8), *366*
Zhukov, P. N., 207(33), *223*
Zieminski, S. A., 232(50), *264*
Zorin, Z. M., 80(87), *97*, 228(28), *263*

Subject Index

Abrasion of wire, 333
Acid-base theory, 124
Adhesion of soil, 270
Admittance, 153
Adsorbed ion, specially 5, 14
Adsorbed layer 140
Adsorption 326-327
  at oil/water interface, 160-162
  entropy of, 318
  heat of, 318
  ion-exchange, 388-390
  mixed, 164
  of dye, 168
  of phospholipid, 168
  of polymer to red cell, 405-407
  of surfactant, 270, 276
  preferential, 124, 128
  specific, 20, 21, 31, 33
  surface dye, 304, 310
  total dye, 310
Adsorption potential, specific
    chemical, 31
Aerosol perspirant, 383
Aggregate, reversible 241
Aggregation
  of bacteria, 410-411

[Aggregation]
  of biological cell, 407-411
  pair formation in, 244
  of white cell, 408-410
Alder transition, 89-91, 93
Antirusting, 387-396

Bjerrum theory, 120
Blood cell, calcium ion binding
    to 404-405
Bridging attraction, 247
Builder, 270, 279
Burton tube, 184

Capillary pressure, 82
C-E curve, 148
Cell, 219, 222
Chelating agent, 376
Chemical affinity, 7
Chemical potential, 8
CMC, 316-320
Charge carrier, electronic 120

Charge
  control agent, 420, 430
  in organic solvent, 360
  of pigment, 356-363
  per particle, 221
Coagulation
  critical zeta potential for, 131
  fast, 61, 238
  rapid (see fast cogaulation)
  slow, 62
Coalescence, 370, 371
  critical potential of, 229-231
  of droplet, 225-233
Coflocculation, 362, 363, 365
Co-ion, 27
Collector, 285-289
Colloidal pitch, 333
Competing binding, 164
Composite plating, 392-394
Conductometric titration curve, 253
Constant surface charge model, 70
Constant surface potential model, 70-74
Cosmetic cream, 348
Coulombic interaction, 131
Coulombic law, 129
Cream, 369, 370
Creaming, 370, 378
Critical flocculation (coagulation) concentration, 59, 233
Critical micelle concentration (see CMC)
Cross-linking agent, 303
Counterion, 25, 27
  binding, 162

Debye-Hückel approximation, 26 28, 29
Debye length, 26, 129
Debye reciprocal length parameter (see Debye length)
Depressant, 285

Derjaguin-Landau-Verwey-Oberbeek theory (see DLVO theory)
Detergency, 269-283, 318
Dialysis, 85, 87
Differential capacity, 22, 29, 37, 40
  measurement, 149
Diffuse layer, 27
  thickness of, 29
Dipole, orientation of, 5
Disjoining pressure, 80
Disordered state, 90
Dispersion
  concentrated, 83-93, 133
  in polar organic solvent, 141
Display
  electrophoretic, 424-428
  photoelectrophoretic, 433-434
Dissipation function, 115
DLVO theory, 57-61, 130, 270, 273, 275, 280, 376
Double layer, 3, 84, 123, 148, 271-274, 280, 376
  asymmetrical, 273, 275, 277
  capacity, 153-158
  curvature of, 208-209
  diffuse, 3, 154
  dynamic, 294
  formation, 154
  interaction energy, 130
  overlapping of, 134, 208-209
  parameter, 148-153
  potential, 10, 128
  symmetrical, 275
  thickness, 300, 371
Drainage, 327-330
Droplet growth, 370
Dropping-time method, 149
Dye:
  acid, 305, 313
  cationic, 305, 309, 315
  direct, 305, 311
  reactive, 315
Dyeability, 303
  surface, 305-311

Dyeing, 303-305
  free energy of, 304
  of paper, 333
  solvent, 345-348
Dynamic jar, 328, 330

Effluent, 333
Electrical charge on particle, 421
Electrical double layer (see double layer)
Electrification of solid, 174
Electroadsorption, 161
Electrocapillarometer, 19
Electrocapillary
  adsorption, 161
  curve, 19, 149, 291
    at oil/water interface, 159-160
  equation, 17, 18
  maximum(ECM), 20
  measurement at oil/water interface, 160
  phenomenon at oil/water interface, 231
  spinning, 350-353
Electrochemical potential, 9
Electrode:
  Ag-AgCl, 101
  dropping mercury(DME), 148
  ideally polarized, 148
  mercury, 148
  polarization of, 204
  reaction, 154
  reversible, 200
  solid, 155
  Zn-$ZnSO_4$, 101
Electrodeposition method, 219-222
Electrokinetic phenomenon, 4, 99, 169
Electrokinetic potential (see zeta potential)
Electrokinetic property, 324, 330

Electroosmosis, 4, 100, 103-111, 134, 199-200
Electrophoresis, 4, 100, 102, 108, 134, 184
  Doppler, 193
Electrophoretic deposition, 394-395
Electrophoretic distribution, 401
Electrophoretic light scattering, 401
Electrophoretic mobility, 220, 376, 378, 398
  distribution of, 198
  in nonaqueous system, 219
  of streptococcus faecalis, 399
Electrophoretic retardation, 109
Electrophotography, 414-424
Electroviscous effect, 207-208
Emulsification:
  electric, 168
  electrocapillary, 339
  nozzle for, 344
Emulsion:
  effect of surface potential on stability of, 376-378
  oil in water, 371
Entropy production, 115
Equi-acid-base point, 388
Escaping tendency, 7
Eversole and Boardman equation, 300

Face powder, 369
FFT spectrum analyser, 194
Filler, 330-332
  treatment of, 333
Fiber:
  amphoteric, 317
  secondary, 333
Floc:
  hard, 328
  soft, 328
Flocculating rate, 233

Flocculation, 326-328, 370, 371, 376
  dual system in, 328
  index, 330
  maximum, 171
  shape effect on, 278
  two-step system in, 328, 332
  value, 233
Flotation, 285-298
  carrier, 295-296
Formation factor in surface conductivity, 215
Free energy of monodisperse latex, 91
Fuoss and Kraus, experiment by, 120

Galvani potential, 5
Gelling agent, 379-383
Geometrical mean law, 78
Gibbs adsorption isotherm, 15, 17, 148
Gibbs free energy, 92
Gilbert-Rideal theory for protein titration, 401
Glass transition point, 301
Gouy-Chapman model, 25-30
Gouy-Chapman theory, 300
Graft copolymerization, 308
Granulocyte, 398

Hard-sphere system, 84, 90, 93
Hamaker constant, 55, 57, 77, 82, 84, 131, 376
Helmholtz free energy, 92
Helmholtz layer, 102
Helmholtz plane:
  inner, 32
  outer, 32
Helmholtz-Smoluchowski equation, 106, 107, 202, 371
Hemi-micelle, 287, 288
Henry equation, 371

Heterocoagulation, 65-80, 270-279, 392
  stability constant for, 281-283
Heterodying method, 194
Hückel equation, 219
Hydration capacity, 301
Hydrophilic or oleophilic property, 125

Impedance bridge, 155
Interface
  binding at oil/water, 162-164
  effect, 85
  nonpolarizable, 17
  polarizable, 15
  viscosity at, 207

Interfacial electrical effect, 278-281
Interfacial impedance, 153
Integral capacity, 25
Iridescence, 85, 87
Iridescent state, 84
Irreversible process, 115
Isoelectric point(IEP), 14, 36, 286, 300, 390-392
  interfacial, 163, 171

Kirkwood-Alder transition, 90
Kolthoff reaction, 249

Latex
  deposition of, 333
  emulsifier-free, 249
  monodisperse, 83, 85, 90, 248
  polystyrene, 84
Lambert-Beer law, 234
Laser doppler method, 193-195

Subject Index

Lippmann electrometer, 149
Lippmann equation, 17, 19
Liquid crystal, lyotropic 370, 378
Liquid development, mechanism of 415-419
Liquid film:
 on capillary wall, 201
 thin, 80-83
Liquidlike structure, 84
London-van der Waals force, 55, 84, 270-272, 275, 376
Lymphocyte, 398
 leukemic, 401
 normal, 401

Macroflocculation, 330
Makeup foundation, 369
Micelle:
 reversed, 120
 swollen, 140
Microcapsule, 348-350
Microelectrophoresis, 188, 371
 commercial apparatus for, 219
 measurement in nonaqueous system, 218-219
 two-tube method for, 190-192
Microemulsion, 140
Microflocculation, 330
Mode mobility, 401
Moving boundary method in electrophoresis, 222

Nail lacquer, 370, 381, 383
Network structure, 140
Nonequilibrium thermodynamics, 115

Ohm's law, 116
Onsager's law, 116
Ordered formation, 87-89

Ordered state, 90
Ordered structure, 85
Organophilic bentonite, 383

Paint, 363
Pallman effect, 169
Patch model, 328
Phenomenological coefficient, 115
Phenomenological equation, 115
Pigment, 355, 420, 429
 coating, 333
 dispersion of, 355
 inorganic, 357, 360, 361, 364
 organic, 361
Platelet, 398
Point of zero charge(PZC), 12-14, 20, 169, 171, 175
Poisson-Boltzmann equation, 26, 29, 49
Poisson's equation, 102
Polarization, 111
Potential
 barrier, 58
 χ-, 5-12
 inner, 5
 of zero charge, 149
 outer, 5
 ϕ-, 5-12
 ψ-, 5-12
 rational, 20
Potential-determining ion, 5, 7, 42
Potentiometric titration, 170
Preservative, 376
Primary minimum, 61
Printing ink, 363
Proton donor and acceptor, idea of, 124
Proton transfer mechanism, 141

Red cell, 398
Regulation model, 74-77

Redeposition, 277-283
Relaxation, 111
 of adsorption-desorption, 154
Relaxation-time effect, 112
Repeptization, 65
Retardation and relaxation
 effect, 219
Retardation factor, 62
Retention, 327-330
 first-pass, 330
 one-pass, 330
 overall, 330
Retention tester, Minidrinier, 330
Rotating grating method, 196-199
Rotating prism method, 195

Schlieren method, 187
Schulze-Hardy rule, 59
Secondary minimum, 61, 84, 241, 255, 262, 376
Sedimentation, 379
 potential, 102, 107, 108
Settling, 370
Shampoo, 370
Size, rosin 332
Sizing, 332-333
 agent, 333
 surface, 333
Slime coating, 295-296
Slipping plane, 103
Soil, removal of, 270-272, 277-283
Sol:
 AgI, 3
 concentration effect, 169
Specialty paper, 333
Stability constant, 282
Stability of colloid, 4, 226-229, 270, 422, 431
Stability ratio, 62, 79, 131, 237, 408
Stability of nonaqueous dispersion, 130-133
 effect of water on, 138-140

Stabilization, exceptional, 232-233
Stabilizing agent, 430
Steric repulsion, 247
Stationary level, 189-190
Stern and Grahame model, 30-36
Stern effect, 164
Stern layer, 30, 103
Stern plane, 30
Stern potential, 304
Stern's adsorption model, 404
Stokes' law, 377, 379
Streaming potential, 102, 107, 134, 202-213, 390, 401
Stopped-flow spectrophotometer, 235
Surface charge, 27, 43
 density, 303, 317
 total, 129, 221
Surface conductance, 106, 111, 206-207, 214-217
 activation energy for, 217
Surface conductivity, 4
 of capillary or flat plate, 214
 or powder sample, 214-215
Surface excess, 148
 relative, 17
Surface pK value of bacteria, 399
Surface polarization, 5
Surface potential, 273, 274, 281
Suspension:
 effect, 169
 capacitance of, 175
Synthetic hectorite, 382-383

Tensametry at oil/water interface, 165-167
Thixotropy, 378, 379, 381
Tiselius electrophoresis cell, 188
Toothpaste, 370
Transient method, 158
Transition, 84
 of first order, 89
Turbidity, 234

Subject Index

Ultraflotation, 295

Van der Waals attractive force
 (see London-van der Waals force)
Voltammetry, phase sensitive AC 156
Volta potential, 5
V value, 301

White water, 333
Walden's law, 123

White cell, cation binding to 403-404

Zero point of charge(ZPC) (see PZC)
Zeta potential (ζ potential), 274-275, 299, 310, 371, 376, 388
 effect of water on, 138-140
 for negative fiber, 300
 of human hair, 401
 of platinum, gold and silver, 392
 of stainless steel, 392
 of titania, 393